Optical 3D-Spectroscopy for Astronomy

Optical 3D-Spectroscopy for Astronomy

Roland Bacon and Guy Monnet

Authors

Roland Bacon
CRAL – Observatoire de Lyon
9, avenue Charles André
69230 Saint-Genis-Laval
France

Guy Monnet
CRAL – Observatoire de Lyon
9, avenue Charles André
69230 Saint-Genis-Laval
France

Cover
Galaxy: (c) NASA

All books published by Wiley-VCH are carefully produced. Nevertheless, authors, editors, and publisher do not warrant the information contained in these books, including this book, to be free of errors. Readers are advised to keep in mind that statements, data, illustrations, procedural details or other items may inadvertently be inaccurate.

Library of Congress Card No.: applied for

British Library Cataloguing-in-Publication Data
A catalogue record for this book is available from the British Library.

Bibliographic information published by the Deutsche Nationalbibliothek
Die Deutsche Nationalbibliothek lists this publication in the Deutsche Nationalbibliografie; detailed bibliographic data are available on the Internet at http://dnb.d-nb.de.

© 2017 Wiley-VCH Verlag GmbH & Co. KGaA, Boschstr. 12, 69469 Weinheim, Germany

All rights reserved (including those of translation into other languages). No part of this book may be reproduced in any form – by photoprinting, microfilm, or any other means – nor transmitted or translated into a machine language without written permission from the publishers. Registered names, trademarks, etc. used in this book, even when not specifically marked as such, are not to be considered unprotected by law.

Print ISBN: 978-3-527-41202-0
ePDF ISBN: 978-3-527-67485-5
ePub ISBN: 978-3-527-67484-8
Mobi ISBN: 978-3-527-67483-1
oBook ISBN: 978-3-527-67482-4

Cover Design Schulz Grafik-Design, Fußgönheim, Germany
Typesetting SPi Global, Chennai, India
Printing and Binding Markono Print Media Pte Ltd, Singapore

Printed on acid-free paper

Contents

Foreword *xi*
Acknowledgments *xiii*

The Emergence of 3D Spectroscopy in Astronomy *1*
Scientific Rationale   *1*
3D History   *4*
3D Technology   *9*

Part I   3D Instrumentation *11*

1	**The Spectroscopic Toolbox**	*13*
1.1	Introduction	*13*
1.1.1	Geometrical Optics #101	*13*
1.1.2	Etendue Conservation	*15*
1.2	Basic Spectroscopic Principles	*18*
1.2.1	The Spectroscopic Case	*18*
1.3	Scanning Filters	*20*
1.3.1	Introduction	*20*
1.3.2	Interference Filters	*22*
1.3.3	Fabry–Pérot Filter	*24*
1.4	Dispersers	*25*
1.4.1	Prisms	*25*
1.4.2	Grating Principle	*27*
1.4.3	The Grating Spectrograph	*28*
1.4.4	Grating Species	*29*
1.4.5	Grating Etendue	*30*
1.4.6	Conclusion	*31*
1.5	2D Detectors	*31*
1.5.1	Introduction	*31*
1.5.2	The Photographic Plate	*32*
1.5.3	2D Optical Detectors	*32*
1.5.4	2D Infrared Arrays	*35*
1.5.5	Conclusion	*35*

1.6	Optics and Coatings 36	
1.6.1	Introduction to Optics 36	
1.6.2	Optical Computation 37	
1.6.3	Optical Fabrication 40	
1.6.4	Anti-Reflection Coatings 42	
1.6.5	High Reflectivity Coatings 43	
1.6.6	Conclusions 44	
1.7	Mechanics, Cryogenics and Electronics 45	
1.7.1	Mechanical Design 45	
1.7.2	Alignments 48	
1.7.3	Cryogenics 48	
1.7.4	Electronics and Control System 49	
1.8	Management, Timeline, and Cost 50	
1.9	Conclusion 52	
2	**Multiobject Spectroscopy** 61	
2.1	Introduction 61	
2.1.1	MOS History: The Pioneers 61	
2.1.2	MOS History: The Digital Age 62	
2.1.3	MOS Flavors 62	
2.2	Slitless Based Multi-Object Spectroscopy 62	
2.2.1	Slitless Spectroscopy Concept 62	
2.2.2	Slitless Spectroscopic Systems 64	
2.3	Multislit-Based Multiobject Spectroscopy 64	
2.3.1	Multislit Concept 64	
2.3.2	Multislit Holders 66	
2.3.3	Multislit Systems 69	
2.3.4	Multislit Instruments 70	
2.4	Fiber-Based Multiobject Spectroscopy 70	
2.4.1	Multifiber Concept 70	
2.4.2	Positioning Systems 71	
2.4.3	Fiber-Based Spectrograph 75	
2.4.4	Fiber Systems Performance 75	
2.4.5	Present Multifiber Facilities 76	
2.4.6	Conclusion 77	
3	**Scanning Imaging Spectroscopy** 81	
3.1	Introduction 81	
3.2	Scanning Long-Slit Spectroscopy 81	
3.2.1	The Scanning Long-Slit Spectroscopy Concept 81	
3.2.2	Astronomical Use 82	
3.3	Scanning Fabry–Pérot Spectroscopy 83	
3.3.1	Introduction 83	
3.3.2	Fixed Fabry–Pérot Concept 83	
3.3.3	Scanning Fabry–Pérot 85	
3.4	Scanning Fourier Transform Spectroscopy 88	
3.4.1	Fourier Transform Spectrometer 88	

3.4.2	Fourier Transform Spectrograph	*90*
3.5	Conclusion: Comparing the Different Scanning Flavors	*91*

4	**Integral Field Spectroscopy**	*95*
4.1	Introduction	*95*
4.2	Lenslet-Based Integral Field Spectrometer	*95*
4.3	Fiber-Based Integral Field Spectrometer	*102*
4.3.1	The Fiber-Based IFS Concept	*102*
4.3.2	The Fiber-Based IFS Development	*103*
4.3.3	Conclusion	*103*
4.4	Slicer-Based Integral Field Spectrograph	*104*
4.4.1	Introduction	*104*
4.4.2	Integral Field Spectroscopy from Space	*107*
4.5	Conclusion: Comparing the Different IFS Flavors	*108*

5	**Recent Trends in Integral Field Spectroscopy**	*115*
5.1	Introduction	*115*
5.2	High-Contrast Integral Field Spectrometer	*115*
5.2.1	Exoplanet Detection	*115*
5.2.2	High-Contrast Integral Field Spectrometer	*116*
5.3	Wide-Field Integral Field Spectroscopy	*117*
5.3.1	The Rationale for Wide-Field Integral Field Spectroscopy	*117*
5.3.2	Current Wide-Field Projects	*117*
5.3.3	Wide-Field Systems 3D Format	*119*
5.4	An Example: Autopsy of the MUSE Wide-Field Instrument	*120*
5.4.1	MUSE Concept	*120*
5.4.2	MUSE Approach	*120*
5.4.3	MUSE Conclusions	*122*
5.4.4	Validity of the Multi-instrument Approach	*123*
5.5	Deployable Multiobject Integral Field Spectroscopy	*123*
5.5.1	Concept	*123*
5.5.2	The First Deployable Integral Field Units System	*124*
5.5.3	Near Infra-Red Deployable Integral Field Units	*124*
5.5.4	Deployable Multi-Integral Field Systems: Conclusion	*126*

6	**Comparing the Various 3D Techniques**	*129*
6.1	Introduction	*129*
6.2	3D Spectroscopy Grasp Invariant Principle	*129*
6.3	3-D Techniques Practical Differences	*130*
6.3.1	Packing Efficiency	*130*
6.3.2	Observational Efficiency	*131*
6.4	A Tentative Rating	*133*

7	**Future Trends in 3D Spectroscopy**	*137*
7.1	3D Instrumentation for the ELTs	*137*
7.2	Photonics-Based Spectrograph	*138*

7.2.1	OH Suppression Filter	*138*
7.2.2	Photonics Dispersers	*141*
7.2.3	Photonics Fourier Transform Spectrometer	*141*
7.2.4	Analysis	*142*
7.3	Quest for the Grail: Toward 3D Detectors?	*144*
7.3.1	Introduction	*144*
7.3.2	Photon-Counting 3D Detectors	*144*
7.3.3	Integrating 3D Detector	*145*
7.4	Conclusion	*146*
7.5	For Further Reading	*146*
Part II	**Using 3D Spectroscopy**	*151*
8	**Data Properties**	*153*
8.1	Introduction	*153*
8.2	Data Sampling and Resolution	*153*
8.2.1	Spatial Sampling and Resolution	*154*
8.2.2	Spectral Sampling and Resolution	*155*
8.3	Noise Properties	*158*
9	**Impact of Atmosphere**	*167*
9.1	Introduction	*167*
9.2	Basic Seeing Principles	*168*
9.2.1	What is Astronomical Seeing?	*168*
9.2.2	Seeing Properties	*170*
9.3	Seeing-Limited Observations	*172*
9.3.1	Seeing Impact on 3D Instruments	*172*
9.4	Adaptive Optics Corrected Observations	*173*
9.4.1	The Need for Overcoming Atmospheric Turbulence	*173*
9.4.2	Adaptive Optics Correction Principle	*173*
9.4.3	Adaptive Optics Components	*176*
9.4.4	Adaptive Optics: The Optical Domain Curse	*178*
9.4.5	Addressing the Lack of Reference Stars	*179*
9.4.6	Addressing the Small Field Limitation	*182*
9.4.7	Large Field Partial AO Correction	*183*
9.4.8	AO-Based Scanning Interferometers	*184*
9.4.9	AO-Based Slit Spectrographs	*185*
9.4.10	AO-Based Integral Field Spectrographs	*185*
9.4.11	AO-Based Near-IR Multiobject Integral Field Spectrographs	*187*
9.4.12	Deriving AO-Corrected Point-Spread Functions	*188*
9.4.13	Conclusion	*188*
9.4.14	For Further Reading	*189*
9.5	Other Atmosphere Impacts	*189*
9.5.1	Atmospheric Extinction	*189*
9.5.2	Atmospheric Refraction	*189*

9.5.3	Night Sky Emission	*191*
9.6	Space-Based Observations	*192*
9.6.1	The Case for Space-Based Observations	*192*
9.6.2	Why all Telescopes are not Space Telescopes	*193*
9.7	Conclusion	*194*

10	**Data Gathering**	*199*
10.1	Introduction	*199*
10.2	Planning Observations	*199*
10.3	Estimating Observing Time	*200*
10.4	Observing Strategy	*204*
10.5	At the Telescope	*206*
10.6	Conclusion	*209*

11	**Data Reduction**	*213*
11.1	Introduction	*213*
11.2	Basics	*214*
11.3	Specific Cases	*216*
11.3.1	Slitless Multiobject Spectrograph	*216*
11.3.2	Scanning Fabry–Pérot Spectrograph	*216*
11.3.3	Scanning Fourier Transform Spectrograph	*217*
11.3.4	Getting Noise Variance Estimation	*217*
11.3.5	Minimizing Systematics	*218*
11.4	Data Reduction Example: The MUSE Scheme	*219*
11.4.1	Detector Calibration	*221*
11.4.2	Flat-Field Calibrations and Trace Mask	*222*
11.4.3	Wavelength Calibrations	*224*
11.4.4	Geometrical Calibration	*225*
11.4.5	Basic Science Extraction and Pixel Tables	*226*
11.4.6	Differential Atmospheric Correction	*226*
11.4.7	Sky Subtraction	*228*
11.4.8	Spectrophotometric and Astrometric Calibrations	*229*
11.4.9	Data-Cube Creation	*232*
11.4.10	Data Quality	*233*
11.5	Conclusion	*236*

12	**Data Analysis**	*237*
12.1	Introduction	*237*
12.2	Handling Data Cubes	*237*
12.2.1	The Spectral View	*238*
12.2.2	The Spatial View	*239*
12.2.3	The 3D View	*239*
12.3	Viewing Data Cubes	*240*
12.4	Conclusion	*241*
12.5	Further Reading	*243*

13		**Conclusions** *245*	
	13.1	Conclusions *245*	
	13.2	General-Use Instruments *245*	
	13.3	Team-Use Instruments *250*	
	13.4	The Bumpy Road to Success *251*	

References *253*

Index *269*

Foreword

The development of astronomy has gone hand in hand with advances in technology. A celebrated step was taken by Galileo Galilei in 1609 when he used the newly invented telescope to observe the night sky and discovered mountains on the Moon, the phases of Venus, and four moons orbiting Jupiter. This revolutionized our world view, and was made possible by the increased light-gathering power and image sharpness provided by the 5 cm telescope lens compared to using the naked eye.

Since then, a number of such transformational steps enabled by new technology have occurred. The move from the lenses of refracting telescopes to the mirrors of reflecting telescopes allowed those telescopes to be of much larger diameter. Replacing the human eye as the detector behind the telescope, first with photographic plates and subsequently with almost perfectly sensitive electronic detectors, made it possible to collect light over time and hence observe much fainter objects. Dispersing the light into a spectrum revealed the physical nature of objects through the study of absorption and emission lines. The detection of infrared light and radio waves from the ground expanded astronomers' view beyond the wavelengths of visible light, and the launch of telescopes into space gave access to the entire electromagnetic spectrum. With the ability to detect particles and, most recently, gravitational waves emitted by celestial objects, astronomers now have even more ways of observing the Universe.

Spectroscopy in the optical and near-infrared regions was initially possible with a single aperture, which was adequate for observing stars. The development of spectroscopy with a slit allowed a more efficient study of extended objects and, more recently, the ability to perform spectroscopy over an extended area has once again provided an enormous jump in capabilities. This latest revolution is the topic of this book. The techniques for integral-field spectroscopy in the visual and infrared wavelength region have now matured to a level where the angular resolution of the spectroscopic observations can be as high as is achievable in direct imaging, and many telescopes have been equipped with such integral-field spectrographs, with others under development for the next generation of giant telescopes. A comprehensive overview is hence timely.

The authors are world-renowned experts who have had a major role in driving the development of integral-field spectroscopy from initial prototypes such as TIGER on the Canada France Hawaii Telescope and SAURON on the William

Herschel Telescope to the transformational MUSE instrument on ESO's Very Large Telescope. In this book, they cover the relevant principles of optics, discuss the advantages and disadvantages of different technical solutions, and link these to existing and future instruments on a variety of telescopes around the world. They also address the role and challenges of data reduction, the importance of adaptive optics to provide an ultra-sharp view and the relative merits of observations by telescopes on the ground and in space, and include illuminating science examples as well as a glimpse of potential future steps. The result is an eminently readable book that contains much helpful information for instrument builders and is highly recommended for any astronomer interested in spectroscopy.

Tim de Zeeuw

Acknowledgments

The authors warmly thank all those who have helped to produce this book, and in particular Martin Roth for his initial help in defining its content and for his enthusiastic support, Norbert Hubin for many cogent corrections on the adaptive optics chapter, Ghaouti Hansali for creating many good-looking figures for this book, and Tim de Zeeuw for his time spent to comment on a complete draft of the book and add a nice foreword.

Thanks also to all those who gave us permission to use figures to illustrate this book, namely: Joss Bland-Hawthorn, Michael Blanton, Richard Bower, Harvey Butcher, Warrick Couch, Jean-Baptiste Courbot, Laurent Drissen, Frank Eisenhauer, Paul Jordan, Reinhard Genzel, John Hill, Norbert Hubin, Sebastian Kamann, Etienne LeCoarer, Oliver LeFèvre, Simon Morris, Sebastian Sanchez, Brent Tully, Pieter Vandokkum, Peter Weilbacher. Special thanks to Davor Krajnović, Edmund Cheung, and Enrico Marchetti who have reformatted their figures specifically for us.

A warm mention to Wiley, our Editor, who have been more than patient along many months. This book's writing has been a long story by itself, full of twists, turns, and suspense.

Last, but not least, thanks to all our present and past collaborators who have shared with us their creativity and their excitement along the road of development of a variety of 3D spectroscopic instruments, and more generally to the many academic and industrial minds over the last four decades or so who have made these high-tech innovative devices possible. Without them, this book would have been no more than an empty box.

The Emergence of 3D Spectroscopy in Astronomy

Scientific Rationale

On a global scale, the "visible" Universe appears as a tapestry of galaxies, ranging from our own (the Milky Way) to billions of remote ones (Figure 1). Individual galaxy light comes from some 10^{11} stars–essentially fusion reactors confined by gravitation–either directly or reprocessed by surrounding ionized gas clouds in discrete narrow emission lines. A large fraction of astronomical investigations is thus directed toward getting detailed spectroscopic data (photon flux versus wavelength) on these faint two-dimensional light distributions over the sky in the most efficient way. These data give a wealth of dynamical, physical, and chemical information on individual stars, star clusters, individual galaxies, groups and clusters of galaxies, and finally on the structure of our whole Universe itself. Note that in this book, we use arc minutes (') or arc seconds (") as units of angular size on the sky and microns (μm) or nanometers (nm) as wavelength units, with $1' = 60''$ and $1\,\mu m = 1000\,nm$.

The most fundamental limitation in these spatiospectral investigations is of a dimensional nature. Let us take as a frequent line of study star/gas motions inside a nearby galaxy, such as that in Figure 2. This is intrinsically a six-dimensional(6-D) dynamical object, that is, with six degrees of freedom: three for the spatial position of stars or ionized gas clouds inside the galaxy and three for their mean velocity vector. In any galaxy except our own,[1] of these six kinematical components, only three can be measured: two spatial components from an image of the object, that is, emitted light integrated along the line of view, and one spectral component, the radial velocity V_r, that is, the spatial velocity projected and integrated along the line of view. The latter is measured from spectral line shifts $\Delta\lambda$, coming from stars or ionized gas clouds, from their laboratory rest wavelength λ: this is the Doppler–Fizeau effect, with $\Delta\lambda/\lambda = V_r/c$, c being, as usual, the speed of light in vacuum.

Note that the finite spatial resolution of the observations – most often in the 0.5"–1.5" range on a ground-based telescope due to atmospheric turbulence – means that any data point results from the combined light of a large

---

1 In our Galaxy, positions of individual stars change over a few years, mostly by minute but now measurable amounts owing to the GAIA space facility's superb position accuracy, well below 10^{-3} arcsec or 1 mas. This will soon add the two lacking velocity components for a billion stars.

Optical 3D-Spectroscopy for Astronomy, First Edition. Roland Bacon and Guy Monnet.
© 2017 Wiley-VCH Verlag GmbH & Co. KGaA. Published 2017 by Wiley-VCH Verlag GmbH & Co. KGaA.

Figure 1 HST Ultra Deep Field (credit AURA/STScI). This is a small (2.4′ across) "randomly" located multiband image taken with an unprecedented depth by the Hubble Space Telescope over the 0.4–0.95 μm wavelength domain. Apart for a handful of forefront Milky Way stars, the field is dotted by many galaxies, some dating back to a mere ∼ 700 million years after the birth of our 13.6 billion-year old Universe. (http://hubblesite.org/gallery/album/galaxy/pr2016028a/.)

number of stars or from a whole ionized cloud, except for a handful of extremely close by galaxies for which a few brightest stars can be individually studied. To add insult to injury, this spatially blurred 3-D only information has perforce to be mapped on a 2-D detector, the best available tool at present: for a long time, this topological mismatch was solved by losing one of the three measurable components, an easy way out, but giving an even more degraded data set. This book's main aim is to cover a full zoo of the so-called 3D spectroscopic facilities that have been recently developed to fully utilize the available information.

Until the mid-1970s, spectrographic instruments were thus providing 2D information only on their 2D detectors (usually a photographic plate), with two main species, the long-slit spectrograph (covering one of the two spatial dimensions and the spectral ones) and the Fabry–Pérot interferometer (covering the two spatial dimensions but only one spectral value). The art of getting the full 3D information (two spatial and one spectral) in a squarish sky field then started, driven by the arrival of the first 2D digital detectors and affordable

Figure 2 The pinwheel galaxy Messier 101 (credit European Space Agency & NASA). This Hubble space telescope image shows a prototypal spiral galaxy, made of about 100 billion stars orbiting around the galaxy center and surrounded by a huge dark matter halo. The *fat* central bulge holds the oldest stars dating back from the galaxy formation some 13 billion years ago. It is surrounded by a *flat* rotating disk of stars, gas, and dust with continuous star formation till now. The UV-bright young disk star clusters are enveloped by large ionized gas regions, easily studied spectroscopically from their bright narrow emission lines. Typical radial velocity range covered by the ionized gas clouds and the stars is $\sim \pm 150$ km s^{-1}.

computers. It has been christened 3D spectroscopy in the astrophysical domain, although it was later called hyperspectral imaging when it was independently derived for other kinds of observation, in particular of planet Earth from space. There is no single way to do it. On the contrary, there is by now a full zoo of instrumental species, detailed in this book, coming from different – and unavoidable – tradeoffs between the relative amount of spatial and spectral information gathered by the detector. Each species is actually optimized for some collection of astrophysical objects, for example nearby, intermediate, or remote galaxies; sparse versus dense stellar/galaxy fields. Note that this book, being astrophysically centered, mainly refers to the astronomical domain, but the need for 3D spectrography is equally present in other domains, such as Earth observation from above and medical imaging.

This instrumental diversification has been cost-efficiency driven: The consolidated cost of telescope observing time is not cheap – a few USD/Euros per second on a large telescope, with typical individual observations lasting 1–4 h. This puts a large premium on the so-called highly multiplexed spectrographic instruments, that is, which observe many different objects at once (multiobject instruments) or deliver many spatio/spectral bins at once on a single object (bona fide 3-D instruments), in spite of their high cost, easily in the 10^7 USD/Euro range. Multiplexes provided by these specialized species are greater by at the very least a factor 20 than those of their venerable ancestor (the long-slit spectrograph), provided they are used well inside their respective "ecological" niches. On top of this, getting the best possible light efficiency and detector sensitivity is mandatory, again in spite of significant cost increase.

3D History

As pointed out above, astronomical spectroscopy was for a long time at best a 2D affair, with the overwhelming majority of observed data coming from the ubiquitous long-slit spectrograph, the dominant species of that time. On this instrument, the 2D detector format is filled by 1D spatial data along the narrow (typically 2–4 detector pixels wide) slit width and 1D spectral data over a large spectral range (typically 1 octave, that is, a factor 2 in wavelength) in the orthogonal direction. The long-slit spectrograph was then a natural match to the available dispersers (a prism or a ruled grating) and the large 2D (photographic plate) detector. Its simple "Cartesian" spatiospectral data format was relatively easy to extract, a crucial point in this nondigital era, and it gave physical information on any type of target, small or extended, with absorption lines, emission lines, or both. This is illustrated in Figure 3, which shows an early (1908) galaxy spectrum [1].

During the same period, a significantly lower life form, the Fabry–Pérot interferometer, was used by a number of teams worldwide. It was also a 2D only

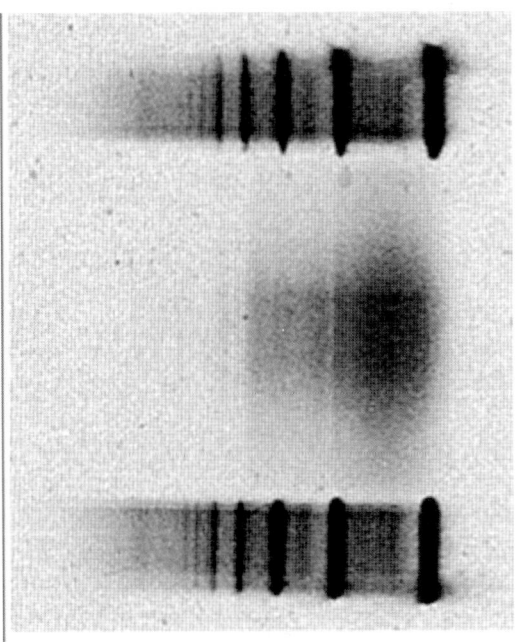

Figure 3 Long-slit spectrum of our nearby galaxy cousin, the Andromeda Nebula (Messier 31). The slit is vertical, with spectral dispersion in the horizontal direction. *Central part*: galaxy spectrum with strong calcium absorption lines, coming from the integrated light of a vast number of stars. *Upper and lower part*: comparison emission lines spectrum of a gas discharge hydrogen lamp, used to precisely derive the nebula absorption line wavelengths. This 8.8 h-long observation was done by E. Fath in 1908 on the 38 in. diameter Lick telescope. (Credit Lick Observatory, USA.)

Figure 4 Fabry–Pérot rings illuminated by the 372.7 nm singly ionized oxygen doublet, projected on an image of the Orion Nebula, a nearby (1350 light-years away) gas cloud ionized by hot blue stars. Ring radii, compared to those given by a laboratory gas discharge lamp, directly gave the gas radial velocity. This 40 min integration was performed by Fabry et al. in 1914 at the 80 cm telescope at Marseilles Observatory. Credit Observatoire de Marseille.

system, covering a full 2D spatial field at the cost of getting only one wavelength on each spatial pixel. As such, its use was strictly restricted to extended emission line regions. Its first astronomical result in 1914 on the Orion nebula [2] is shown in Figure 4.

Starting in the mid-1970s, this sedate phase was replaced over the next two decades or so by a Cambrian-type evolution explosion, with a host of new phyla, adapting the spectrographic data formats to the spatiospectral shapes of different astronomical targets. True to its namesake Earth's geological era, a large fraction of these new instrumental life forms quickly vanished, but three robust 3-D phyla remain in use today and actually have all but outlived their two venerable 2D ancestors (Figure 5).

The first new successful data reshuffling approach, cross-dispersed echelle spectroscopy, is actually still a 2D mode, and thus formally lies outside the scope of this book. It covers the need for getting the spectrum of a point-like object – typically a star or a quasar (the nucleus of an ultra-active galaxy) – over a wide spectral band (about 1 octave) with a high to very high spectral resolution (3.10^4 to 2.10^5). This requires at least 10^5 spectral bins or, for a classical slit spectrograph, an awkward 3-m long spectrum for typical 15 μm detector pixels. To solve that conundrum, a high-order echelle grating gives tens of overlapping

Figure 5 Schematic three-dimensional (sky coordinates α, δ; wavelength λ) illustration of the current three main 3D phyla spatiospectral coverage. These are multiobject spectrography (MOS, Section 2.1), scanning Fabry–Perot spectrography (SFPS, Section 3.3.3) or Fourier transform spectrography (Section 3.4), and finally integral field spectrography (IFS, Section 4.1). Note the very different shapes of the data "cubes" delivered by the instruments.

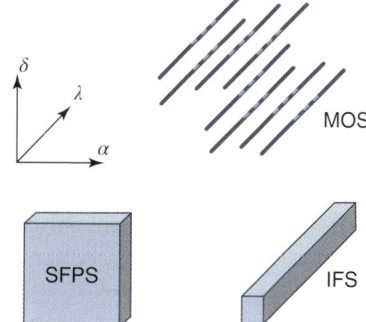

Figure 6 First 1971 echelle spectrum (order 67 to 57 from top to bottom), obtained at the Pine Bluff 36″ Telescope on the very bright star Capella. It roughly covers the 360–430 nm UV-violet spectral range. Wavelength increases from left to right and from top to bottom. Many absorption lines, coming in particular from Ca, K, and Fe atoms, can be seen. Credit University of Wisconsin, USA.

high-resolution spectra that are separated in the orthogonal direction by low-resolution cross-dispersion with a prism or a grating. In that way, the full spectrum is cut into about 20–40 pieces stacked on the 2D detector. The resulting spectral format neatly covers a squarish detector with a hundred thousand contiguous wavelength bins or more. See Figure 6 for the first use of this approach for astronomical purposes by Schroeder and Anderson [3]. This phylum is by now highly successful in its own ecological pasture.

The first true 3D system, the scanning Fabry–Pérot spectrograph concept, a brainchild of Brent Tully, had first light in 1974 [4], as shown in Section 3.3.3, Figure 3.3. It works exactly like its 2D ancestor with still only one wavelength bin per spatial pixel for any given exposure. However, now a small format (typically 15–40 elements) spectrum is built at each spatial pixel by moving the interferometer's optical cavity by a small fraction of wavelength from one exposure to the next. This approach proved efficient, but still mainly limited for the study of emission line regions in relatively closeby galaxies. Together with its more complex but also more versatile Fourier transform spectrograph cousin fathered by Maillard [5] in 1992, this approach remains as of today a relatively narrow niche capability.

A typical sky image – as in Figure 1 – is sparsely filled by small (typically less than 1″ across) ultra-faint galaxies. Obtaining their spectra one at a time (two by orienting adequately the spectrograph slit) with a classical long-slit spectrograph would take forever. Hence, the spectacular success of the so-called multiobject spectroscopy (MOS), with the spectra of many (thousands at last count) objects in a relatively large patrol field (many arc minutes across) obtained in a single observation. There are two variants of this often dubbed 2 1/2-D technique: (i) the multifiber mode, with individual optical fibers put on the targets and reformatted in a long-slit to feed a classical long-slit spectrograph, and (ii) the multislit mode, with many short slits put on as many targets and feeding directly the spectrograph. See Figures 7 and 8 for their respective first lights

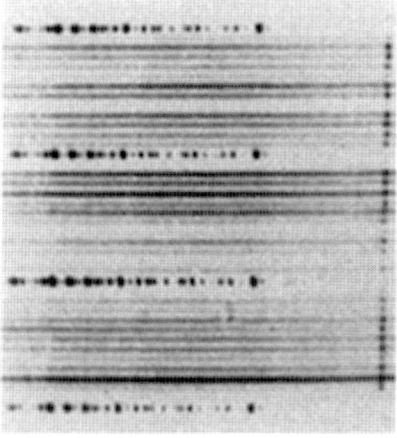

Figure 7 First multifiber data in the optical range within the galaxy cluster Abell 754 by Hill *et al.* in 1979 at the Steward Observatory 2.3 m telescope. Spectral dispersion is horizontal, with 37 fibers distributed along the vertical axis. Of these, 32 fibers were put on the sky objects in the 0.5° patrol field, and 5 were locally illuminated by a calibration HeAr gas discharge lamp. (Hill *et al.* [6]. Reproduced with permission of SPIE.)

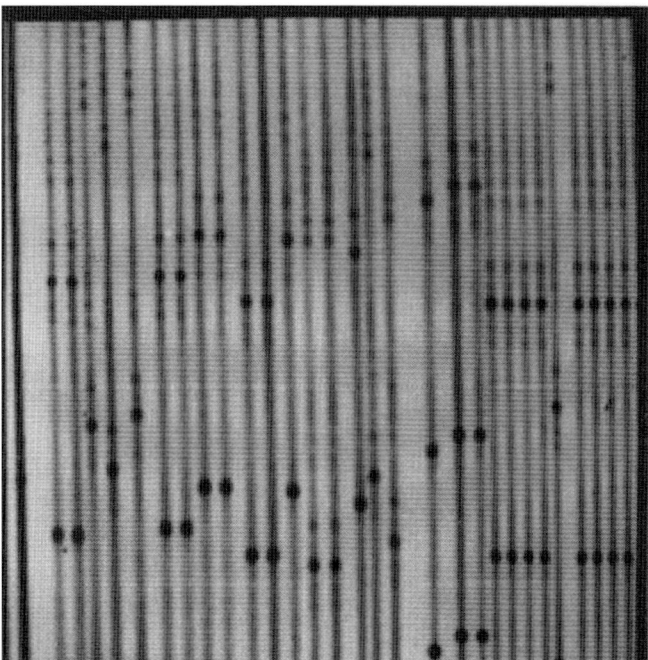

Figure 8 First multislit (actually multiholes) data in the optical range within the galaxy cluster 0054+1654 by Butcher *et al.* in 1982 on the Mayall 4-m telescope at Kitt Peak National Observatory. The spectra cover the 500–900 nm range with a spectral resolution ~ 300. The picture shows a portion of the full data set. Spectral dispersion is vertical. The bright spectrum on the left is of an alignment star. (Butcher *et al.* [7]. Reproduced with permission of SPIE.)

by Hill *et al.* [6] and Butcher *et al.* [7]. The success of that approach has been phenomenal, with this phylum's ubiquitous presence at all large telescopes, plus a number of standalone multiobject facilities on moderate size (2–4 m) telescopes in order to conduct multiyear surveys of up to a million or more objects.

The opposite case of full spatiospectral coverage of a small field (arc seconds across) densely covered by a compact object of any kind (solar system objects, star clusters, planetary nebulae, galaxy nuclei, distant galaxies, …) in a single exposure was solved around 1980 with the so-called integral field spectrographs (IFS), fathered by G. Courtès and C. Vanderriest at the concept level. The data cube

Figure 9 The 1980 first light of a fiber-based integral field spectrograph demonstrator at the Mauna Kea 2.2 m UH telescope by C. Vanderriest [11] on the active galaxy 3C 120 nucleus. The hexagonal science field is paved by 169 fibers, with four additional rows of 9 fibers each to monitor an empty sky region. Each 0.8″ diameter fiber gives a spectrum over the 390–690 nm spectrograph range with ∼ 330 spectral resolution. (a) Fiber-based image dissector input. (b) A few successive spectra from a 20-min exposure on the object, including one spectrum (bottom) from an empty sky location. (Vanderriest [11]. Reproduced with permission of IOP.)

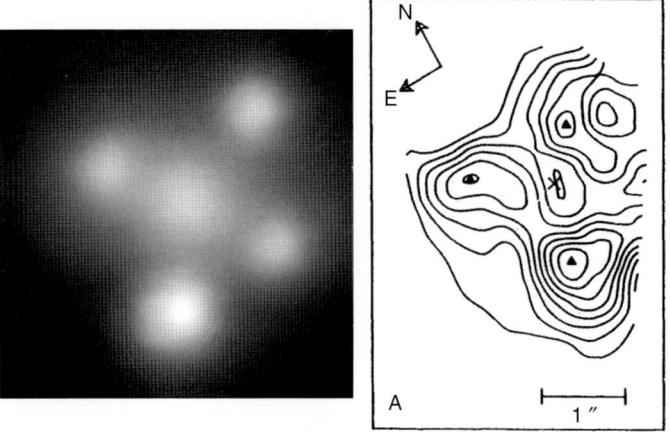

Figure 10 First light of the lenslet-based TIGER integral field spectrograph by Adam *et al.* 1989 [8]. The 5″ × 5″ field is paved by 70 lenses, giving 70 spectra in the 370–689 nm range with ∼ 1000 spectral resolution. (Left) Direct visible image of the Einstein Cross, the four-element gravitational mirage of a distant quasar by the nearby galaxy located at the center. (Right) Reconstituted image from the 70 spectra in the redshifted 190.9 nm doubly ionized carbon emission line, showing three of the four mirages. Credit Observatoire de Lyon, France. (Reproduced with permission of Astronomy & Astrophysics.)

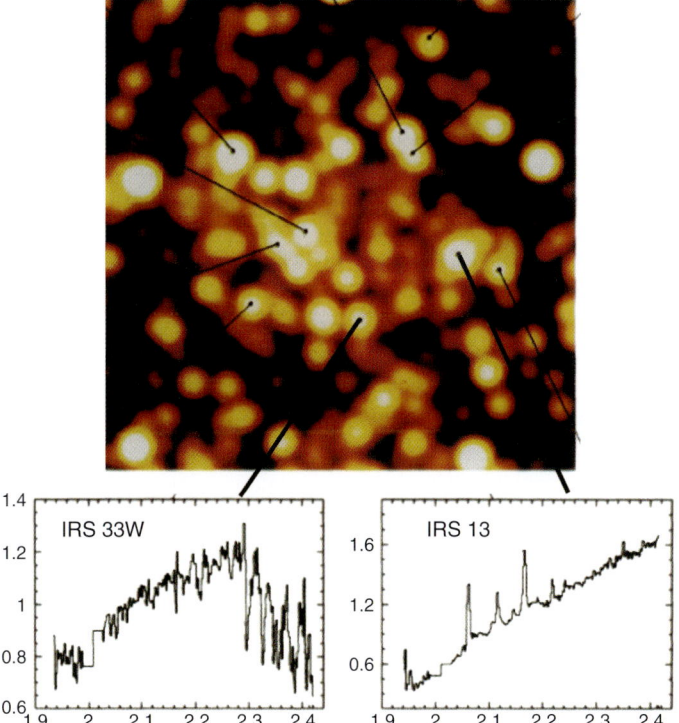

Figure 11 First light of the SPIFFI slicer-based integral field spectrograph by Krabbe *et al.* [12] at the 2.2 m La Silla telescope. The 8″ × 8″ field of view is sliced in 256 slits, each giving K-band spectra at ~ 1000 spectral resolution. Central square: Reconstituted K-band (1.85–2.4 µm) image of the central star cluster in our Galaxy. Surrounding boxes: Extracted K-band spectra of stars, orbiting around our Galaxy 3.5 million solar masscentral black hole. (Krabbe [12]. Reproduced with permission of IOP.)

produced by an IFS in a single exposure consists of one spectrum, typically of one octave width, for every spatial pixel in the squarish small field of view. Three operational variants were eventually developed, the 1989 lenslet-based TIGER IFS [8], the 1990 fiber-based SILFID IFS [9], and finally the 1993 slicer-based SPIFFI IFS [10]. See Figure 9 for first result of the 1980 fiber-based IFS demonstrator and Figures 10 and 11 for the first scientific use of respectively the lenslet-based and the slicer-based variants. After a slow start, this last phylum is now thriving, with a strong presence at most large telescopes, and an even larger one foreseen aboard the next generation of extremely large telescopes currently being built.

3D Technology

Successive generations of astronomical observational tools – telescopes and their instrument suites–always critically required access to some of the latest cutting-edge technologies of their time, sometimes beyond what was then available. Yet, corresponding financial resources always fell much beyond the kind of 'astronomical' size that would have been needed to develop such technologies from scratch. The remarkable progress in the observing power of observational techniques since Galileo's time (a factor 10^{12}–10^{15}) came mainly from co-opting technological advances made for the two "real" markets in our western-style society, the military domain and mass production for the layman. This was already the case at the very first stage of telescope-aided observation: Galileo's 1609 refracting telescope was made possible by the thriving development of mass-produced lenses to correct people's impaired vision, and the significant cost of building the first telescopes justified by their military potential for combat area surveillance, or more mundanely expressed, detection, at a larger distance than from the unaided eye, of a column of soldiers moving to invade your beloved Italian City State.

In a similar vein, in the last quarter of the twentieth century, the phenomenal observing gains offered by brand new two-dimensional digital detectors stemmed from their development in the optical region for the consumer market and in the infra-red region for military purposes. 3D spectrography has neatly surfed in the last four decades on market availability of its key components. That includes not only the crucial large 2D digital detectors and powerful computers, but also exotic glasses (developed for integrated circuit manufacturing), diamond-turned lens surfacing (surveillance aircrafts cameras), volume phase holographic gratings (jet fighters head-up displays), laser cutters (jet plane parts manufacturing), and high-efficiency light-transmitting fibers (telecommunication). In most cases, significant tinkering has been needed to adapt these technologies to their purported astronomical applications. There has also been some useful return, as in turn these astronomy-specific developments found new "useful" applications: one example is the development 30 years ago of zero expansion glasses for the minuscule market of large telescope mirrors, which soon after has been redirected toward the huge market of cooking induction tops.

Part I

3D Instrumentation

1

The Spectroscopic Toolbox

1.1 Introduction

Present spectroscopic instruments use essentially bulk optics, that is, with sizes much greater than the wavelengths at which the instruments are used (0.3–2.4 µm in this book). Diffraction effects – due to the finite size of light wavelength – are then usually negligible and light propagation follows the simple precepts of geometrical optics. A special case is the disperser or interferometer that provides the spectral information: these are also large size optical devices (say 5 cm to 1 m across), but which exhibit periodic structures commensurate with the working wavelengths. The next subsection gives a short reminder of geometrical optics formulae that dictate how light beams propagate in bulk optical systems. It is followed by an introduction of a fundamental global invariant (the optical etendue) that governs the 4D geometrical extent of the light beams that any kind of optical system can accept: it is particularly useful to derive what an instrument can – and cannot – offer in terms of 3D coverage (2D of space and 1 of wavelength).

1.1.1 Geometrical Optics #101

As a short reminder of geometrical optics, that is, again the rules that apply to light propagation when all optical elements (lenses, mirrors, stops) have no features at scales comparable or smaller than light wavelength are listed:

1) Light beams propagate in straight lines in any homogeneous medium (spatially constant index of refraction n).
2) When a beam of light crosses from a dielectric medium of index of refraction n_1 (e.g., air with n close to 1) to another dielectric of refractive index n_2 (e.g., an optical glass with refractive index roughly in the 1.5–1.75 range), part of the beam is transmitted (refracted) and the remainder is reflected. The normal to the surface, the input ray, and the reflected and transmitted rays are in the same plane, called the incidence plane. For a ray at incidence angle (the angle between the ray and the normal to the surface) θ_i, the transmitted angle θ_t and the reflected angle θ_r are given by the very simple formulae: $\theta_r = \theta_i$ and $n_2 \sin \theta_t = n_1 \sin \theta_i$ (see Figure 1.1).

As indexes of refraction vary with wavelength, a multi-wavelength single light ray is transmitted as a multicolored fan. Note that the transmission

Optical 3D-Spectroscopy for Astronomy, First Edition. Roland Bacon and Guy Monnet.
© 2017 Wiley-VCH Verlag GmbH & Co. KGaA. Published 2017 by Wiley-VCH Verlag GmbH & Co. KGaA.

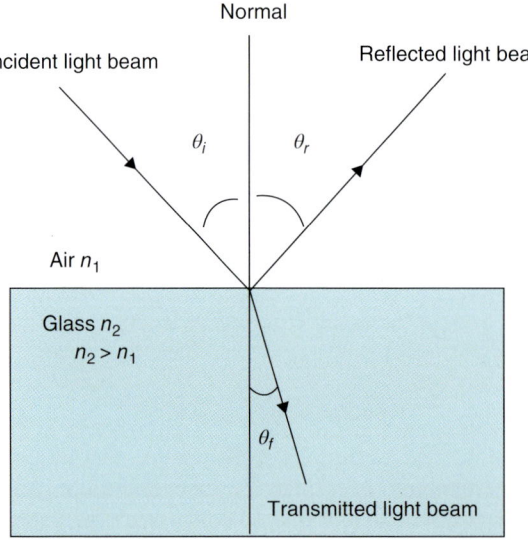

Figure 1.1 Light propagation at an air–glass interface.

formula above does not give a real θ_t value when $(n_1/n_2) \sin \theta_i$ is greater than 1. Hence, for an incidence angle greater than $\arcsin(n_2/n_1)$, there is no transmitted light and the beam undergoes total reflection inside the high refractive index medium n_1. This is a useful trick when applicable, since this is the only way to get reflection of a beam of light with 100% efficiency, provided all rays have incidence angles greater than the critical value and the glass surface is superclean.

To extend the above formulas to mirrors in an index of refraction n_1 medium, one just uses n_1 before the mirror and $n_2 = -n_1$ after (1 and -1 when the mirror is in vacuum).

3) Light is actually an electromagnetic wave that carries two orthogonal so-called polarization states, the p-state with the electric field parallel to the incidence plane and the s-state with the electric field perpendicular to the incidence plane. The laws of geometrical reflection and transmission of light are exactly the same for both polarizations, except when using the few so-called anisotropic crystals. On the other hand, the reflection coefficients R and the transmission coefficients T at the interface between two dielectrics are different for the p and the s components except for normal incidence, that is, for $\theta_i = \theta_t = 0$. They are given by the Fresnel equations:

$$R_p = \frac{(n_1 \cos \theta_t - n_2 \cos \theta_i)^2}{(n_1 \cos \theta_t + n_2 \cos \theta_i)^2} \quad R_s = \frac{(n_1 \cos \theta_i - n_2 \cos \theta_t)^2}{(n_1 \cos \theta_i + n_2 \cos \theta_t)^2} \quad (1.1)$$

From energy conservation, the transmission coefficients are $T_p = 1 - R_p$ and $T_s = 1 - R_s$.

Nearly unpolarized input light, that is, with an equal mix of p and s states is actually the most common case for artificial light sources, with the notable exception of many lasers. This is true also for most natural astrophysical sources with a few exceptions (active galactic nuclei in particular). In the unpolarized case, the reflection coefficient is the mean value of R_s and R_p. For common optical glasses and small incident angles, this gives an about 4% light loss (percentage of reflected light) when crossing from air to glass. Many IR glasses or crystals, however, have indexes of refraction as high as 2.5, giving much higher reflection losses (~ 18%). For reasonable angles, light beams inside spectrographs remain largely unpolarized, except when a high blaze angle grating is used as seen in Section 1.4.5, unless the instrument is dedicated to spectropolarimetric investigations, using its own internal polarization device to separate the p and s beams.

It is easy to see that R_s is never equal to zero; on the other hand, $R_p = 0$ at the so-called Brewster incidence angle θ_B given by $\tan\theta_B = n_2/n_1$. For $n_1 = 1$ (air) and $n_2 = 1.5$ (typical low index glass), this gives $\theta_B = 56°$. Light rays striking a glass at Brewster incidence angle are thus fully s-polarized. Finally, at the critical incidence angle $\arcsin(n_2/n_1)$, with $\cos\theta_t = 0$, Fresnel equations give $R_p = R_s = 1$, indicating indeed total reflection of the rays for the two polarizations.

1.1.2 Etendue Conservation

Let us remind first that the solid angle of a cone of any shape is the area it subtends on a sphere of unit radius; it is thus a dimensionless quantity. In particular, the solid angle of a circular cone of light with half apex angle α is $\Omega = 2\pi(1 - \cos\alpha)$. For small values of α, this gives approximately $\Omega = \pi\alpha^2$.

The etendue or optical throughput expresses quantitatively how much a beam of light is spread out in area and solid angle. Taking an infinitesimal surface element dS immersed in a medium with refractive index n, emitting light inside an infinitesimal solid angle $d\Omega$ and at an angle θ from the normal to the surface, the resulting etendue is $d^2E = n^2 dS \cos\theta \, d\Omega$ (see Figure 1.2). Solid angles being dimensionless quantities, the etendue has the dimension of area. For a full light beam, it is obtained by integrating d^2E over area and solid angle, giving

$$E = n^2 S\Omega \tag{1.2}$$

In particular, for a light cone of half-angle α orthogonal to the surface S, we get

$$E = n^2 S \int_0^\alpha \cos\theta \, d\Omega = n^2 \pi S \sin^2\alpha \tag{1.3}$$

This computation can be carried out in principle at any location along the optical path; in practice, for imaging systems, it is usually done either at the level of the light source itself (or any of its image) or at the level of the pupil (or any of its image). Seen from the source of light (e.g., from the telescope focal plane for astronomical purposes), this is essentially the product of the sky field area by

Figure 1.2 Visualization of the infinitesimal etendue component $d^2E = n^2 dS \cos\theta \, d\Omega$.

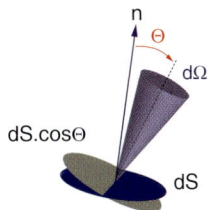

the solid angle subtended by the pupil (telescope primary mirror). Seen instead from the pupil, this is as well the product of the pupil (telescope mirror) area by the solid angle subtended by the sky field. Note that the light flux Φ carried by a beam of radiance L and etendue $n^2 S\Omega$ is $\Phi = LS\Omega$: consequently, while it is often useful to consider optical systems illuminated by a point light source ($S = 0$) or a parallel beam ($\Omega = 0$), both have no physical meaning as they carry zero energy. Actually, there is a minimum beam etendue that is set by finite light (wave)length λ. For a circular source of diameter d in air, the minimum beam half-angle set by diffraction is $\alpha \sim \lambda/d$, or a minimum etendue:

$$E = \left(\pi^2 \lambda^2\right)/4 \qquad (1.4)$$

Note that this is the etendue of a diameter λ disk uniformly emitting in half-space.

The etendue concept is a fundamental and highly useful tool, because as a beam propagates inside any optical system, its etendue never decreases, being at best constant, the so-called $S\Omega$ conservation. The crucial point here is "any optical system": the light beam can be, for example, transmitted through a bundle of tapered (conical) optical fibers, sliced with multi-mirrors and then recombined; in fact it can go through any imaging and non-imaging combination you care to consider, and still at best its original etendue is conserved. For an extreme example, launch a low-etendue beam from a He–Ne laser and put a diffuser on the beam trajectory: the etendue can easily increase by a factor of 10^6 or more as almost fully collimated laser light is diffused almost uniformly over a whole half-space (a 2π solid angle). On the other hand, for imaging systems with small optical aberrations both in the source and pupil planes, the etendue computed either from pupil images or from source images is the same and nearly conserved, easily by better than one part in one thousand, as the light beams propagate inside the optical system: following the etendue along the light path, ultimately down to the detector plane, is thus a simple and powerful way to size up the optical components and the detector.

Derivation of this fundamental invariance from the two principles of thermodynamics is quite straightforward from the following thought experiment: A blackbody source of area S and radiance L is immersed in a medium of index of refraction n and emits light in a solid angle Ω, as per Figure 1.3, upper part. Light goes through a non-absorbing (perfect light transmission) arbitrary optical system and emerges through a surface S' immersed in a medium of index of refraction n', with a solid angle Ω' and radiance L'. The total flux

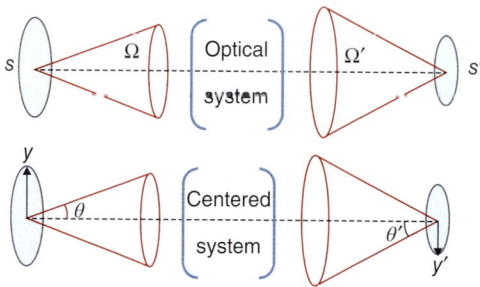

Figure 1.3 Schematic illustration of the 2D etendue conservation $S\Omega = S'\Omega'$ for any optical system and the 1D etendue conservation $y \sin \theta = y' \sin \theta'$ for any centered system.

emitted by the source is $\Phi = LS\Omega$. The total flux collected at the output is $\Phi' = L'S'\Omega'$. From the first principle of thermodynamics (flux conservation), $L'S'\Omega' = LS\Omega$. From the second principle of thermodynamics (non-decreasing entropy), $L'/n'^2 < L/n^2$; otherwise, for example, a thermocouple connecting S and S' would give an electric current with only one source of heat (the blackbody source), in clear violation of the second law. Finally, for any optics $n'^2 S'\Omega' > n^2 S\Omega$. Q.E.D.

Most optical imaging systems actually use centered optics, that is, with all powered (non-flat) optical surfaces of lenses and mirrors having the centers of curvature aligned along a common axis, called the optical axis. Often, all surfaces are on top rotationally symmetric along this axis, but this is not the case when, for example, astigmatic or toroidal lenses and/or mirrors are used, as for prescription glasses used to correct eye's astigmatism. For such centered optical systems and negligible aberrations in the pupil and field images, the etendue conservation actually works in two dimensions in any plane section containing the optical axis, as established below. This can be derived from Fermat's principle, namely, that light follows trajectories for which the optical path $\int n \, dl$ is an extremum: the end result is that for a small 1D source of half-length y perpendicular to the optical axis emitting light in a cone of half apex angle θ (which, on the other hand, can be very large), and any centered optical system with small aberrations, the image of the source has a half length y' and emits light in a cone of half angle θ' with $ny \sin \theta = n'y' \sin \theta'$ (the so-called Abbe's sine condition). Given that the general 2-D etendue conservation gives in that case $n^2 y^2 \sin^2 \theta = n'^2 y'^2 \sin^2 \theta'$, one sees that for centered systems, there is, in addition, conservation of the 1-D "linear" etendue $y \sin \theta$ in any section along the optical axis (see Figure 1.3, lower part).

OPTICAL ETENDUE #101 TOOLBOX

Optical medium refraction index n
Source circular area $S = \pi y^2$
Light cone solid angle Ω, half-angle θ.

Follow beam propagation over each source/pupil image

- For any optical system: $n^2 S\Omega$ (at best) invariant
- For any centered optical system: $ny \sin \theta$ (at best) invariant

---

One telling illustration of etendue conservation concerns optical fibers (Figure 1.4). They are commercially available as almost unlimited length cables with three cylindrical components from center to edge: a high-index glass core of up to a few 100 µm diameter, a lower index glass cladding, and a protecting envelope. For a beam angle smaller than the fiber critical angle, θ_M, light injected in the core is trapped by total reflection and propagates to the other end, with essentially zero energy loss from the many reflections at the core-cladding interface. In practice, this angle limitation is expressed by the fiber maximum

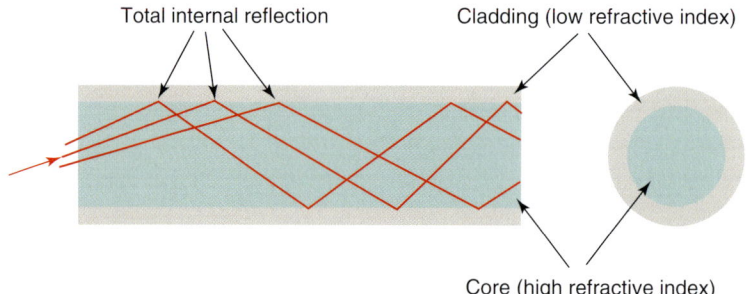

Figure 1.4 Principle of the optical fibers. Light entering the fiber core is trapped by total reflection at the core-cladding interface and propagates to the fiber end.

numerical aperture (NA) $n_1 \sin \theta_M$. It is easy to show that NA $= \sqrt{n_1^2 - n_2^2}$, where n_1 is the core refraction index and n_2 the cladding refraction index. A typical value is NA ~ 0.22, corresponding to an acceptance light cone of half-angle 12.7° in air. For astronomical applications, fiber length is relatively short, 50 m at most, and the fiber's internal light transmission is extremely good, from roughly 0.4 to 1.7 µm.

For fiber core diameters greater than ~ 10 µm, geometrical optics applies, and owing to etendue conservation, beam linear etendue $n \sin \theta$ is in principle perfectly conserved as the beam propagates through and exits the fiber. In real life, it increases in case of even low fiber stress due to cable handling, and/or even gentle bending applied to carry the light beams to their required location. One typical application uses hundreds of 30-m long fibers to pick astronomical objects at the moving prime focus of a telescope some 15 m up and carry the light to a number of spectrographs conveniently located on the floor. A rule of thumb is then a $\sim 15\%$ linear etendue degradation for an input beam close to maximum acceptance angle, and more for smaller angles.

1.2 Basic Spectroscopic Principles

1.2.1 The Spectroscopic Case

In the IR-optical domain covered in this book, individual photon frequencies ν are much too high (3×10^{14} Hz at 1 µm wavelength) for present technology to be directly measured with a coherent detector, as routinely done in the radio to far infrared domain. A separate coherent device is thus required to sort out the photons according to their frequency before they are sent to a non-coherent 2D detector. The detector then registers the total number of photoelectrons generated at each of its pixels during the exposure. The main figure of merit of such a spectrographic instrument is its spectral resolution $\mathcal{R} = \lambda/\delta\lambda$, where $\lambda = 1/\nu$ is the wavelength and $\delta\lambda$ is the smallest wavelength variation that can be detected by the instrument.

In practice, this sorting out can be done either by using a filter or by using a disperser. As the name implies, a filter lets out only one wavelength slice at a time; to get a number of wavelength bins, it is thus necessary to use a filter whose bandpass can be shifted at will (not a simple endeavor though) and make successive exposures. As the name also implies, a disperser (a grating or a prism) receives, say, a parallel beam of light and sends back dispersed parallel beams (i.e., with different inclinations for different wavelengths), which are imaged on the detector. In the astronomical domain, exchangeable interference filters coupled to an imager are widely used for multi-wavelength imaging with spectral resolutions of at most ~ 50. On the other hand, dispersers are by far the most common device used for bona fide spectrography, loosely defined as delivering a minimum spectral resolution of ~ 300.

Irrespective of their design, spectral properties of spectrographic instruments are characterized by a set of three generic values, their central wavelength λ_c, spectral range $\Delta\lambda$, and resolved spectral width $\delta\lambda$. This set gives two unitless parameters defining the instrument spectral grasp, namely, its free spectral range $R_c = \lambda_c/\Delta\lambda$ and its mean spectral resolution $\mathfrak{R}_c = \lambda_c/\delta\lambda$. The wavelength domain covered by large 'optical' telescopes (as opposed to radio-telescopes), that is, a whopping 0.3–24 µm range, is usually split in four domains: the so-called optical domain (0.3–0.95 µm); the near-IR (0.95–2.4 µm); the thermal IR (2.4–7 µm); and the medium IR (7–24 µm). They correspond to quite different instrument technologies and even science goals, with most ground-based astronomical observations performed in the first two spectral regions. In terms of spectral resolution, there are essentially four regimes: low spectral resolution (500–1500) for large surveys of distant galaxies; medium resolution (3,000 – 6,000) for most galaxy studies, high resolution (15,000–30,000) for precise radial velocities and/or abundance studies of individual stars or ionized gas regions, and very high spectral resolution (>100,000) for ultra-precise abundance determination in stars or in the interstellar/intergalactic medium, plus search for exoplanets. The 3D line of work explored in this book is mainly concerned with the first two regimes, that is, low and medium spectral resolution.

It is essentially impossible to build a single spectrograph that could cover efficiently the full 0.3–2.4 µm optical to near-infrared range, and most spectrographs are in fact limited to one octave at best, that is, a factor of 2 in wavelength breadth. This corresponds to a maximum free spectral range $R_c = 1.5$. Nevertheless, to cover the full optical-near infrared spectral range simultaneously, one can build, for example, a three-arm instrument with a combination of two dichroic beamsplitters[1] sending three selected spectral windows of manageable widths (e.g., 0.3–0.5 µm, 0.5–1 µm, 1–2 µm) to three optimized spectrographs; one example is the X-shooter at the European Southern Observatory (see the corresponding ESO web pages). One advantage of that multiarm approach is that short-lived phenomenas, such as γ-ray burst remnants (resulting from one of the most powerful known explosions in the Universe), can be identified over this wide spectral range in, say, a single 30-min exposure, before fading below detectivity limit.

---

1 A plane parallel plate with a complex set of dielectric coatings on one surface, which reflects the short wavelengths and transmits the longer ones. It is usually put at a large angle – for example, 45° – with respect to the optical beam to separate the reflected and transmitted components.

1.3 Scanning Filters

1.3.1 Introduction

The most commonly used scanning filter in astronomy is the Fabry–Pérot interferometer. This is a resonant cavity made of two parallel plane glass plates facing each other and with highly reflective ($R \geq 90\%$) internal surfaces. Cavity spacing between the two plates is usually between a few hundred microns and a few millimeters for astrophysical applications. The two outer surfaces of the plates are antireflection coated and with a $\sim 0.5°$ wedge to prevent troublesome artifacts. As shown below, the depth of the cavity must be controlled to a very small fraction, easily 1% of the light wavelength λ, by no means a trivial endeavor at optical to near-IR wavelengths. Plates are usually made of fused silica to take advantage of its very low thermal expansion coefficient and ability to be accurately figured and exquisitely polished. Very high reflectivity coatings with low absorption are obtained in the optical region above ~ 450 nm by vacuum deposition of alternative high and low refractive index interference layers on the plates' inner surfaces; a thin gold layer is used instead above ~ 1000 nm.

A light ray of wavelength λ entering the Fabry–Pérot (FP) cavity – also called etalon – at an angle θ with respect to the normal to the cavity undergoes multiple reflections inside its cavity of optical length ne (cavity depth e & refractive index n), as shown in Figure 1.5. Parallel rays exiting the etalon interfere with each other. In the idealized case of perfect parallel plates with 100% reflectivity, that interference is perfectly constructive when the optical path difference between two successive rays is any multiple of the wavelength. This gives the canonical FP phasing equation:

$$2ne \cos \theta = p\lambda \tag{1.5}$$

Here p, the etalon order, is a positive integer, usually in the 200–2000 range. For such rays, 100% of the light is transmitted, while in all other cases the rays are reflected back to the light source. Transmitted light versus wavelength for any fixed angle θ is then a Dirac comb function with zero width peaks separated by the (slowly varying) etalon free spectral range $\Delta \lambda = \lambda/p$. In most cases, one needs to get only one wavelength peak, and the many others transmitted by the cavity need to be filtered out.

When illuminated uniformly by a monochromatic source, transmitted light appears as a series of concentric rings at infinity (Figure 1.6) of angular radii θ_p given by the phasing equation. Different wavelengths give ring patterns of different sizes; this is the basis of the two most common astronomical filters, the

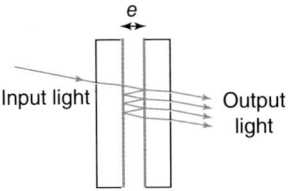

Figure 1.5 Principle of the Fabry–Pérot interferometer. Light rays are trapped by multiple reflections inside a cavity of depth e. The cavity acts as a spectral filter as only rays undergoing wavelength-dependent constructive interference are transmitted; all the others being reflected by the etalon.

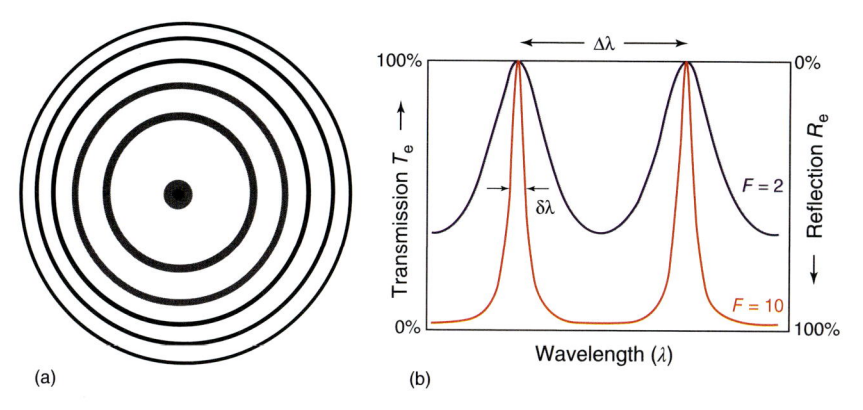

Figure 1.6 (a) Set of k rings ($k = 0$ to 5) from monochromatic light uniformly illuminating a Fabry–Pérot etalon, with exact phasing at the center ($k = 0$). The successive ring radii obey the canonical equation $2ne \cos \theta_p = p\lambda$, which for small angles θ_p translates into ring radii $r_k \propto \sqrt{k}$. With rings FWHM $\propto 1/\sqrt{k}$, rings etendues are all the same. (b) Transmission function for two etalons of respective finesses 2 (blue) and 10 (red). (Credit DrBob, Wikipedia.)

non-scanning interference filter with a fixed optical depth cavity and the scanning Fabry–Pérot interferometer with a variable depth cavity.

In real life, however, interferences are not perfectly constructive and the crucial etalon quality factor is its finesse N, the optical equivalent of the resonant cavity quality factor Q in the radio domain. This is the unitless ratio between the transmission peaks separation (aka the free spectral range) and their full width at half maximum (FWHM). At the cavity level, N can be seen as the effective number of parallel exit rays coming from a single entrance ray and interfering with each other. At the data set level, this is the number of independent spectral bins that can be distinguished. Ideally, N would be set solely by the reflectance factor R of the plates, with $N_r \sim \pi\sqrt{R}/(1-R)$. In that case, normalized transmitted light intensity T versus wavelength λ has an Airy shape (Figure 1.6), with

$$T_\lambda = \frac{1}{1 + N_r^2 \sin^2 \pi \frac{(\lambda - \lambda_0)}{\Delta\lambda}} \tag{1.6}$$

In this formula, $\Delta\lambda$ is the free spectral range and λ_0 the peak wavelength.

Moreover, reflective coatings slightly absorb light (absorption factor A), and the normalized peak transmission is not 1 any more, but $\sim (1 - N_r A)$ instead. For reasonable values of the reflective finesse ($N_r \leq 50$), this gives a few percentage light loss above ~ 480 nm. Moreover, the effective finesse N is always smaller than N_r, owing to the cavity residual optical defects: this gives a small range for e in the canonical equation above, and hence again non-perfect constructive interference. It is also necessary to allow for a finite range of θ to fall on any detector pixel, since a beam with a zero width θ angle would have zero etendue, and hence would carry a zero photon flux. This again spoils the light beam's constructive interference and lowers N.

The end result is an overall finesse N and a global transmission τ, with a comb-like spectral transmission curve $T \sim \tau/[1 + N^2 \sin^2 \pi(\lambda - \lambda_0)/\Delta\lambda]$.

$\lambda - \lambda_0 = 0$ corresponds to the transmission peak $T = \tau$; $\lambda - \lambda_0 = 1/N$ to a halved transmission $T = \tau/2$; $\lambda - \lambda_0 = \Delta\lambda/N$ to the minimum transmission $T = \tau/(1 + N^2)$.

The insert below gives quantitative estimates of the overall finesse, spectral resolution, and transmission of a Fabry–Pérot etalon.

FABRY–PÉROT ETALON COOKING BOOK

Plates reflectivity R, absorption A; cavity depth e & r.m.s. defects d
Beam incidence angle θ & radial angular width $\delta\theta$
Cavity order $p = 2e \cos\theta/\lambda_0$, free spectral range $\Delta\lambda = \lambda_0/p$

Reflective finesse $N_r \sim \pi\sqrt{R}/(1-R)$; efficiency $\tau_r \sim 1 - N_r A$
Defect finesse $N_d \sim \lambda/2d$; efficiency $\tau_d \sim (N_d/N_r) \arctan(N_r/N_d)$
Beam finesse $N_\theta \sim 1/p\theta\,\delta\theta$; efficiency $\tau_\theta \sim (N_\theta/N_r) \arctan(N_r/N_\theta)$
$[\theta = 0 \Rightarrow N_\theta \sim 2/p\,(\delta\theta)^2]$

Overall finesse N with: $1/N^2 \sim 1/N_r^2 + 1/N_d^2 + 1/N_\theta^2$

Spectral resolution $\mathfrak{R} = pN$; efficiency $\tau = \tau_r \times \tau_d \times \tau_\theta$

Transmission $T_\lambda \sim \tau/[1 + N^2 \sin^2\pi(\lambda - \lambda_0)/\Delta\lambda]$

To get a feel of what this threatening formulae mean, let us take a generic order p etalon with the reflective finesse of plate coatings N_r, and hence with a potential spectral resolution $\mathfrak{R}_r = pN_r$. It is always a good idea to get the cavity root mean square defects (due to plates figuring, polishing, and parallelism errors) as small as possible, with at least $N_d = 2N_r$. Similarly, the working field and the resolved angular size (set up by the detector pixel size) should be small enough to get at least $N_\theta = 2N_r$. With those minimum values, the insert formulas give $\mathfrak{R} \sim 0.82\,\mathfrak{R}_r$ and $\tau \sim 0.85\,\tau_r$. These are reasonably good results, which justifies a $N_d \geq 2N_r$ & $N_i \geq 2N_r$ rule of thumb to get a decent etalon performance. In practice, this means matching the etalon figuring and adjustment requirements with the plates reflectivity specification, and limiting the field of view to the maximum value compatible with the required beam finesse, and possibly less.

1.3.2 Interference Filters

Many spectroscopic systems require a spectral filter that lets through light in a fixed spectral band toward the instrument. A classical Fabry–Pérot etalon can provide any required spectral band, even exceedingly narrow ones, but cannot foot the bill because of the many other spectral bands getting through the different etalon orders. What is used instead is an interference filter, an avatar of the classical etalon, with an extremely low order ($p = 1$ or 2) solid cavity sandwiched between multilayer dielectric stacks, all created by vacuum deposition. A huge reflective finesse of up to ~ 800 can be obtained. The resulting spectral resolution can then be up to ∼ 1000 in the optical range above about 480 nm, with a minimum transmission peak around 50%. These performances would be all but impossible to attain with the classical etalon and its maximum defect finesse ∼ 100. Such high performance interference filters are commercially available from a number of vendors and are widely used in a large variety of optoelectronics systems.

These are on top very rugged and highly stable devices, with lifetimes of usually over a decade, provided they are kept in a reasonably dry environment. One proviso is their significant temperature-related bandpass shift. The temperature coefficient is usually positive, that is, the central wavelength transmitted by the filter shifts to a higher value when temperature increases, by very roughly +0.01 nm per degree: this could in theory be used to shift the filter bandpass at will; this is, however, not practical because of the many hours required to change the filter temperature significantly in a homogeneous manner. Performance (spectral resolution and peak transmission) drops rapidly below ∼ 480 nm, because of rising internal absorption from any known high-index material. Interference filters above ∼ 1800 nm central wavelength are easily subject to delamination of the thicker coating layers, especially as they are usually used in a cryogenic environment (operating temperature of ∼ 77°K) to avoid excessive thermal emission from the filter itself.

The spectral bandpass of a simple interference filter, made with a central $\lambda/2$ optical depth cavity, sandwiched between alternative $\lambda/4$ optical depth high and low index layers, has the classical strongly peaked etalon Airy shape. By using a more complex layer set, it is in fact possible to almost get the optimum square shape, if at the expense of some peak transmission loss. Note that unwanted light, that is, outside of the filter spectral bandpass, is reflected rather than absorbed as with color filters: on the negative side, extra care must be taken to avoid that it is reflected back toward the instrument by any optical surface behind the filter; on the positive side this light might be used if needed, for example, to monitor atmospheric transmission changes.

Interference filters owe their usefulness to their large accepting etendue. Their linear optical etendue in any direction is $y \sin\theta$, where y is a circular filter half-size, and θ the maximum beam cone half-angle. Standard interference filter sizes are usually up to 2 in. in diameter (50.8 mm), but a few vendors can provide up to ∼ 185 × 185 mm filters, which can be mosaicked to get even larger areas. The beam half-angle θ for a filter with spectral resolution (for a parallel orthogonal beam) $\mathfrak{R}$ is readily computed from the cooking book insert as $n/\sqrt{\mathfrak{R}}$. Here, n is the cavity refractive index, about 1.5 if it is made of a low index material, and 2.4 if it is made of a high index one. Even at its maximum spectral resolution of ∼ 1000, a high-index interference filter can still accept a rather big cone of light, with a half-angle $\theta \sim 0.1$ rad. or 5.7°. For filters with a wider bandpass, the acceptance angle is larger, but limited anyway to roughly 12° because of polarization effects in the coating layers at too large angles. Note that the acceptance angle decreases only with the square root of the spectral resolution, and not with the spectral resolution when using a grating to filter the light.

There are two basic ways to insert an interference filter in an astronomical instrument, namely, either on an image of the pupil (pupil mounting) or on an image of the sky field (field mounting).

a) *Pupil mounting.* This gives the best peak transmission at the cost of a gradual shift of the central wavelength to lower values for field positions away from the field center. It is in particular preferable when the input light etendue is much smaller than the available filter etendue (say, by at least a factor of 3 in linear etendue), since the shift effect above becomes fairly negligible. Note that the optical quality of the filter then needs to be as good as that of the main instrument optics, with the corresponding technical specification as required to the vendor.

b) *Field mounting.* This is often the preferred choice since it gives the same bandpass for all points in the field (provided the filter bandpass is the same over the useful area of the whole filter, an important technical specification), if at the cost of significant light loss, $\sim 30\%$ when the input etendue is equal to the filter etendue. With this mounting, the filter figuring error budget is much relaxed; on the other hand, extra care must be taken to minimize dust particles on the two outside faces of the filter that would give artifacts on the final field image. To minimize this effect, the filters are generally put slightly out of the exact field position in order to defocus any remaining dust image on the detector.

1.3.3 Fabry–Pérot Filter

As for an interference filter, there are two ways to insert an etalon, the pupil mounting and the field mounting. Typical useful size of an etalon is 50 mm diameter, but up to 150 mm diameter plates can be manufactured.

a) *pupil mounting.* This leads to the classical Fabry–Pérot spectrograph, presented in Section 3.3, which gives spectral information over a wide field of view. Its accepting etendue is indeed extremely large, according to the cooking book recipe $\theta\delta\theta = 1/2\mathfrak{R}$ (to get the beam finesse N_θ twice larger than the final finesse N). Here, θ is the beam cone half angle and $\delta\theta$ the ring width at θ, which must cover at least 1 pixel on the detector. Taking a typical 50 mm diameter etalon, an F/1.4 camera, and a 4k × 4k detector with 12.5 μ pixels, this gives a huge beam cone half angle $\theta = 21°$ for an already large maximum spectral resolution $\mathfrak{R} = 7,656$. For a large 8-m diameter telescope, this means a quite sizeable 7.9′ diameter working field. On the other hand, requirements in terms of plates figuring and parallelism quality are tough, since the $N_d \geq 2N_r$ cooking book recipe applies to the full pupil area.

b) *Field mounting.* This is rarely used, as the beam half-angle on the etalon must be at most $\sim 1/\sqrt{\mathfrak{R}}$. This leads to a much smaller field of view than with pupil mounting. Accepting linear etendue of a $2y$ diameter etalon is $y/\sqrt{\mathfrak{R}}$, obviously that of a same resolution and same diameter interference filter with cavity refractive index equal to 1. On the other hand, plates figuring and parallelism requirements are much relaxed since they now apply at the very small size of a detector pixel projected back on the etalon ($y/2048$ when using a 4k × 4k detector).

1.4 Dispersers

As their name implies, dispersers act by dispersing, that is, by changing the inclination of incoming light beams as a function of light wavelength in the so-called dispersion plane. To separate the output beams according to wavelength, the incoming beam must have a narrow angular width in the dispersion direction. On the other hand, the angular length in the orthogonal direction can be very large, as it is only limited by the field of view of the camera. The usual arrangement is then to limit the beam with a narrow, but possibly long, slit located at infinity with respect to the disperser. Note that while the slit is nearly always put on a sky image and the disperser on an image of the telescope primary or secondary mirror, the opposite combination works too. Actually, this exotic variant is used for the lenslet-based integral field spectrograph (Section 4.2).

1.4.1 Prisms

Since Isaac Newton's seminal experiences and for about two centuries, prisms have been *the* disperser used for all spectroscopic observations, including of course astronomical ones. Prisms' principle is the essence of simplicity: since the index of refraction n of all transparent materials varies with wavelength λ (actually decreasing for increasing wavelengths), the output light from a glass wedge of apex angle A illuminated with a polychromatic parallel beam (incidence angle θ) consists of a fan of parallel monochromatic beams deviated toward the prism's base and with emergent angles θ' nicely sorted out according to λ.

As illustrated in Figure 1.7, prisms are typically used in their minimum deviation (symmetrical) configuration, with the emergent beam angle θ'_c at the central wavelength λ_c equal to the entrance angle θ. This happens for $\sin\theta = n_c \sin(A/2)$, where n_c is the glass refractive index at λ_c. It is easy to show that the central angular deviation $\delta\theta'/\delta\lambda$ is then equal to $K\delta n/\delta\lambda$, with the constant K given by

$$K = \frac{2\sin(A/2)}{\sqrt{1 - n_c^2 \sin^2(A/2)}} \tag{1.7}$$

K can be quite high, for example, $K \sim 1.7$ for a quite typical apex angle of 60° and central index $n_c = 1.62$, but the glass dispersion $\delta n/\delta\lambda$ is always small for transparent glasses, with typically only a 2% refractive index variation from the blue (486 nm) to the red (656 nm). Near strong absorption bands, glasses might in fact exhibit large dispersions, but then would be hugely variable and associated with large light losses.

Figure 1.7 Prism's principle: The figure shows a prism used at minimum deviation for the central wavelength (green ray). This is a symmetrical configuration with similar beam incident and emergent angles θ. The extreme wavelength beams are shown in red and blue.

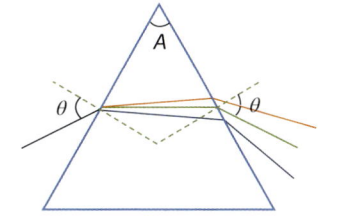

Linear etendue conservation in the dispersion direction between the grating plane and the detector plane can readily be used to set up the basic spectrograph parameters as per the following insert. In the orthogonal direction, the available etendue is very large and the slit length limited only by the spectrograph optics field and/or the detector length. In the dispersion direction, on the other hand, it is limited by the prism's low dispersion for any required spectral resolution $\Re = \lambda/\delta\lambda$. Linear etendue conservation between the prism plane and the detector plane gives the insert formulae. The prism figure of merit $K\lambda\, \delta n/\delta\lambda$ is a dimensionless number that expresses its dispersion efficiency. For the typical glass selected above, the figure of merit is ∼ 0.11, an order of magnitude less than the corresponding figure for a grating (see Section 1.4.2).

Prism Spectrograph Cooking Book

Prism apex angle A; diameter d; glass relative dispersion $\Delta = \lambda\, \delta n/\delta\lambda$
$K = 2\sin(A/2)/\sqrt{1 - n^2\sin^2(A/2)}$; unitless figure of merit: $K\Delta$

Telescope diameter D, on-sky slit width α

Spectral resolution $\Re = (K\Delta)\, d/D\, \alpha$ (at minimum deviation)

Prism diameters can be very large, 1 m or more for a very few common optical glasses including fused silica. For diameters ≤ 30 cm, a large palette of glasses can be produced, including expensive UV and/or IR transparent crystals. Care must be taken to subject prisms to only slow homogeneous temperature changes, as glass refractive indexes are temperature dependent.[2] This can be important for large prisms, given their huge thermal inertia.

Prism transmissions are usually quite high, 90% or more if its two surfaces are antireflection coated for the mean optical ray's incidence/reflective angle. Furthermore, the spectral range potentially covered is in principle limited only by the transparency of the prism material. This means that it is possible to cover easily an 1 octave spectral range (i.e., a factor of 2 between the lowest and the highest wavelength) in one go, or even much more, if at the cost of significant reflection losses. For example, most glasses are transparent from about 0.4 to 1.5 μm, and some crystals from, for example, 0.15 to 9 μm, a whopping factor of 60. Another very attractive property is that all the light is concentrated in a single spectrum, not in a number of different "orders" as for diffraction gratings (see Section 1.4.2).

Nevertheless, as pointed out above, a big limitation of prisms is their very small angular dispersion $\delta\theta'/\delta\lambda$ (see Exercise 3): to get the relatively high spectral resolutions (a few thousands) most often required for astronomical purposes, one would need to put an exceedingly narrow slit on the object under study, thus rejecting most of its light. In addition, glass dispersion significantly decreases with increasing λ, typically by a factor of 5 over 1 octave in wavelength, with a nonideal corresponding variation of 2.5 in spectral resolution. As a result of

---

2 At room temperature, fused silica 633 nm refractive index temperature dependence is $+10^{-5}/°K$, equivalent to a 0.4 nm wavelength shift per °K.

these two shortcomings, prisms are now used only for a few specific cases, and diffraction gratings with much higher and more uniform angular dispersion (see Exercise 4) are chosen instead for most spectrographic applications.

1.4.2 Grating Principle

A diffraction grating is a mirror or window with a periodic structure (often called grooves), which diffracts polychromatic (i.e., made of many wavelengths) input light into monochromatic beams traveling in different directions. In the canonical case of a plane parallel beam of wavelength λ impacting at incidence θ a plane diffraction grating with a parallel straight grooves per unit length, the various output light directions θ'_λ are restricted to the values for which light scattered from adjacent elements of the grating are in phase (see Figure 1.8). The relationship between the incidence and diffracted angles in media of respective refractive indexes n and n' can be obtained easily. This is the grating fundamental equation:

$$n \sin \theta + n' \sin \theta' = ka\lambda \qquad (1.8)$$

Here, the order k is any integer – positive, negative, or null.

For $k = 0$, we recover the classical reflection law for a normal mirror: this nondispersed zero order "white" light beam is mostly a nuisance, giving bright parasitic light on the detector. In a few cases, however, it is useful, for example, to monitor sky transparency in real-time.

At optical-NIR wavelengths, it is quite convenient to express λ in micrometer: a is then the number of grooves per micrometer, with $a\lambda$, as it must be, a dimensionless number. The grating equation shows that the groove period $1/a$ cannot be smaller than $\lambda/2$: for this limiting value, θ and θ' are respectively equal to $+90°$ and $-90°$, that is, the incident and the order 1 diffracted beams are both parallel to the grating surface.

The periodic structure (or grooves) that transforms a mirror in a reflection grating is made by modulating its surface shape (amplitude grating). This can be done by pushing aside a metal coating on a glass surface (surface relief gratings) or by etching a light-sensitive material illuminated by interfering laser beams (holographic gratings). Efficient transmission gratings are made by modulating the refractive index of a thin gelatine layer on top of an optical window (volume phase holographic gratings, VPHG), also through illumination by interfering laser beams.

Figure 1.8 Grating's Principle's: A parallel polychromatic light beam (black rays) falls on a plane reflection grating at incidence angle θ. In that illustration, first order green rays (not shown) are diffracted back at the same angle θ along the input beam (Littrow condition), while extreme wavelength beams are shown respectively in red and blue.

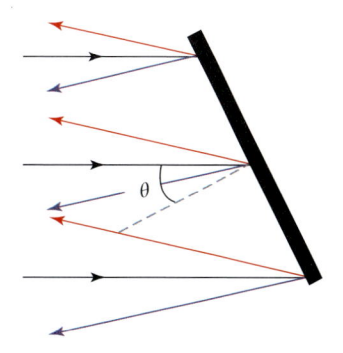

Most astronomical applications require covering a wide spectral range at once. This usually leads to using diffraction gratings in their first order ($k = +1$): the grating equation above shows that one can then cover almost one octave without mixing first order diffracted light with any of the other orders, including the non-dispersed, thus always bright, zero order. Of paramount importance is then getting the best possible efficiency, that is, with most of the diffracted light concentrated in the first order. For amplitude gratings, this can be obtained by manufacturing nearly triangular grooves with facets at the so-called blaze angle ϕ with respect to the normal to the grating. Peak efficiency close to 1 is then obtained at the blaze wavelength λ_c, given by $2 \sin \phi = k a \lambda_c$. At this wavelength, the incidence angle and the first order diffraction angle are the same (Littrow condition) and equal to the blaze angle. In practice, for blaze angles up to ∼ 30°, first order efficiency is still good (say > 70%) over 1 octave. For higher blaze angles, groove size becomes comparable to the light wavelength: this results in efficiency curves for the two polarizations that are still highly peaked, but at different wavelengths. The end result for natural (unpolarized) light is smaller efficiencies, say, below 50% for a 60° blaze angle over 1 octave.

"Normal" optical systems, that is, combinations of mirrors and lenses, obey the principle of inverse return of light, both for light path and light efficiency, and that for each of the two linear polarizations. This is still true with a grating along the optical path.

1.4.3 The Grating Spectrograph

The canonical plane grating long-slit spectrograph uses the following components (see Figure 1.9): (i) a long but exceedingly narrow width slit, (ii) a collimator imaging the slit of at infinity (angular width α), (iii) a plane grating located on the exit pupil (diameter d), and (iv) a camera (aperture ratio Ω) to image the dispersed slit on the detector (projected slit width w). Note that an image of the sky is usually put on the slit with an image of the telescope primary (or secondary) mirror

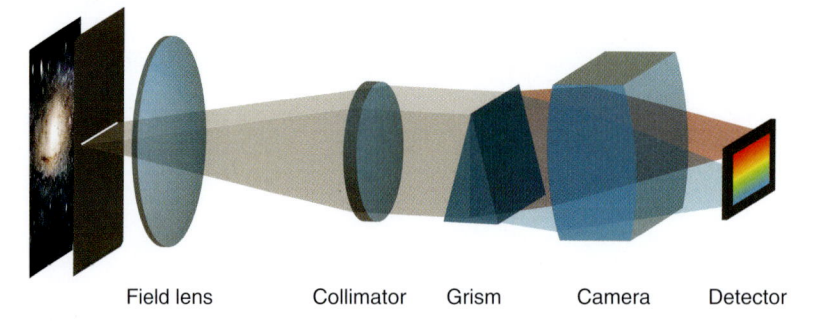

Field lens   Collimator   Grism   Camera   Detector

Figure 1.9 3D view of the long-slit grating spectrograph concept. (a) The 2D image at telescope focus is sliced by a long narrow horizontal slit. (b) Light on the detector is dispersed in the vertical direction. Note that we follow in this book the two usual conventions for figures showing light propagation inside an optical system: (i) whenever possible, light goes from left to right (top to bottom for a vertical optical axis) and (ii) the horizontal and vertical scales are generally not the same, usually exaggerating the angles of the optical beams for better clarity.

on the grating, but the other way would work too and has been occasionally done.

The long narrow slit, typically 3×4096 pixels at the detector level, or $0.6'' \times 15'$ on the sky or with a 4-m class telescope, is dictated by the dispersive geometry of the grating. This extremely thin shape is far from optimum though, as astronomical objects other than stars basically always have much more roundish shapes, and most of their light is just wasted when cut by the slit. On large telescopes, the light of even originally point-like stars is not always fully collected by long-slit spectrographs: star focal images have a disk shape of roughly $0.4''$ to $1.8''$ diameter depending on atmospheric turbulence, with a median value around $0.7''$ at the best sites in the world. In the example given above, more than half of the time a significant fraction of the precious starlight would be wasted.

2D etendue conservation does not however prevent sending an initially round object through a narrow rectangular field, while keeping a round pupil at infinity, provided the slit and object areas are equal, or the object is of diameter up to $24''$ for the $0.6'' \times 15'$ slit referred above. Unfortunately, this transmogrification does not conserve the 1D linear etendues and thus cannot be done with simple centered optical systems, for example, by using a set of cylindrical lenses. This necessitates developing instead complex multimirror/lenses systems that are notoriously difficult to fabricate and align (see Chapter 4 for much more on these so-called image slicers).

As seen above, for an input plane parallel light beam at angle θ with respect to the normal of the grating, an exit plane parallel beam at angle θ' for the wavelength λ will be in phase if and only if $\sin\theta + \sin\theta' = ka\lambda$; k, the spectrum order, is by principle an integral number, and a is the number of lines per micrometer on the grating plane, with λ the wavelength in micrometer.

In the $k = 0$ case, we have $\theta = -\theta'$ independent of the wavelength of light. This so-called zero-order or white light thus follows a simple reflection on the grating plane, collecting light at all wavelengths. This is not only a waste of light, but also adds a bright line imprinted on the detector. It is thus particularly important to get a grating with no more than a small percentage of light in zero order.

1.4.4 Grating Species

There are essentially two grating species, the surface relief grating (mechanically ruled or holographically imprinted) for which the periodic wavefront modulation is governed by periodic grooves depth, and the VPHG with periodic modulation of the index of refraction of a thin gelatine dichromate layer deposited on an optical plate. The first variant is now used solely in reflection, with an aluminum or gold (for the NIR) layer deposited on the front surface. The second works in transmission only, with a better efficiency than the equivalent surface relief transmission grating, $\sim 85\%$ for a $45°$ blaze angle instead of $\sim 30\%$ only. VPHGs are often sandwiched between two identical prisms so that light at the central wavelength goes directly through the disperser with zero deviation.

1.4.5 Grating Etendue

For optimum sensitivity, gratings are used mostly at or close to zero deviation (Littrow condition), that is, with $i = i' = \phi$ (blaze angle) for the central wavelength λ_c. In order to fully accept the usually circular pupil, they have in general a rectangular shape with an aspect ratio (length L over height d) at least equal to $1/\cos\phi$. Differentiating the canonical equation with respect to wavelength gives the angular width of the slit β as $\beta = \delta i = 2\tan\phi / \mathfrak{R}$, where $\mathfrak{R}$ is the spectral resolution $\lambda_c/\delta\lambda$. Note that in first approximation and according to the canonical equation, the spectral resolution over the spectral domain covered on the detector is proportional to λ/λ_c. This is not perfectly constant, but better than when using prisms.

Linear etendue conservation in the dispersion direction between the grating plane and the sky plane can be used to set up the basic spectrograph parameters as per the following insert. In the orthogonal direction, the available etendue is very large and the slit length limited only by the spectrograph optical field and/or the detector length. Note that the fundamental relationship between the spectral resolution and the slit linear etendue has exactly the same shape as for a prism, the only difference being the dimensionless figures of merit of different dispersers ($2\tan\phi$ for a grating).

Grating Spectrograph Cook Book

Grating height d (pupil diameter); blaze angle ϕ
Telescope diameter D, on-sky slit width α

Spectral resolution $\mathfrak{R} = 2d\tan\phi/D\alpha$ (at zero deviation)

A number of manufacturers provide off-the-shelf reflective gratings with a large variety of groove periods/blaze angles and sizes, a few up to about 30 cm height. They all are actually replicas molded in the thousands from very expensive ruled masters. Larger sizes can be achieved by mosaicking a few identical gratings, if with stringent alignment requirements. Reflective gratings can be produced for virtually any working wavelength.

VPHGs are on the other hand mostly custom made, with heights (size perpendicular to dispersion) up to about 25 cm. Mosaicking is rarely done since it is very difficult to produce two closely identical VPHG. Due to the gelatine dichromate bandpass, VPHGs can work roughly from 370 to 1700 nm. They can be operated at cryogenic temperatures when needed.

Maximum blaze angle contribution $\tan\phi$[3] is usually ~ 0.7 for a VPHG covering one octave (with simple inline optics); ~ 1.7 for a ruled or holographic reflection grating (with larger, more complicated optics however); ~ 4 for a high-order echelle grating (which however requires an additional disperser used as an order sorter). The bottom line is that the only free construction parameter is the grating diameter d, leading automatically to very large gratings (hence big

---

[3] Higher values are technically feasible, but entail up to 50% light loss as the peak efficiency wavelengths as the p and s polarization components become widely separated.

expensive instruments) when wanting either a large spectral resolution or a substantial on-sky slit width and especially when requiring both.

w, the width of the instrument slit projected on the detector by the camera of aperture ratio Ω, usually corresponds to ~ 2 detector pixels (so-called Nyquist condition); it can occasionally be set up as large as 3–4 pixels when a high signal-to-noise ratio is required: this is in practice the province of high ($\mathfrak{R} \sim 3.10^4$) and very high spectral resolution ($\mathfrak{R} \sim 10^5$) astrophysics for which reaching minimum signal to noise ratios of 50 and 500 per resolved spectral pixel respectively is the norm. Note that because of linear etendue conservation between the grating plane and the detector plane, w is given by $w\,\Omega = D\,\alpha$.

In most cases, the spectral resolution is thus directly set up by the slit width, itself imposed by the need for collecting enough of the science object and/or the use of a high aperture ratio on the detector to get enough sensitivity. In the case of bright objects, a very narrow slit can be used in principle: getting a substantial spectral resolution with a small grating (hence a small instrument) then becomes easy. There is a limit, however, as part of the light going through the narrow slit is diffracted and begins to overfill the instrument pupil. The theoretical limit is given by a simple formula, namely, $\mathfrak{R}_M = kN$, where k is the grating order and N is the total number of grooves covered by the pupil. Settling at that limit though would result in $\sim 50\%$ light loss and a practical upper limit for the actual spectral resolution of a spectrographic instrument is in fact $\sim kN/2$, unless the grating/camera combination is enlarged (at a cost) to collect a significant part of the diffracted light.

The optimum spectral resolution $\mathfrak{R}$ for an astronomical spectrograph is very much science dependent, for example, varying from 10^3 or so for optimum detection of extremely faint objects in the near-UV to yellow region to 4.10^4 for determining abundances of key chemical elements in Galactic stars, to more than 10^5 for the indirect detection of exoplanets from minute radial velocity variations of their parent stars. See Exercise 12 for the specific analysis of absorption line radial velocity accuracy versus signal to noise ratio.

1.4.6 Conclusion

The vast majority of astronomical spectrographic instruments uses the disperser approach, almost always a plane grating, either in transmission or in reflection. This is the basic building block for the many variants presented later in this book, which differ in the shape of the sky field selected at the telescope focal plane. This ranges from a simple narrow slit (long-slit spectrography) to multislits/holes on well-separated targets (multiobject spectrography) to a single squarish field (integral field spectrography).

1.5 2D Detectors

1.5.1 Introduction

A suitable 2-D digital detector (for lack of an even better 3-D ones, see Section 7.3) is the key element of any imager and particularly the spectroimagers covered in this book. There is a long list of perquisites for such detectors, and particularly, (i) high, ideally up to 100% quantum efficiency, that is, one electron "created" by each incoming photon, in a large wavelength range; (ii) very low dark noise (spurious electrons created during integration time) and readout noise (spurious electrons created by the detector readout electronics); (iii) large format, up to billions of detector pixels, and (iv) high linearity and high dynamical range, that is, output signals that are precisely proportional to the incoming photon flux over a wide flux range. There are also qualitative, but still crucially needed, additional features, such as being rugged (idiot/astronomer-proof), easy to use, and with few if any troublesome artifacts, such as charge blooming around overexposed pixels. This section covers the phenomenal progress enjoyed during the last 50 years and the current state-of-the-art 2D detectors landscape.

1.5.2 The Photographic Plate

Very large 2D integrating detectors were already available at the turn of the nineteenth century, namely, photographic plates or films, sensitive from the near-UV to the red domain. A single 50 cm × 50 cm photographic plate with about 12 µm spatial resolution (a typical value for plates optimized for low-level flux measurements) actually offered a whopping 1.6 billion spatial pixels, a format now just attained by the largest detector mosaics being built; see below. Besides, photographic plates are cheap, rugged, easy to use (just open and then close a shutter for an exposure followed by a few hours to develop, fix, wash, and dry), are operated at room temperature, and have the nice feature of doubling as their own data archive. Unfortunately, they also get a long list of dire shortcomings: extremely low quantum efficiency, possibly as "high" as a few 10^{-3}, but then with a very small dynamical range, plus extreme nonlinear behavior not only relative to the object flux but also with respect to integration time. Data extraction was terribly slow and of limited accuracy, adding to the overall spectacular inefficiency of this purely analog device.

Digital detectors with a much higher quantum efficiency (up to 10% in the blue), near perfect linearity, and high dynamical range were actually available soon after World War II, with the photomultiplier tubes. An incoming photon strikes a photocathode, ejecting one electron through photoelectric effect. These electrons are accelerated inside a vacuum tube, striking multiple dynodes. As a result, a single primary electron ultimately gives a hundred million electrons, which are easily detected at the end of the tube as a very short current pulse. Note that this is not an integrating detector but works by counting incoming photons on the fly one by one. Unfortunately, photomultipliers are essentially 0-D detectors, that is, offer only one spatial pixel, and thus took only a small part of the astronomical pastures, mostly to measure the integrated light flux of stars or central parts of galaxies.

1.5.3 2D Optical Detectors

The first 2D electronic optical detectors (as well as some 1-D ones) were introduced in the late 1960s, with many variants of video cameras of increasing sensitivity, ultimately up to individual photons detection. Essentially 2D versions of the photomultiplier, the latter featured single photoelectrons counting, quantum peak efficiency ~ 10% in the blue (typically 4% in the red) domain, and respectable, but not huge, dynamics with recordable fluxes from around 1 to a few thousand photoelectron per pixel per hour. Despite their small formats (a few 10^5, typically 40 × 40 µm, pixels), they quickly displaced photographic plates for all astronomical work, except for completing the few long-term imaging surveys of large fractions of the sky carried out at the time.

However, soon after, in the late 1970s, these 2D photon counting detectors were themselves quickly and almost entirely replaced by the charge coupling devices, or in short CCDs, invented a decade earlier [13]. These are essentially integrating detectors, such as photographic plates but unlike photon counting ones. They thus suffer not only from dark noise but also from readout noise. Initially, CCD formats were also quite small, quite comparable to photon counting formats at the time.

Somewhat ironically, their introduction resulted initially in a significant loss in low photon flux detectivity compared to photon counting detectors, as CCDs then featured huge readout noises, around 100 electron r.m.s. per pixel (versus essentially zero with photon counting!). But, they already have much better quantum efficiencies, peaking around 50% in the blue–green, high dynamics from a few to up to 32,000 electrons per pixel, and most importantly turned out to be extremely rugged and easy to use. Their sensitivity loss for extremely low photon fluxes had actually a much smaller impact than could have been imagined: the vast majority of astronomical (and non-astronomical) observations relies on detecting a small variation in a high photon flux, rather than a very small photon flux over a negligible background. Canonical examples of the former are (i) all wide spectral band imaging for which the photon flux is dominated by the night sky background and the goal is to detect the very small flux variation due to, for example, a distant galaxy, and (ii) most absorption-line spectroscopy of galaxies for which the goal is to measure relatively slight flux dimming (due to atomic or molecular absorption) of the strong object stellar continuum. On the other hand, the canonical example of the extremely low photon flux case is high spectral resolution (typically 12,000) spectroscopy of emission lines in extremely faint ionized gas regions, a respectable scientific domain, but representing a very small fraction of all astronomical observations.

A mid-2010s top-of-the-art CCD (see Figure 1.10) is an almost ideal detector, featuring >80% QE over one wavelength octave, very small dark current ($1e^-$/px/hr or less), and readout noise ($1.5 - 3e^-$/px/readout). The dynamical range is huge – a factor of 30,000 or so, and data output is immediately available in a digital format. Note that while off-the-shelf CCDs for laymen applications (video cameras) are dirt cheap, a so-called science-grade 4096 × 4096 (4k × 4k in detectors lingo) 12.5 µm pixel CCD costs ~ 50,000 Euro/USD. A major cost escalation driver is the thinning of the detector: this is a time-consuming low-yield process, but one which effectively doubles the detector QE. Another is the on-chip implementation of extremely low-noise electron amplifiers to drastically reduce the readout noise of standard CCDs. Also, while commercial CCDs operate at room temperature with very short integration times (50–100 Hz

Figure 1.10 This shows the state-of-the-art 4096 × 4112, 15 μm pixels CCD231-84 from e2v, one of the leaders in the field. Since this high performance device is four-side buttable, it can be used as a building block for the development of extremely large mosaics. Credit Paul Jordan [14], e2v, the UK. (Reproduced with permission of Paul Jordan.)

video frame rate), astronomical CCDs are operated with typically 30–45 min integration times (a factor of 2.10^5!). To lower dark current to at most a few electrons per pixel per hour, detectors are cooled to around −90 °C, a significant design and operating complication: in particular, the detector is housed in vacuum inside a cryostat, with a thick entrance window, often doubling as the last lens element of the camera. A gentle dry nitrogen flow is often used to prevent frost formation on the window front face.

CCDs can be fully buttable on their four sides with very little dead space between them, and mosaics of more than a billon pixels have been built, with 4k × 4k single CCDs used as building blocks. This is essential for most imaging instruments, covering large fields on the sky, but is also increasingly required for spectrographic 3D instruments, as builder teams get more and more ambitious in terms of simultaneous spatial and spectral coverage.

CCDs "natural" spectral coverage starts from the near-UV to roughly 0.8 μm: as the light wavelength increases, the detector thin silicon material gets transparent, leading to free-falling quantum efficiency and parasitic fringing at the level of a small percentage as part of the light going through the detector is reflected back and interferes with itself. The so-called red-optimized thicker CCDs retain good quantum efficiencies up to ∼ 0.98 μm and exhibit much less fringing (∼ 0.4%). Note that most fringes arise from the highly variable upper atmosphere emission lines strongly prominent at these wavelengths and are not easily calibrated out. Another annoying feature comes from cosmic rays impacting the detectors and creating point-like bright spots on the final data: typical spot counts are ∼ 4 per second integration time with a standard 4k × 4k CCD, ∼ 20 for the thicker red-optimized one. This means some 15,000–75,000 impacts recorded for a typical 1-h exposure. Fortunately, their distinct highly peaked shape helps much to get highly efficient rejection algorithms in the subsequent data processing phase (see Section 8.3).

1.5.4 2D Infrared Arrays

Until the early 1980s, for lack of the equivalent of photographic plates in the optical range, near IR astronomical observations above 1 µm were painstakingly performed with single-pixel semiconductor or bolometer detectors. Development for the US Air Force of 2D IR Arrays based on HgCdTe or InSb semiconductors that now feature up to 4k × 4k pixels (e.g., the Teledyne H4RG array) has been a tremendous bonus, as soon as these (at first much smaller) detectors hit the civilian market. They remain relatively expensive, with a cost per pixel one order of magnitude higher than for optical CCDs. Like CCDs, they are fully buttable and large arrays featuring up to 100 million pixels are currently in operation. Unlike CCDs, it is possible – and almost always advisable – to perform multiple nondestructive readouts during the integration time as the IR photons slowly build the spectra (or the image) on the detector.

Performance is splendid with, in particular, about 80% quantum efficiency in a large domain, starting around 0.9 µm and reaching up to 5.5 µm. Actually, with recent improvements in the manufacturing of the HgCdTe material, the long wavelength cutoff can be tailored anywhere between 1.6 and 5.5 µm by tweaking the chemical element ratios. Near IR arrays are generally cooled around 70 °K to reduce integration noise to about 0.02 e^-/px/s; this relatively high level compared to CCDs usually limits individual integration times to a few minutes at most. With low speed readout around 50,000 pixels per amplifier per second and special reading tricks, readout noise is of the order of 7 e^-/px (again almost an order of magnitude higher than with CCDs): to avoid excessive total reading time, an individual 4k × 4k array holds up to 64 amplifiers working in parallel and the detector controller continuously performs low-speed nondestructive readout of the array until the end of the integration time.

IR arrays are less sensitive than CCDs to cosmic ray impact and, besides, their multiple nondestructive readout schemes can be used for real-time rejection. On the other hand, they are more "touchy" than CCDs, with possible artifacts, such as hot (high-noise) pixels, or parasitic light emission from the amplifiers that can reach the array corners.

1.5.5 Conclusion

The optical/NIR 2D-detector astronomical landscape is currently dominated by two commercial solid-state integrating detectors, CCDs for the optical range (up to 0.98 µm for the so-called deep depletion CCDs) and charge injection devices (CIDs) for the NIR, with long wavelength cutoffs that can be tailored to the instrument requirements in the 1.6–5.5 µm range. Pixel size is usually around 12 µm for CCDs and 15 µm for near IR arrays. Very large mosaics can be built, covering any spectrographic need. Their low readout and integration noise and high quantum efficiency (actually close to 100%) make them almost ideal detectors, except for a few photon-starved cases such as high-resolution spectroscopy of fast transient sources or real-time measure of atmospheric turbulence to correct image blurring (Section 9.4).

One worrisome recent commercial development is the increasing replacement of CCDs by cheaper optical charge injection devices for most laymen applications (surveillance, mobile phones, etc.), except the really high-end ones. This trend might well someday make science-grade CCDS no more available for astronomical purposes, at the cost of a significant hit in detectivity, owing to the intrinsically higher CID readout noise.

A recent emerging detector is the avalanche photodiode 2D array. This is a return to the short-lived photon-counting era, but with much higher QE (typically 60% peak), more rugged devices, and a larger wavelength range covering both the optical and NIR domains, for example, up to 1.65 µm, but with extremely small formats (typically up to 8 × 8, 50 µm, pixels) and non-negligible dark (integration) noise. Much larger format devices have been built for defense purposes, but are not yet available for the civilian market. They are well adapted to wavefront sensing for adaptive optics systems with their typical sub-millisecond integration times (see Section 9.4), but not yet – if ever – for hour-long integrations as the main science detector for spectrographic instruments. In the same vein, large format CCDs, which, owing to internal electron multiplication, reach full photon-counting capability, have been developed by an e2v-University of Montréal collaboration [14].

Finally, all 2D detectors – including photographic plates – currently feature plane light-sensitive surfaces. This does not mean that there is any strong technical difficulty in developing detectors with any other sensing surface curvature, just that zero curvature is by default *the* commercial standard. This is a real limitation, as spectrographic optics designers would love getting concave detectors when opting for refractive cameras and, conversely, convex detectors to match reflective optics. Developing 2D detectors is however an extremely expensive endeavor (hundreds of million USD/Euros), and there is little hope to ever get off-the-shelf curved detectors, except with the kind of massive financial investment that ground-based astronomy, and even space-based astronomy, could hardly afford.

1.6 Optics and Coatings

1.6.1 Introduction to Optics

As discussed above, the main optical path for the various spectrographic modes involves two basic subsystems, namely, a collimator to image either a full 2D field or a 1D field (an input slit) at infinity, and a camera to image back the field or the spectra on a 2D detector. Cameras and collimators can be based on lenses (dioptric systems), or mirrors (catadioptric systems), or a combination of both. The spectral range to be covered is usually one octave at most, unless multiple cameras/gratings are used in parallel.

It would be nice to use off-the-shelf optical subsystems, gaining very much in project cost and timeline, but that is not generally possible while retaining high light transmission and excellent image quality. In particular, most spectrographic cameras necessarily feature an entrance pupil some 5–10 cm *before* their first lens, in order to insert their transmission grating; this is a very significant constraint on the camera optical system, and all off-the-shelf cameras are designed instead with the entrance pupil well inside their optical body. Main astronomical instrument optical systems are thus usually expensive One Off prototypes that are in-house designed and then contracted to industrial optical companies. Optical cost can then easily be in hundreds of thousand USD/Euros, with the total development time around 2 years. In this chapter, we will look at a few basic principles on how these complex optical systems are designed, fabricated, and tested.

1.6.2 Optical Computation

Optical computation of the various optical subsystems is the first development step. This endeavor remains much of an art, even if the market offers (generally at a cost, one of the very few exceptions in the mid-2010s being WinLens 3D basic for Windows) remarkably efficient optical design programs that are extensively used to optimize delivered image quality, taking into account the many constraints inserted by the designer. Besides elementary design constraints such as spectral range, entrance pupil position (significantly before the camera first lens for grating and Fabry–Pérot based spectrographs), and the collimator and camera focal lengths and aperture ratios, there are many other less obvious ones, for example: (i) near telecentricity,[4] that is, with the output pupil as seen by the detector located near infinity, when precise measurement of object locations or spectral lines positions is required; (ii) use of cheap glasses when budget is limited; (iii) reasonable glass lengths; (iv) small excursions of the final image position with temperature when the instrument is not kept at constant temperature; (v) achievable optical tolerances, that is, reasonable required precision on optical component parameters (refractive index, thickness, radius of curvature, tilt and centering, inter-lens/mirror separations); (vi) no exposed glass surface too close from an image of the field (as any dust particles would then create an artifact on the final image); (vii) no harmful parasitic images, with particularly no glass/mirror center of curvature close to the conjugate of the detector plane, as any bright point in the field would then get a small halo around it, the so-called Narcissus effect from the eponymous Greek demigod; (viii) no radioactive glasses or coatings, especially if close to the detector, in order to avoid extra detector noise, and so on.

One major difficulty in getting high-performance dioptric systems is in canceling their inherent chromatic aberrations: this requires at least two kinds of transmitting materials with different dispersions (relative variation of refraction index over the working wavelength range). This is relatively easy for 1D long-slit spectrographs since a large part of the chromatic effects can be easily offset by tilting slightly the detector. When working in the main part of the optical spectrum, say between 450 and 950 nm, cheap optical glasses can then be used (as an example, see the MUSE instrument case in Section 5.4.2). For the extreme

---

4 See Figure 1.11 for a visual appraisal of non-telecentricity effects.

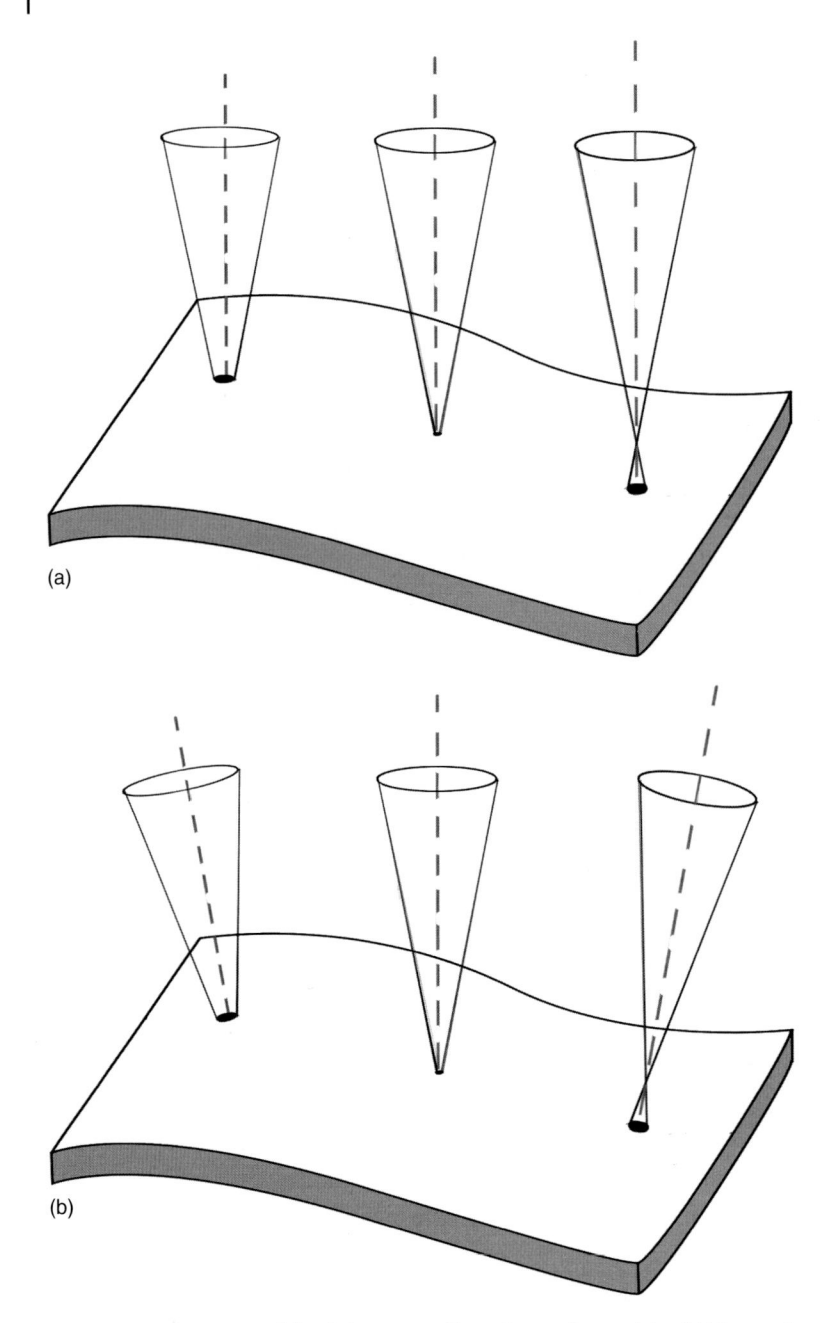

Figure 1.11 Illustration of the deleterious effect of non-telecentricity. (a) Three telecentric beams (with parallel optical axes) fall on a detector. Non-perfect flatness of the detector degrades the images, but does not move their centers of gravity with respect to each other. (b) The same, but for non-telecentric beams (optical axes not parallel). There is a similar image degradation, but now their centers of gravity are displaced with respect to each other. This leads to significant measuring errors, typically a few micrometers for, say, 10–20 µm flatness deviation. (Credit Colombine Majou.)

blue and near-ultraviolet, as well as the near-infrared domains, standard optical glasses, except the low-index low-dispersion fused silica, are no more transparent and more expensive and fragile crystalline glasses are used instead. One special difficulty for IR lenses is that the whole optical train is then usually cooled to cryogenic temperatures (say 77 °K for the near IR): knowledge of refractive indexes at these temperatures is quite sketchy, which makes optical optimization difficult. Also a remotely controlled cryogenic motorized system for accurate focusing on the detector surface is then required on such instruments.

A much better chromatic correction, called apochromatism, is needed for 2D field (multislit or slitless) spectrographs, for which the above tilting trick cannot work by design. In the optical domain this usually requires using fluoride glasses that are more expensive, difficult to polish, and easily cracked during the antireflection coating process: this results in steep cost escalation (say ×4) and long delays, easily 2 years for getting a full optical subsystem.

Mirrors have the huge advantage of no chromatic aberration at all, and on top give less geometrical aberrations than single lenses of equivalent converging/diverging power and aperture ratio. Their big disadvantage is geometrical in nature: light bounces back from mirrors and in most cases, especially for 2D fields, the output light beam is entangled with the input beam. Nevertheless, one popular solution for high-aperture cameras is the Schmidt design with a spherical mirror and the detector cum cryostat located at the mirror focal plane (see Figure 1.12). A different approach has been chosen for the JWST near-infrared multislit spectrograph NIRSpec: given its enormous spectral range, 0.6–5 μm, a train of three strongly aspherical off-axis mirrors (see Figure 1.13) is used for

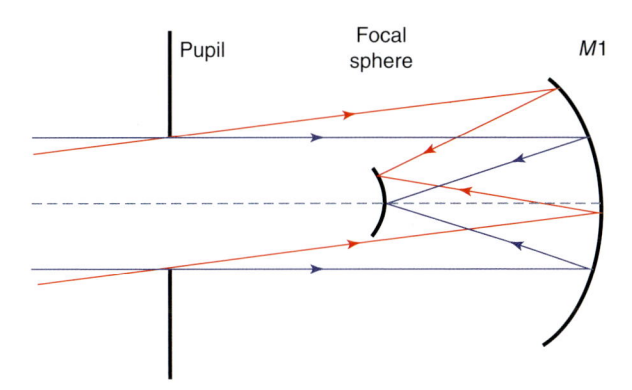

Figure 1.12 Illustration of the Schmidt mounting, in essence a spherical mirror with the entrance pupil located at its center of curvature. Two parallel light beams at two different inclinations are shown. Owing to the system's full rotational invariance, all input parallel beams are imaged with the same (small) aberration irrespective of their 3D inclination, a trick first discovered by the philosopher (and lens maker) B. Spinoza in 1600 and implemented by B. Schmidt in the 1930s. This field-invariant aberration can, for example, be canceled by adding an aspheric window on the pupil; its correction effect then varies over the field, but only as a cosine function, which in most cases is good enough. Again, because of rotational invariance, the images are located on a spherical segment with its center of curvature on the pupil. A field flattener (a thick convergent lens possibly doubling as the detector entrance window) can also be placed just before the focus.

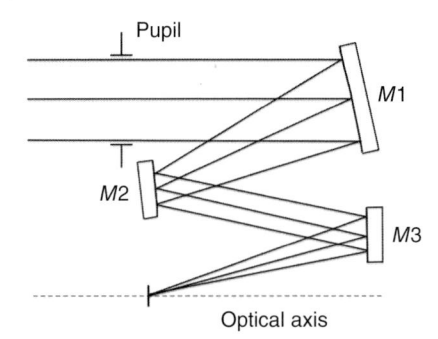

Figure 1.13 The so-called three mirror anastigmat (TMA) is a centered optical system (i.e., with a common optical axis) made of three highly aspheric mirrors, the shapes and positions of which are tuned to give extremely good images on a flat focal plane over a wide field of view. In real life, to avoid 100% beam obscuration by M_2, only off-axis cuts of the three mirrors are used. The TMA can be used as a camera with light reflected successively by M_1, M_2, and M_3, or as a collimator when used in reverse.

both the collimator and the camera. As can be seen on the figure, the off-axis part neatly solves the disentangling beam problem, at the (huge) cost of large diamond-machined monolithic optomechanical systems.

One integral part of the optical computation effort is to derive the manufacturing tolerances compatible with the required optical quality, so that the manufacturer can plan and test accordingly. Like for Goldilocks and the three bears, it is essential to design and fabricate the optics with the right tolerances, no more, no less: too loose ones than needed would give poor quality images and too tight ones, overly expensive or even unfeasible optics.

1.6.3 Optical Fabrication

Optical fabrication spans a huge range of technical procedures, adapted to the many different substrates (glass, crystal, metal, plastic, ceramic), sizes (from about 0.1 mm to 8.4 m diameter in the astrophysical domain), shapes (spherical, mildly or hugely aspherical), and required surface qualities (from a few λ for auxiliary lenses to $\lambda/4$ for precision lenses to $\lambda/100$ for interferometric components, where λ is the shorter wavelength at which the component is used).

For mirrors and lenses alike, the first step is usually to grind the substrate to get an approximate shape or even an accurate one, but with still rough surfaces at the micrometer size level. The next step is to lap/polish the surface to get a local smooth finish at close to the nanometer level, or even an extra-smooth one, for , for Fabry–Pérot or Michelson plates, or when the optics must image faint structures near an intensely bright one (the so-called coronagraphic grade optics).

Here are a few tidbits connected to optical surfacing:

- One might think that the most important manufacturing requirement is to get the exact theoretical shapes of the lenses/mirrors. Actually, this is not generally the most difficult part, as shape tolerances are often as "large" as a few micrometers, or even more, especially on large (meter-size) optics. What is more difficult is to avoid introducing significant slope deviations at all scales from the full diameter of the optical piece to about half a wavelength. Penalty for not reaching the required smoothness level is a significant fraction of scattered light, especially when working at short wavelengths. This means significant light losses and problematic artifacts around bright objects in the field. Slope requirements are especially tighter for the coronagraphic grade optics required to observe very faint sources near highly bright ones.

- Classical optical figuring uses a statistical process that automatically generates smooth spherical surfaces, including plane ones: the glass piece or blank, ground to the global shape, is put in contact with a same-size matching tool, with a mixture of abrasives and water in between. The blank and the tool rotate and oscillate with respect to each other in a pseudorandom manner. Over hours while using increasingly finer abrasives, this process automatically generates two spherical surfaces of opposite curvatures, since only two such spherical elements can remain in full contact for any orientation and lateral displacement. To produce plane surfaces, two blanks are lapped over a roughly plane tool, and the two blanks are also lapped against each other. This statistical process can be altered to get instead aspheric surfaces, for example, by stressing the optical surface during and/or after polishing, but these are slow, expensive processes, especially when one wants to avoid significant scattered light (say, no more than 1%). This is too bad because even only a few aspheric surfaces often allow the design of performant optical systems with much less lenses than their all-spherical variants.
- Molding techniques produce large quantities of dirt-cheap spherical or aspheric optics of quite reasonable optical quality. Molds are expensive to fabricate (say typically ~ 50 kEuros for a 10 cm diameter component), but can then produce tens of thousands of identical components at very low added cost. Owing to significant post-molding shrinking of the plastic materials, this process is however not well suited for production of high precision optical components. Molding is thus generally used for mass production of relatively low-tech optical components, for example, microlens arrays, cameras for smart phones, and so on. One nice feature is that a single mold can directly produce a full subcomponent, for example, a mirror with its mounting cell and reference alignment points, saving significant fabrication, integration, and adjustment time.
- Diamond-turning is a process of direct mechanical machining of precision optics using a computer-controlled lathe equipped with a diamond-tipped tool. Diamond-turning is used to manufacture spherical or aspheric lenses or mirrors alike from a number of crystals, metals, and plastics. Note that it is only since the beginning of the 2010s that it has been possible to get high-quality low-scatter diamond-turned optics good enough for the optical domain. One nice feature of this technique is that it is relatively easy to produce (at a cost) a monolithic all-mirror piece, for example, all aluminum or copper, incorporating, for example, three off-axis aspheric mirrors located at their precise theoretical locations within one micrometer or so. Such a built-in subsystem permits to evade having to perform tricky high-precision mechanical adjustments, and besides cannot subsequently become misaligned. Moreover, such subsystems can be put directly in a cryogenic environment; metal (isotropic) shrinkage will slightly alter the optics prescription in a homothetic way, but with the optics still aligned and focused. This does not work for lenses though, since optical train changes are then dominated by temperature shift of the refraction indexes of glasses.
- Optical testing during the fabrication process is essential for high-quality components and actually the only way to converge to the desired shape and surface

finish. This can easily be a full subproject in itself, with the study and fabrication of often huge testing devices. The usual rule for optical manufacturing is "if you can test it good enough and fast enough, you can get it good enough and fast enough." And the corollary, "if you cannot test it right, you will never get it right."
- For major subsystems (collimators, cameras, etc.) with exacting mounting tolerances, it is almost always better to get (at a cost of course) the optical components fully integrated inside their mechanical body from the industrial manufacturer. The client must nevertheless plan for independent end-to-end testing before closing the contract: image quality testing is generally quite easy in the optical domain, but takes much longer – usually days – for NIR optics working at cryogenic temperature. Checking light throughput is always difficult, and even more in the UV and NIR domains.

1.6.4 Anti-Reflection Coatings

When a beam of light crosses from a dielectric medium of index of refraction n_1 (e.g., air whose n is very close to 1) to another dielectric of index n_2 (e.g., an optical glass with index roughly in the 1.5–1.75 range), part of the light is transmitted (refracted) and the remainder is reflected. For optics made of transmitting elements – lenses, prisms, transmission gratings, beam splitters, Fabry–Pérot plates, and so on – the transmitted part is the useful one, while the reflected part is essentially a big nuisance, reducing overall light efficiency and possibly bouncing back from the other optical surfaces to finally create unwanted artifacts on the detector. The intensity of the reflected (R) and transmitted (T) components is given by the Fresnel equations as given in Section 1.1.1. In particular, the relative loss for light at normal incidence on a surface of refractive index n_2 immersed in air is $R = (n_2 - 1)^2/(n_2 + 1)^2$. This is about 4% loss for standard optical glasses, and is much higher with more exotic materials such as diamond at all wavelengths, or silicon, zinc sulfide, and zinc selenide in the near-IR range.

For astronomical optics, antireflection (AR) coatings are normally applied to the surface of all transmitting elements, including the front surface of transmission gratings and detectors. The simplest theoretical interference AR coating would consist of a single quarter-wave layer (meaning that its optical thickness ne is equal to $\lambda_c/4$) of a transparent material whose refractive index is the square root of that of the lens. This gives zero reflectance at λ_c for normal incidence light (of whatever polarization) and typically less than 2% reflectance over one octave in wavelength and for incidence angles less than 15°. In the optical and UV domains, but not the NIR, this is somewhat theoretical though as there are no two transparent and durable materials with such a wide refraction index contrast. Cheap AR coatings for cameras and prescription glasses are usually overcoated with a quarter wave layer of MgF_2 whose refractive index is 1.386 at 500 nm, versus typically 1.5–1.65 for cheap glasses.

On the opposite side, a perfect AR coating would consist of a material whose refractive index would continuously vary from the bulk material value on the inside to the air index ($n = 1$) on the outside. In electrical terms, this would give perfect impedance matching between the surrounding air and the lens material, and hence no light loss at all. Incredibly enough, this can be rather closely achieved by spin-coating lens surfaces with variable porosity silicon prepared through a sol–gel process: see Cleveland Crystals Inc. for commercial availability up to 300 mm diameter, in particular for coating highly fragile crystals since the process is very gentle. Another process to cover a wide wavelength band, still at the research laboratory stage, is to print periodic nanostructures on the surface, mimicking biostructures on moth's eyes. With top-structured pyramids of about 1 μm size, better than 1% reflection has been obtained over 0.5–2.5 μm, or 2.3 octaves.

Most coatings for astronomical purposes consist instead of a sandwich of dozens of thin multilayers made of alternating high and low refractive index materials. Layer thicknesses of, for example, MgF_2 and ZnSe thin films are tailored to produce destructive interference in the beams reflected from the glass–air interfaces, and reciprocally constructive interference in the corresponding transmitted beams. In practice, one can get better than 1% reflectance over one octave wavelength and for incidence angles less than 15°. The layers are deposited one by one in a vacuum chamber. Whenever possible, it is better to get hard coatings that can be easily dusted off and cleaned. The hardening part involves baking the coated lenses at high temperature though and only soft coatings can be applied to fragile materials such as crystals and fluoride glasses. It is usually possible to get such complex vacuum-deposited coatings from commercial firms over about half a meter in diameter lenses (or mirrors).

1.6.5 High Reflectivity Coatings

Mirrors present a somewhat different set of challenges, whether working in the NIR or in the optical to near-UV domain.

For the NIR domain, beyond about 1 μm, an extremely durable very high reflectivity ($\geq 99\%$) can be readily obtained by overcoating the glass or metal mirror with a layer of vacuum-deposited gold. From Kirchhoff's law, this also means that the thermal emissivity of a such a mirror is below 1%, meaning that even at room temperature its thermal IR emission will usually be negligible.

For the optical domain, a protected silver coating gives good reflectivity (90–98%) in the 0.5–1 μm range, but does not cover the blue and near-UV domains. To do that, multilayer dielectric coatings are needed and can be obtained from a few industrial firms, very much like for AR coatings. Extremely high reflectivity $\geq 99.99\%$ can even be attained, but only for a single wavelength, a single incidence angle, and a single polarization to boot. For the usual astronomical requirements of 1 octave wavelength range, reasonable incidence angles, and unpolarized light, very good reflectivity $\sim 99\%$ is achievable, if by stacking up to ~ 100 coating layers.

One difficult case is the coating of the large telescope glass mirrors. They usually work from the atmospheric UV cutoff (0.31 μm) to the atmospheric IR cutoff at 24 μm, a huge spectral range for which there is no good solution, only trade-offs. For almost all telescopes, the lesser bad choice is a thin (~ 100 nm) vacuum deposited aluminum layer, spontaneously overcoated by a very thin Al_2O_3 layer as the vacuum chamber is opened to outside air. During the first few weeks of operation, this gives a respectable >92% reflectivity over the whole range (except for an 86% dip around 0.85 µm). Quite unfortunately, aluminum coatings do age though and, even with mirror cleaning every month or so to remove dust particles, reflectivity in the visible range drops to maybe 80% within 18 months, and recoating, usually with a custom plant at the telescope premises, needs to be performed. Alternatively, the Gemini-South 8-m diameter telescope mirror is currently overcoated with a custom-protected silver coating: it cannot be used below 0.4 µm, but is more durable and gives better and more stable reflectivity than even fresh aluminum for all wavelengths beyond 0.45 µm, also provided it is regularly cleaned.

1.6.6 Conclusions

By way of summary, here is the typical, if somewhat convoluted, way in which an optical subsystem for a major astronomical instrument is developed over possibly a 3–5 year time span:

- The instrument's main optical train is defined by the user, with all fundamental parameters (field, focal length, wavelength range, pupils, disperser, detector), and special requirements clearly set up.
- Detailed optical computation, including optical and mechanical tolerances, is performed by an optical engineer, with some iterations back and forth to the previous step and to the mechanical engineer in charge of the instrument design. A detailed specification and requirement document is then sent out for competitive tendering.
- Optical fabrication is contracted out to an optical firm, with some iterations back and forth to the previous step, for example, to refine glasses index of refraction to that of already available blanks, the radiuses of curvature to those of tools already available, etc. Acceptance tests both at the manufacturer premises and in-house are clearly spelled out. For ultra-high precision optics, it is not uncommon (at a cost though) that the most critical lens is produced first, and its actual parameters accurately measured and used for a next iteration of the whole optical system design.
- Coatings are usually subcontracted by the optical firm. Their specification and progress must be closely followed, because of the high potential for bad performance – for many possible reasons such as improper lens or mirror cleaning before deposition, deposition on the wrong surface, and even use of radioactive materials (!) that would saturate the detector. Besides, vacuum deposition of hard layers is not a gentle process, and lens/mirror cracking or permanent surface distortion is quite common, especially when exotic glasses are used. Any such event can easily delay the project by a year.
- Progress meetings/reports are regularly performed as contractually agreed upon.
- Subsystems delivery and acceptance are performed as contractually agreed upon.

1.7 Mechanics, Cryogenics and Electronics

1.7.1 Mechanical Design

Mechanical design of an astronomical instrument is a complex venture. The primary role of mechanics is to house the optical components of the instrument with accuracies/stabilities that can range from 1 mm to 1 nm (a factor of 1 million range!), depending on the component functions. It must allow for easy, precise, and stable adjustments, at first for the initial instrument assembly, but also during its whole lifetime. It almost always provides for the motion/exchange of key optical components (filters, gratings, scanning interferometric system, etc.), again with highly variable requirements in terms of accuracy and stability. Note that accuracy and stability are two separate issues; for example, exchanging one plane reflective grating with another for, say, modifying the spectral resolution calls for a modest accuracy (a 1 mm centering error for a 15 cm diameter grating would not make any sizeable harm), but requires an extremely good stability during a whole 1-h exposure, usually within a few micrometers. It is important to analyze quantitatively all these requirements before starting the mechanical design of the instrument.

Proper housing of optical components means putting them firmly and accurately in place in their holders, yet without exerting any strong mechanical constraint that would distort their shape, or even break the glass. One important mounting concept here is that of mechanical degrees of freedom: any optomechanical component has six degrees of freedom that fully define its position in space and must be all constrained to get a highly stable and repetitive positioning, possibly down to a fraction of a micron; see Figure 1.14 for an archetypal example using three spheres in contact with three grooves. In the laboratory, one can rely on gravity to ensure that components stay on their contact points (except during a strong earthquake!). For telescope-mounted instruments, one adds springs exerting forces orthogonal to the glass surface and directed toward the support points: any mismatch here would create constraints that would distort or even break the glass.

For less demanding applications, say at the 10 µm repetitiveness level, one can slightly overconstrain the system, for example, with extended soft contact points. For even less precise requirements, one can opt out entirely of the kinematic mounting business and overdefine the component positions, for example, insert a thin elastic rubber band between a lens and its housing: this is easier to install and much more gentle for a glass component, but also less stable and much less repetitive. For more on the fascinating issue of kinematic mountings, you may look at the nice University of Arizona tutorial at http://fp.optics.arizona.edu/.

Mechanical development also comprises housing of electronics systems (instrument and detector control), fluids cabling (electric power, data stream, cooling fluid, dry gas lines), and various calibration systems. Add that mechanical design complexity is more than often underappreciated and systematically "slaved" to the more glamorous optical design effort to boot, it is no wonder that this is historically **the** main source of frustration and delays in (astronomical) instruments development.

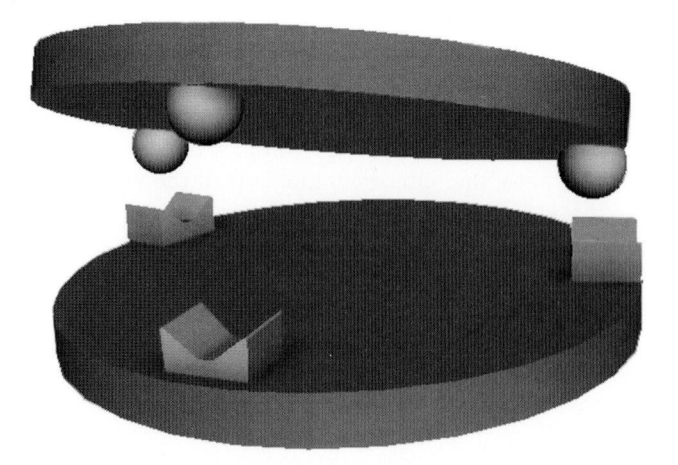

Figure 1.14 This classical kinematic mounting features three optically polished sapphire spheres glued at 120° to the underside of a component and three right-angle hardened ground steel grooves at 120° on top of its mounting plate. This gives the required six contact points, ensuring incredibly stable and repetitive positioning. The upper component can be removed and then put back (gently) in place, and without any adjustment repositions itself within a fraction of a micron. Note that this requires only very lax absolute accuracy (say only at the millimeter level) for the relative positions of the spheres and grooves. (Reproduced with permission of Colombine Majou.)

One significant difficulty is that most astronomical instruments call for extreme stability, yet are more than often moving a lot with respect to gravity with a typical angular speed of 15° per hour as the telescope tracks the sky: that is a lot when a few micron stability per hour is required, for instance, between the optics output and the detector. There are quite a number of ways for a telescope to feed its instruments (Figure 1.15): Prime Focus and Cassegrain instruments are actually moving in two dimensions; Nasmyth instruments are rotating orthogonally to gravity at the cost of one more mirror; folded Nasmyth are rotating along gravity (hence with no differential flexures) at the cost of two; finally coudé instruments are fully stable at a minimum additional cost of three mirrors (or alternatively tens of meters of optical fibers). It is usually a good idea to opt for a nonmoving instrument when stability requirements are stringent and the field of view is small enough to make it possible. Note that for moderately large fields (say 1′ diameter for a 10-m telescope), a so-called field derotator (e.g., a three-mirror combination, see Figure 1.16) can be inserted in the optical beam in order to feed a fixed-orientation instrument on a Nasmyth platform: the instrument still rotates slowly as the alt-azimuth telescope tracks an object on the sky, but only along gravity.

Still, many instruments are actually operated under big gravity changes and must carefully be designed to get negligible flexures, and, even more important, correctly built to avoid mechanical instabilities. It is not uncommon to find that a given instrument, which according to finite elements analysis should not flex

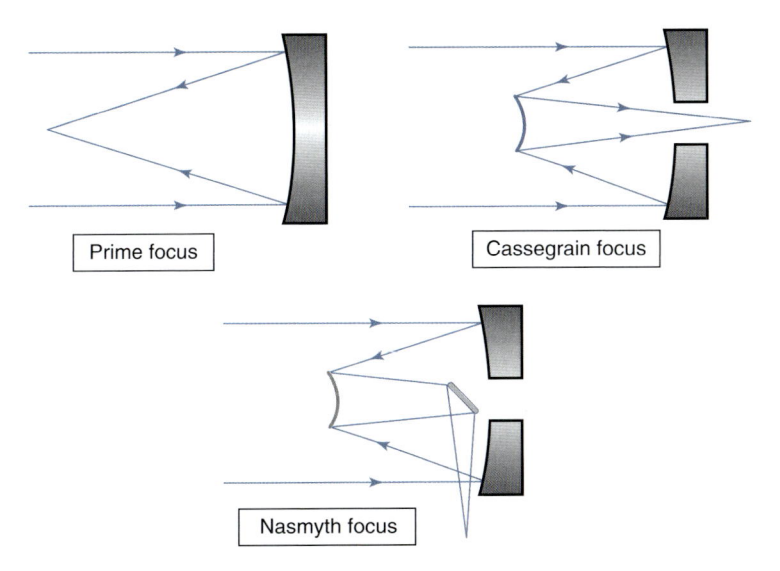

Figure 1.15 The three main telescope foci are shown, namely prime focus, Cassegrain focus, and Nasmyth focus. Additional mirrors are needed for the folded Nasmyth and coudé foci. As the telescope tracks during the night along two orthogonal axis, much like a warship turret, instruments at prime focus and Cassegrain focus move along, and on top usually rotate to cancel field rotation. At Nasmyth focus, owing to the rotating tertiary mirror, light is sent to a horizontal rotating platform when the instrument sits; field rotation has still to be canceled, though.

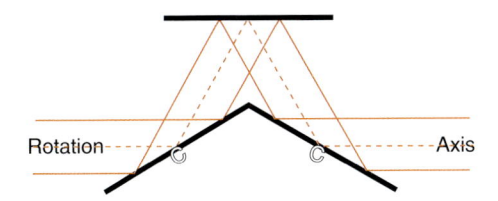

Figure 1.16 This is a schematic view of the classical three-mirror derotator in the case of a parallel beam input. It works also with a convergent beam, for example, with an image of the field on mirror #2. Field rotation is nulled by counterrotating the derotator around its horizontal axis. For small enough light beams, a prism with three internal reflections can be used instead.

by more than a few micrometer, actually internally moves by millimeters, due to, for example, an improperly tightened nut: this is a rather trivial example, but a number of "completed" instruments have never been operated because of unmanageable flexures. In some cases, flexure requirements are just too harsh to be attained with purely passive means; it is then necessary to incorporate active flexure compensating systems, despite the added complexity: as an example, no DVD player would work at its required submicron accuracy level without its many internal control loops.

1.7.2 Alignments

In any instrument design, and not only for astronomical ones, building and documenting an alignment/adjustment strategy is essential. Here are a few cardinal rules:

- Like for any crucial element of a project, if you fail to plan, you plan to fail.
- Start on it as soon as you are at the instrument concept level, certainly not as an afterthought at the end of instrument design. An alignment impossibility is arguably one of the most likely hidden traps that might spell doom for your project right from its start.
- Define carefully the instrument elements that are to be adjusted and how you are going to do it. If you do not put enough degrees of freedom, the instrument will never be aligned and so will never work. If you put more adjustments than needed, you might ultimately succeed, but that will take more efforts and might cause significant delays.
- Evaluate correctly the various adjustment ranges/accuracies required. If you set them too small/too loose, the instrument will never be adjusted correctly. If they are overly large/tight, the instrument might ultimately be adjusted, but with an impact on cost and timeline.
- Design and build adjustment devices that are repetitive, highly stable, and most preferably equipped with digital or analogic encoders. Manual adjustments are much cheaper to develop, but more than often could not be accessed safely, at least on large telescopes: fully motorized and encoded adjustment systems are then required. All that design and implementation effort might cause project overcosting and delays in the short term, but chance is that it is going to be recouped many times later on.
- When performing instrument adjustments, do not hesitate to be dumb and lazy and proceed empirically: see, for instance, Exercise 6 for a rather generic "blind" adjustment scheme that avoids figuring out the precise metrology of the adjustment scheme.

1.7.3 Cryogenics

Since the demise of the photographic plate, every astronomical instrument features at least one cryogenic system in order to operate its digital detector at proper temperature, around 150 °K for CCDs and 75 °K for NIR arrays. This requires integrating the detector and its proximity electronics in a cryostat (in essence a magnified thermos bottle) under good vacuum and developing a cooling system, generally either a liquid nitrogen bath or a cryocooler. In the latter case, much care is needed to avoid the vibrations from the cryocooler operation propagating inside the instrument. In the CCD case, the window cryostat is usually the interface between the cooled and uncooled parts of the instrument. To avoid frost formation on the outside face of the window, one can, for example, maintain a gentle nitrogen flow in front of the window. All in all, this requires a significant number of cables, pipes, and regulating mechanisms.

Near-IR spectrometers installed at normal telescope sites (meaning neither in space nor in Antarctica) must be entirely cooled when working above ~ 1.65 μm.

Looking backward along the light path, this must extend up to the entire focal plane of the instrument: the reason is that the disperser is actually "looking back" over a full hemisphere to any thermal radiation from the instrument mechanics and sending a good fraction of it along the main optical path, ultimately up to the detector. Any detector pixel thus sees thermal radiation over the full spectral range, but science light only over one resolved spectral element. NIR instruments are thus installed inside big cryostats, with usually a strong thermal gradient between the instrument entrance at say 150 °K, which is enough to get negligible thermal radiation up to 2.4 μm, and the detector support at 75 °K, about the best operating temperature for the detector array. Apart from the additional cost, this makes initial adjustment and integration, as well as subsequent repairs, excruciatingly slow: a week cycle forth and back to cryogenic operation for a few hours of repair on an open cryostat at room temperature is a frustrating but common occurrence. Also, any motorized motion inside the cryostat is difficult and expensive to develop, as the very few cryocompatible motors tend to exhibit vanishingly small torques and small lifetimes when operated at such low temperatures. It is also not uncommon that a failing motion at cryogenic temperature reworks spontaneously when the cryostat is still closed but already back to less frigid conditions. This makes repairs even more problematic, such as for proverbial car's faults that never occur at dealers' premises.

1.7.4 Electronics and Control System

Like modern cars, most astronomical instruments (including of course 3D-spectrometers) incorporate many motors and encoders and cannot be operated without fully automated control systems. Given that the time to, for example, reconfigure hundreds of fibers in the focal plane of a multiobject instrument to address a different sky field, exchange a grating to modify the spectral resolution, exchange a filter to modify the wavelength range, and so on, is time lost for observing, it is vital to develop optimized automatic sequences in order to perform these functions in parallel. Note that all this requires a lot of cabling, which must be done most professionally. It is not uncommon that analysis of the failures history of an operating instrument shows that more than half are connected to cabling problems, especially for instruments that are regularly taken in/out of their telescope focal locations.

Detector environments too usually require a number of controlled functions, such as a light shutter for all CCDs (to be closed during frame readout) or an internal focusing mechanism for most NIR instruments (because of the huge temperature difference between an opened and closed cryostat). A highly specialized electronic controller is also needed to set up the detector voltages, start/end the exposures and read out the detector pixels. This is often done with in-house built electronic racks; in some cases the required functions are (mostly) provided by an off-the-shelf integrated circuit developed by the detector provider.

Given the number and sophistication of controlled functions in any modern astronomical instrument, operation is now always fully computer controlled and closely integrated with the operation of the telescope. To save precious telescope time, whenever possible the various steps needed are done in parallel, for example, disperser and filter exchange while the telescope moves toward the next target. User's friendly feedback is continuously sent to the observer, usually through a graphical user interface. In the few cases where the observer must actually provide a real-time input, for example, precisely set a narrow slit on the science target, a "super" user's friendly environment is provided. For instance, just clicking on two locations on a sky image taken with the instrument will automatically move the telescope and set its field rotator to place a slit or an integral field unit at the required sky position.

1.8 Management, Timeline, and Cost

As can be gathered from the technical complexity of 3D spectrometers, successful development of such full-scale instruments requires heavy management investment over a long timeline – up to a decade from first concept to start of routine operation – and carries a substantial global cost, easily in the tens of million Euros/USD.

Just to give a flavor of what is generally needed for a successful endeavor, here are a few pragmatic "rules" for the many project stages (see Figure 1.17):

- At start, there is a Concept, a Principle Investigator (P.I.) eager to transform his/her goal ("a goal is a dream with a deadline," Napoleon Hill 1925, The Law of Success) into a Project to be carried out over a 10-year period or so, and at least one Institution eager to get the instrument and ready to support a team of competent people led by the P.I. and find/provide adequate funding. Very often, this very first phase takes actually many years of preliminary studies and intense lobbying before the Project gets its first green light and moves to the definition phase, usually with a flashy acronym attached.
- In every case, at least most subsystems, or even the whole instrument, will be built by high-tech industrial firms. Always remember that they are most likely to know better than you how your specifications can be achieved at minimum cost and/or timeline by their technology: impose only really needed specs, and never how they are going to be met, making industry part of the solution, rather than of the problem. Remember also that tradeoffs are unavoidable, as

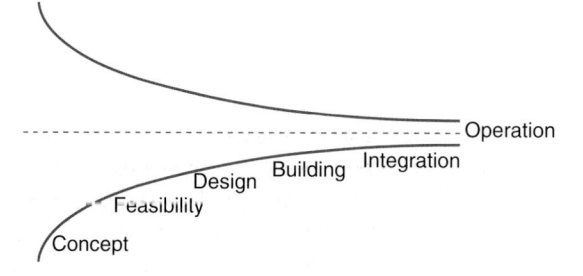

Figure 1.17 Project Funnel. This small cartoon illustrates how starting from a broad concept, any instrumental project becomes more and more tightly defined as it moves through successive stages toward start of operation. Along the way, uncertainties, in terms of cost, timeline, and/or performance drastically decrease, that is for a successful project.

per the (already optimistic) adage "Performance, Cost, Timeline, choose two." Competitive tendering (when at all possible) is a must, as a factor of 4 overall range in the industrial offers for the same optomechanical component is not uncommon. Beware, while you might have got a very cost-effective deal in the tendering process, changing later the specs at the building phase means reopening the whole cost issue, a recipe for heavy cost increases: this is a widely valid warning, not limited to astronomy or even high-tech developments, but which applies to any big project, as attested by many high-profile horror stories of public building developments.

- Any substantial Project goes through a number of phases, namely: Feasibility Study; Preliminary Design; Final Design; Fabrication; Assembly, Integration and Testing (AIT); Telescope Installation and Commissioning. Each step carries the Project to a well-defined level, including detailed planning of the next step, and ends with a Review, preferably led by external consultants. Projects can be deeply modified or even canceled following any Review by whatever organization the builders are working for, in particular, should meeting reasonable performance, cost, and/or timeline appears highly problematic.
- The basic aim of Feasibility Study is to show that the project can be achieved at minimum risk with the required performance at an affordable cost and within a reasonable timeline. This is for a substantial part "just" a paper effort, but with additional in-house and/or external technological developments to prove the validity and cost of the key concepts of the instrument. Preliminary and Final Design are in principle just design phases as their names imply, again producing a lot of (electronic) papers, but are often coupled with procurement in parallel of long-lead subsystems, such as the main optical components. These two phases lead to the Fabrication phase, which includes careful components/subsystems validation, both at the manufacturers' premises and subsequently in the AIT hall.
- AIT is the next crucial step, with the T (Testing) generally the longest and most expensive part. In particular, it usually requires special premises. That may include an integration hall with proper environmental parameters (temperature, hygrometry, dust level, vibration level, etc.) and various equipments (handling tools, vacuum pumps, gas containers, cryogenics, electric power, etc.), a metrology laboratory for subsystem test and acceptance, a clean room for detector integration, and any required custom-made test equipment: this can be a full subproject by itself that needs to be planned and even possibly fabricated well in advance as an integral part of the whole Project. In the 1980s, the then fashionable time- and cost-saving strategy of careful testing only the subcomponents as a way to avoid the higher burden of whole instrument testing resulted in the one billion dollars Hubble Space Telescope near-fatal disaster: a single error in the optical testing of the secondary mirror, compounded by NASA's adamant refusal for any full system testing, resulted in an orbiting telescope delivering strongly aberrated images. This was eventually fixed by adding optical correctors in front of each on-board instrument, but this now proverbial story remains as a reminder to come back to the harsh but much more sensible way of full instrument characterization prior to shipping to a distant mountain and even more to outer space.

- For a really big project there will be many specific team positions: Project Investigator; Project Manager; Instrument Scientist; Control Manager; Mechanical, Optical, Electronics and Software Leaders, and so on. For a smaller one, the same functions are still needed but are concatenated to be filled by fewer people. For both small and large projects, advanced work planning (often on a yearly basis), with at least monthly progress assessments, is a must. Keeping such large teams eager and enthusiastic over a decade or more, while facing unavoidable problems and setbacks, is by no means a small management challenge.
- Delivering proper documentation is a major part of any project, and a vital tool during the instrument operating phase. Yet, it is too often seen as a chore, to be (badly) done after instrument building has been achieved. *Au contraire*, it should be started early, preferably already at the feasibility study level: Early drafts of, for example, the Users manual, the Alignment, Control Software, and the AIT documents are actually a great way to catch well ahead of time intrinsic problems that might later kill the Project as such, or at least cause big delays and/or cost overruns.
- Providing, documenting, and maintaining a near real-time pipeline that extracts and visualizes the 3D data in physical units (intensity versus wavelength at each sky position) at most a few minutes after observation, is no small feat, but essential to evaluate the validity of typically 1-hour long observations. Together with a thorough off-line pipeline for proper data reduction delivering science-ready products, equally maintained during the whole life of the instrument, requires a big manpower investment over decades. This is also one of the best ways to 'sell' the instrument to its potential users, the ultimate touchstone for instrument success.
- Any project has risks. You can strive to minimize them, but you cannot eliminate them altogether. On one side, you can develop your 10-year project using only well-seasoned risk-free technologies right from the start, and chance is that when completed it will be fully uncompetitive. On the other side of the fence, you could redefine your project for any technological advance and/or new science drivers that happen along the line, and chance is that the instrument will never be completed. Staying at the optimum risk level that maximizes the expected scientific value of the instrument is arguably the best winning strategy, even if much easier said than done.

1.9 Conclusion

As can be gathered from this chapter, developing a state of the art (3D) instrument is a long, expensive, complex, and risky high-tech endeavor. It requires in particular building a large competent team in all relevant technical domains, selecting the best high-tech vendors and learning from them, and during the long development phases striking a delicate balance between rigorous long-term planning and creativity.

This long-term effort is not limited solely to the instrument design, building, and installation phases, but extends as well during its whole operating life, either still by the original building team or by the instrument host observatory, or as a combination of both. It is in particular quite common to rejuvenate an ageing instrument, for instance, by implementing a new state-of-the-art detector or changing its spectral domain and/or spectral resolution. On the other hand, as over time an instrument becomes decidedly uncompetitive, it is usually better to reject aggressive and futile therapy, and build instead a worthy successor. In any case, only a cradle to grave investment can give an instrumental facility the chance to achieve its full observing potential over a reasonable length of time.

*** **Exercise 1**   **Trying to beat etendue conservation #1**

You have been tasked with detecting an all-sky night emission line with a ground flux of 20 ph.cm^{-2} s^{-1} sr^{-1}. Back in the early 1960s, your best detector is an 12% quantum efficiency (QE), 7 mm diameter, photomultiplier, accepting light from a 30° half-angle conic beam. Detector r.m.s. noise is 12 counts per second. The line is selected with a 38% peak transmission narrow-band interference filter. Since the detector QE varies over its sensitive area, for better signal stability, you image the pupil on the detector, with telecentric beams (i.e., sky image at infinity). A reminder: the solid angle of a cone of half-angle θ is $\pi \sin^2 \theta$.

1. Shooting the detector directly at the sky, with a baffle somewhere to avoid input light out of the detector acceptance cone, what count number per second do you get? Hint: First compute the detector etendue.
2. To improve the situation, you put a telescope in front of your instrument in order to collect more light. What is the optimum telescope diameter (if any)?
3. Undeterred, you put a tapered cone in front of the photomultiplier with a d mm input diameter and of course a 7 mm diameter output. What is the best d (if any)? What happens to the light rays?
4. As a last resort, you now insert a tapered paraboloid with its focus at the center of the sensitive surface of the detector. With perfect concentration of all light rays parallel to the optical axis at the center of the detector, will you finally collect more light?

Answer of exercise 1

1. Detector maximum etendue: $(\pi/4) \times (0.7)^2 = 0.385$ cm^2 sr. Photons to detector counts efficiency: $0.38 \times 0.12 = 0.0456$. Detected flux: $20 \times 0.385 \times 0.0456 = 0.35$ counts per second. With the 12 counts per second r.m.s. detector noise, it looks like a short integration, ~ 300 s would lead to say a 5σ detection. However, at the time, photomultipliers had highly unstable noise properties, and a signal of about 3 counts per second was already at the detection limit, irrespective of integration time.
2. Light is collected on a much larger pupil area, but with a smaller solid angle on the sky. Because of etendue conservation, the detected flux is the same, actually smaller because of the not perfect light transmission of the telescope.

3. More of the same, again because of etendue conservation. One might wonder why a larger beam etendue than permitted by etendue conservation injected in the tapered fiber does not strike the detector. A careful look shows that these extra rays, after a few reflections on the cone wall, just come back along the cone, ultimately up to the sky.
4. Again no way, for the same reason. Yes, the whole beam parallel to the optical axis entirely strikes the detector. However, with zero etendue, it carries zero energy. Note that one of the authors (GM) has been indeed tasked at the time to find a way to detect such a source, and went through these steps one by one, including showing with a messy computation[5] that the paraboloid off-axis aberrations indeed prevent breaking the etendue limit.

***** **Exercise 2**   **Trying to beat etendue conservation #2**

Let us take a hypothetical 10″ diameter ionized gas cloud at a galaxy center with a uniform brightness narrow emission line at 656 nm. The gas is rotating as a solid body, with constant integrated radial velocity along the sky plane projection of its rotation axis (the minor axis) and a linear radial velocity gradient $G = 20$ km s^{-1} (″)$^{-1}$ along the projected orthogonal axis (the major axis) from one edge to the other. You are using a long-slit spectrograph observing facility with the following parameters: telescope diameter $D = 3.6$ m, grating diameter $d = 150$ mm, and blaze angle ϕ to be derived.

1. Find the grating blaze angle for which a wide, 10″ width, slit parallel to the cloud minor axis (thus collecting the whole cloud light) gives nevertheless a narrow spectral line on the detector.
2. Assuming 100% optics transmission, this spectral line is an order of magnitude brighter at the detector location than on the sky. Now, here is the tough question: this is a clear violation of the sacrosanct second principle of thermodynamics, right?

Answer of exercise 2

1. Spectral resolution $\mathcal{R} = \lambda/\delta\lambda$ for a 1″ wide slit is equal to $10^5 d\,(\tan\phi)/D$. $\mathcal{R}$ is also equal to c/G', where G' is the radial velocity gradient across the 1″ slit. With $G' = -G$, the emission line peak beams are on top of each other and thus fall on the same detector pixels. This requires $\tan\phi = 10^{-5} cD/dG = 3.6$, or a steep but feasible 75° blaze angle high-order echelle grating.
2. Well, yes and no. The second law as expressed so far —never decreasing entropy in a closed system — is indeed spectacularly violated here. However, the fully correct extended 2nd law — never decreasing entropy plus information in a closed system — is not. It is the known a priori information (location, orientation, and magnitude of the object radial velocity gradient) that is used to achieve this seemingly impossible feat. Note that a similar experiment had actually been done successfully in the 1950s (De Vaucouleurs G., private discussion).

---

5 You are of course most welcome to repeat it.

** Exercise 3   Prism etendue

The goal is to illustrate the small linear etendue of the prisms in the direction of dispersion. A spectrograph at the focus of a $D = 4$ m telescope is using a $d = 75$ mm, 60° apex angle A, BaF10 prism at minimum deviation. Slit width projected on the sky: $\alpha = 1''$ ($\sim 5.10^{-6}$ rad); camera aperture ratio $\Omega = 1/1.5$. Glass index of refraction n given by $n = B + C/\lambda^2$, with λ is in μm, $B = 1.67$; $C = 0.00743$. Central wavelength $\lambda_c = 0.55$ μm.

1. Compute the instrument spectral resolution at central wavelength. Hint: use the prism spectrograph cooking book insert.
2. Compute the slit width w projected on the detector.

Answer of exercise 3

1. Index $n = 1.695$ at 0.55 μm. From prism spectrograph insert, spectral resolution is $\mathfrak{R} = (K\Delta)d/D\alpha$. From the prism data: $K = 1.883$ and $\Delta = 2C/\lambda^2 = 0.049$. Finally $\mathfrak{R} = 346$, just a small value at the lower limit of spectrographic resolution.
2. Etendue conservation gives $w\Omega = D\alpha$, hence $w = 30$ μm, or typically 2.4 CCD pixels.

** Exercise 4   Grating etendue

The goal is to compare the linear etendue of gratings versus prisms in the direction of dispersion, with a similar instrumental setting as in Exercise 3. A spectrograph at the focus of a $D = 4$ m telescope is using a $d = 75$ mm diameter transmission grating (first order Littrow mounting) with a blaze angle $\phi = 30°$. Slit width projected on the sky: $\alpha = 1''$ ($\sim 5.10^{-6}$ rad); camera aperture ratio $\Omega = 1/1.5$. Central wavelength $\lambda_c = 0.55$ μm.

1. Compute the spectral resolution at central wavelength. Hint: use the grating spectrograph cooking book insert. Set up the grating ruling (number of grooves per millimeter).
2. Compute the slit width w projected on the detector.

Answer of exercise 4

1. From the grating spectrograph insert, spectral resolution is $\mathfrak{R} = 2(\tan \phi) d/D\alpha$. From the grating data: $2\tan \phi = 1.155$, giving $\mathfrak{R} = 4330$, a comfortable value, 12 times larger than with the equivalent prism-based instrument. Using $2\sin \phi = a\lambda_c$ gives $a = 1082$ grooves per millimeter.
2. Etendue conservation gives $w\Omega = D\alpha$, hence $w = 30$ μm, or typically 2.4 CCDpixels, of course the same as with the equivalent prism spectrograph.

*** Exercise 5   Grism Rotation Invariance

We consider a generic zero deviation grism, made of a blaze angle ϕ (in air at central wavelength λ_c) VPHG sandwiched between two identical apex angle ϕ

prisms of refractive index n_c. Because of their zero mean deviation, different grisms can be exchanged in the instrument, with, for example, different wavelength ranges and/or wavelength dispersion, with the spectra automatically centered on the detector. Given the use of remotely controlled exchange mechanisms, it is difficult to avoid small rotation angle uncertainties ϵ of the grisms in the dispersion plane, the consequences of which are evaluated here.

1. Find the emergent angle $(\phi + \epsilon')$ at central wavelength λ_c for an incident angle $(\phi + \epsilon)$. Derive the zero deviation departure $(\epsilon' + \epsilon)$. Hint: evaluate approximatively ϵ' as a second degree polynomial in ϵ.
2. Compute the corresponding spectral shift $\Delta\lambda$ versus rotation error ϵ. Find the relationship between the grism rotation error ϵ and the resulting spectral shift 'resolution' $R = \lambda_c/\Delta\lambda$. Does that ring a bell?
3. Taking the same spectrograph as in Exercise 4, what rotation angle error ϵ will shift the central wavelength by one slit width? How does it compare with the slit angular width as seen from the grating.

Answer of exercise 5

1. From the classical grating law, $\sin(\phi + \epsilon) + \sin(\phi + \epsilon') = 2\sin\phi$. Taking $\epsilon' = k_1\epsilon + k_2\epsilon^2$ with the well-known approximations $\sin x \sim x$ and $\cos x \sim (1 - x^2/2)$, one gets $k_1 = -1$ and $k_2 = \tan\phi$. The error angle $\epsilon + \epsilon'$ is equal to $\epsilon^2 \tan\phi$.
2. Using again the classical grating law, but now looking at its λ dependence, we find $\epsilon = \sqrt{2/R}$. And, yes, the wavelength shift versus grism tilt angle (in the dispersion plane) and the wavelength shift versus Fabry–Pérot tilt angle (in any plane) obey similar laws.
3. From Exercise 4, $R = 4330$, hence $\epsilon = 0.0215$ or $1.23°$ at the grism level. Slit width is 1" on the sky, hence 1" × (4000/75), or 0.000267 rad at the grism level. Their ratio is ~80, which can be seen as the grism invariance figure of merit.

* Exercise 6   Optical Adjustments for Dummies

To adjust a mirror inclination inside your instrument, you use a point-light source centered on its input field, the image of which on the 15 μm pixel CCD must be exactly centered, that is, at $x_c = 2048, y_c = 2048$ in pixel units. Initially, the image is instead at $x = 2274.4, y = 1852.7$. You are using two rotating linear screws with encoders. From your own calibration, one positive (clockwise) 360° turn of the screws moves the light source on the detector by $\Delta x_1 = 122.7, \Delta y_1 = 85.4$ for screw #1, $\Delta x_2 = 97.1, \Delta y_2 = -101.6$ for screw #2.

1. Assuming fully linear behavior, what are the screw rotations $\Delta\theta_1, \Delta\theta_2$ in degrees required to make the adjustment?
2. Applying the computed corrections, you now find the source image at $x' = 2046.1, y' = 2048.9$. What screw rotations do you apply now?
3. You should be now happily dead centered: why? What can you say about the relative actions of the two screws in length scales and orientations?

Answer of exercise 6

1. From the calibration: $\Delta x = (122.7/360)\,\Delta\theta_1 + (97.1/360)\,\Delta\theta_2$ and $\Delta y = (85.4/360)\,\Delta\theta_1 - (101.6/360)\,\Delta\theta_2$. Inverting the corresponding 2×2 matrix gives $\Delta\theta_1 = 1.7621\,\Delta x + 1.6840\,\Delta y$ and $\Delta\theta_2 = 1.4811\,\Delta x - 2.1280\,\Delta y$. With the required adjustments $\Delta x = -226.4$ pixels and $\Delta y = 195.3$ pixels, this gives $\Delta\theta_1 = -70.05°$ and $\Delta\theta_2 = -750.92°$.
2. Now $\Delta x = 1.9$ pixels and $\Delta y = -0.9$ pixels. This gives $\Delta\theta_1 = +1.75°$ and $\Delta\theta_2 = +4.84°$.
3. The first iteration missed the target at the 1% level. The second one should give a similar accuracy, that is, this time better than 0.1 pixel adjustment: this has been done just from an empirical calibration, without having to figure out the mirror actual motions, a thankless task. Normalizing the two $\Delta\theta$ relationships gives $\Delta\theta_1 = k_1\,(x\cos\alpha_1 + y\sin\alpha_1)$ and $\Delta\theta_2 = k_2\,(x\cos\alpha_2 + y\sin\alpha_2)$, with $k_1 = 2.437$, $\alpha_1 = 43.7°$, $k_2 = 2.593$, $\alpha_2 = -55.2°$. The two screws thus move the light beam by slightly different amounts, their scale ratio being $k_2/k_1 = 1.064$. Their two directions are also not exactly orthogonal, with an angle $\alpha_1 - \alpha_2 = 98.9°$.

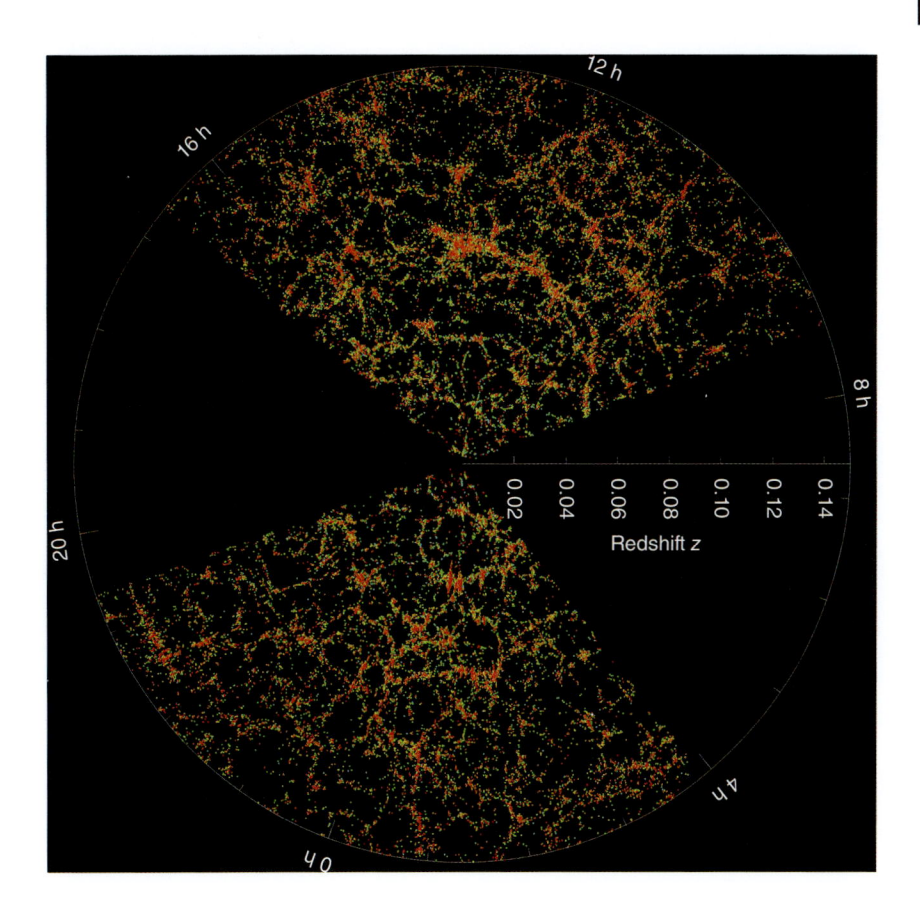

The Sloan Digital Sky Survey has created the most detailed three-dimensional maps of the local Universe ever made, with deep multicolor images of one third of the sky, and spectra for more than three million astronomical objects. The maps show the distribution of galaxies of the local universe [121]. Each dot is a galaxy; the color bar shows the local density. These observations have been obtained with the multiobject fiber-based spectrograph of the SDSS 2.5 m telescope at Apache Point Observatory. ([120]. Reproduced with permission of Michael Blanton.)

2

Multiobject Spectroscopy

2.1 Introduction

A deep sky image of a few square arc minutes with a 4- to 8-m class telescope shows thousands of distant galaxies as small light blobs covering a small percentage of the CCD or NIR array area, with only a few local stars belonging to our Milky Way. To study such a field, one needs to record detailed spectra from as many galaxies as possible, and clearly, taking all these spectra one by one would just take forever. It is thus highly tempting to find a way to cram as many 1D spectra as possible on a large 2D detector. This sometimes labeled 2.5 D (instead of true 3D) technique is called multiobject spectroscopy (MOS) and is presented in this chapter.

2.1.1 MOS History: The Pioneers

MOS is actually an old astronomical technique. As early as 1885, Henry Draper began to use a low dispersion prism in front of an 11-in. diameter photographic refractor to get hundreds of stellar spectra in a single exposure on a large photographic plate. This eventually led to the Draper catalog, classifying the 400,000 brightest galactic stars spread over the whole sky. By the 1950s, the same very simple technique (in terms of hardware, not of data extraction given the many spectra overlaps over the field), installed on telescopes of up to 80 cm diameter, was able to reach thousands of stars in one go over typically a few square degrees field. It was also quite easily installed on large telescopes, using an optical relay with a prism, or more often a grism, put in an intermediate pupil between the optical relay collimator and the camera.

Since the 1960s, however, and for ground-based observations, the scientific usefulness of this approach has been sharply limited by a nagging contrast problem: each useful pixel on the detector gathers star light over one spectral element, that is, a few nanometers wavelength width only, but also the night sky background light integrated over the full spectral range covered by the instrument, for example, 4000 nm wide. Faint objects signal is thus all but swamped out by the background photon noise of the sky, limiting this approach to bright galactic stars only. The "slitless" approach is thus now essentially dead for ground-based observations. As presented in Section 2.2.1, it remains on the other hand very much alive for space-based observations in the UV, optical

Optical 3D-Spectroscopy for Astronomy, First Edition. Roland Bacon and Guy Monnet.
© 2017 Wiley-VCH Verlag GmbH & Co. KGaA. Published 2017 by Wiley-VCH Verlag GmbH & Co. KGaA.

and near-infrared domains, since the night sky seen from above the Earth's atmosphere is extremely dark.

2.1.2 MOS History: The Digital Age

On the ground, the modern MOS concept originated from the never built 15-m diameter new generation telescope (NGT) project put forth by the Kitt Peak National Observatory in the early 1960s. Given the anticipated high cost of the telescope, the project team pointed out that the NGT observing efficiency could be boosted by factors of hundreds by making the spectra of many faint objects in the telescope field in parallel. Implementation was expected to be quite straightforward, at least in principle: a large number of objects are selected in the focal plane of the telescope by inserting either short slits/holes or individual fibers centered on the astronomical targets; light is then directed to a spectrograph that delivers hundreds of spectra in one go. This new approach neatly avoided both the sky background contrast problem and overlap effects nagging the slitless technique.

With the later availability of large digital detectors and the development of sophisticated data extraction and reduction packages, MOS spectroscopy has become arguably the most popular spectroscopic technique on large telescopes. It addresses in particular what is currently one prime observational need for night time astrophysics, namely, huge surveys of millions of galaxies over a sizeable chunk of the sky at small to moderate spectral resolution (a few 10^3). Such surveys have been essential to establish the mass content of the Universe and in particular the existence of dark matter; they are now the main observational way to study the existence and nature of the mysterious dark energy, which apparently provides about 68% of the total mass/energy content of our Universe. Less developed, but coming out fast, is the launching of surveys of millions of stars in our Galaxy at high spectral resolution (a few 10^4) to unravel the complex formation history of its stellar component over its 13-billion year existence.

2.1.3 MOS Flavors

The following three sections deal with the three current MOS flavors, namely, successively slitless spectroscopy, multislit spectroscopy, and multifiber spectroscopy.

2.2 Slitless Based Multi-Object Spectroscopy

2.2.1 Slitless Spectroscopy Concept

As outlined above, this is conceptually the simplest multiobject system, involving just an imaging spectrograph (see Figure 2.1) without the complex focal plane electromechanical systems required by the other MOS flavors (see the next two sections). Any (small size) object in the 2D spatial field acts as a slit and is dispersed by the grating/prism, with the whole spectra bunch imaged on the detector. This means that the position of each object light peak defines its

2.2 Slitless Based Multi-Object Spectroscopy

Slitless spectrograph

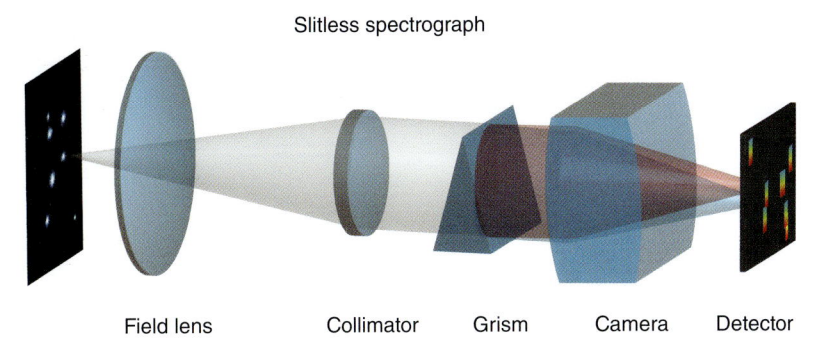

Field lens — Collimator — Grism — Camera — Detector

Figure 2.1 *Slitless spectrograph concept.* An optical relay re-images the sky field at the focal plane of the telescope on the 2D detector at an appropriate focal ratio, typically in the F/1.5 to F/3 range. It also gives an intermediate pupil on which a prism or grism is inserted to provide spectra from any object in the field.

1D spectrum location and that each object extension in the dispersion direction directly impacts (lowers) the spectral resolution of the spectrum. It is thus essential to get an image of each field spectroscopically addressed by the instrument, roughly taken over the same spectral domain, in order to derive absolute wavelengths and mitigate the spectral distortions associated with object size.

One fundamental limitation of this technique is thus that it can address only small size, typically sub-arcsecond, objects: that fortunately includes two prime scientific targets, namely, all galactic stars, but the Sun, and all distant galaxies that light core diameters are typically in the 0.2″ to 0.5″ range. On the other hand, big objects, such as galactic extended ionized gas regions and nearby galaxies do not give usable data, and are actually a nuisance. Another limitation is that spectra of objects roughly aligned along the dispersion direction do overlap, depending on angular separation versus spectral length; this is usually very difficult to disentangle and generally means that such objects must be discarded. One mitigating strategy is to rotate the grating or the field by 90° and take a second exposure for each field, thus recovering a fair fraction of the discarded objects (see e.g., [15]).

Conceptually simple does not necessarily mean technically simple. Optical requirements are more demanding than for a long-slit spectrograph, since the various spectra can originate from any point in the 2D field of view of the instrument. The single collimator mirror routinely used in long-slit spectrographs cannot cover this 3D spatiospectral field because of very strong geometric aberrations, and a refractive (lens-based) collimator is generally chosen instead. Suppressing chromatic aberrations of the collimator and the camera is then especially difficult and usually requires the use of large fluoride glasses and/or large synthetic crystals, which are expensive, fragile, and difficult to polish and antireflection coat. Also, generally, at least one strongly aspheric surface is required in the camera, because of its large FoV. Alternatively, there is an all-reflective solution, free of chromatic aberration, but which require three very large and strongly aspheric mirrors, with drastic centering and stability requirements: in view of its large wavelength range (1–2.5 μm), it is indeed used for the space JWST NIRISS slitless spectrograph.

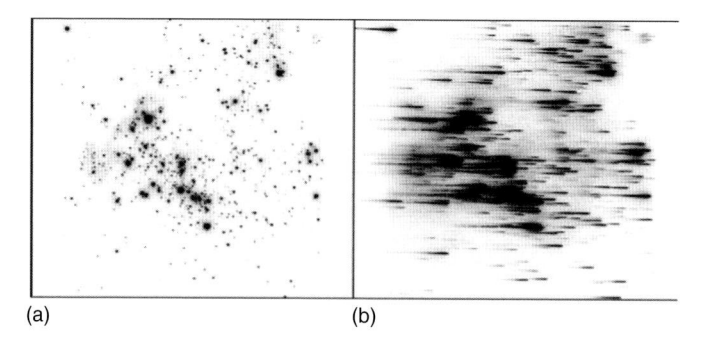

Figure 2.2 (a) 115–362 nm UV image of the young star cluster NGC 604 in the nearby M33 galaxy taken with the STIS instrument aboard the Hubble Space Telescope. (b) STIS slitless prism spectra in the same region. Dispersion is nearly horizontal, and spectral resolutions range from 2500 at 115 nm to 50 at 362 nm. (Credit STScI/NASA.)

2.2.2 Slitless Spectroscopic Systems

As introduced earlier (Section 2.1.1), this approach is now the province of space missions from the UV to the near-infrared domains, as only bright objects can be accessed from the ground owing to the strong night sky background. Figure 2.2, from the Hubble Space Telescope STIS instrument handbook, shows one such use in the UV domain.

For such space applications, it certainly helps that no complex, heavy, and potentially unreliable focal plane target selection hardware is required. Achievable multiplex is generally high, of the order of 1000, even with relatively modest size detectors. Light dispersion is made either with a prism (no parasitic light coming from unwanted orders but low spectral resolutions, usually about 150 or less) or with a grism (much larger spectral resolution, but a number of parasitic orders, including the bright zero order). In large part, the relative simplicity of this hardware is traded off with a number of quite complex software issues in the areas of data extraction and calibration (see Section 11.3.1).

The next such big space-based facility is the EUCLID NISP spectrograph[1], the main goal of which is an all-sky spectrographic survey at $\Re \sim 380$ of at least 50 million relatively bright galaxies in the 1–1.85 µm range, a number quite unthinkable for a ground-based survey. To better disentangle overlapping spectra, each 0.5° field will be observed twice with the grism dispersion in two orthogonal directions. Launch is expected at the start of the 2020s.

2.3 Multislit-Based Multiobject Spectroscopy

2.3.1 Multislit Concept

The Multislit rather straightforward concept is to use a number of short slits (or slitlets), each put on one astronomical target in the 2D field of the instrument, typically in the 5′–10′ diameter range for an 8–10 m class telescope

---

1 See the relevant European Space Agency (ESA) web pages for more.

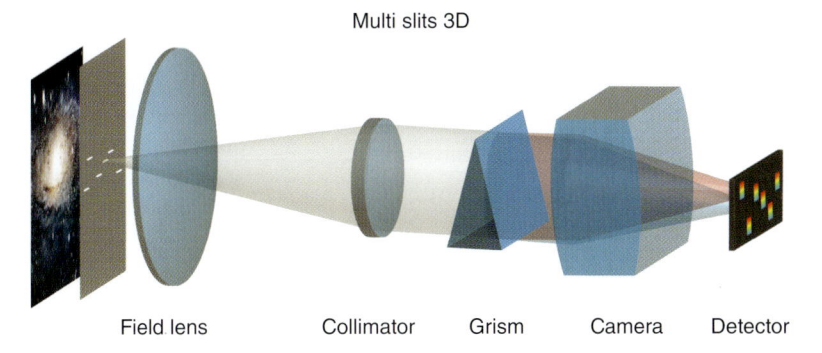

Figure 2.3 Principle of the multislit spectrograph. Many short narrow slitlets are put on selected objects in the field of view. An optical relay re-images the slitlets at the telescope focal plane on the 2D detector at an appropriate focal ratio, typically in the F/1.5 to F/2.7 range. It also gives an intermediate pupil on which a prism or grism is inserted to provide spectra from every slitlet. With no disperser inserted, the instrument gives direct sky images that can be used to measure the accurate positions of potential targets for further multisilt observations.

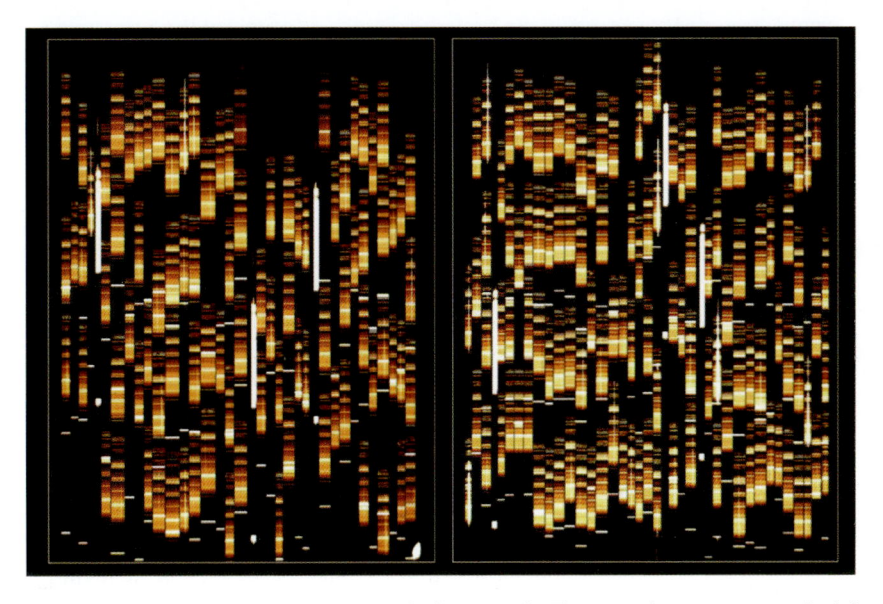

Figure 2.4 First VIMOS exposure in its multiobject mode. This optical instrument was built by a European Consortium led by O. LeFèvre (LAM, Marseille) for the European Southern Observatory (ESO). Two of the four $7' \times 8'$ quadrants are shown. Dispersion is in the vertical direction. A total of 221 low-resolution spectra on as many galaxies have been obtained in this single exposure. ©The European Southern Observatory. (Used Under Creative Commons License : https://creativecommons.org/licenses/by/4.0/.)

(see Figs 2.3 and 2.4). This field can be enlarged by using a number of identical modules, for example, a battery of four identical instruments to get a twice larger field. The instrument multiplex, that is, the maximum number of slits that can be put in its field of view, is typically a few hundreds, up to thousands for the largest instruments presently in operation or in construction.

The multislit systems patrol field (meaning the on-sky area where slits can be put) is always smaller than the one allowed by the CCD size. For example, with a 2k × 4k pixel CCD, the patrol field will usually be limited to, say, the central 2k × 2k pixels square, with the extra CCD space left in the grating dispersion direction Y being used to accommodate spectra coming from objects near the two Y edges of the patrol field. The basic reason for this restriction is that getting spectra for as many objects as possible at once always looks great, but is scientifically useful only insofar as they all cover a common spectral range.

Note that there are additional restrictions on where slits can be put: choosing more than one slit at any given X position (perpendicular to dispersion) runs the risk of overlapping spectra, unless spectra delivered by the instrument are short enough compared to the slits' Y separation. Multislit selection softwares always incorporate automatic overlapping checks, not only between main (first order) spectra but also taking into account light coming from other grating orders, including the small sizes but uncomfortably bright zero orders (which are actually short spectra when using a grating + prism configuration, because of the prism component). For a deep study of specific fields, a rather common observing strategy is to observe the same field at least twice, rotating the field by 90° for the second exposure, exactly as for slitless spectroscopy.

Multislit spectrographs multiplex, namely, the maximum number of objects that can be observed simultaneously depends on many factors including pixel angular size, spectral resolution, and spectral range. On an 8–10 m class telescope, a rule of thumb for 2k-pixel long spectra is about $M \sim 100$ per 2k × 2k detector pixels. Of course, these are only maximum values that can be actually obtained only in sky fields that feature enough putative targets.

2.3.2 Multislit Holders

While the multislit approach is conceptually simple, actually putting a huge number of slits with a few micron accuracy on, say, a 40 cm × 40 cm plate is no small feat. Ideally, we would love to get a digital focal plane display with millions of pixels that can be individually switched electronically, without any cumbersome mechanical motion. Such a component is in fact ubiquitous and very cheap, with the thin film transistor liquid crystal displays widely used in electronic goods. Unfortunately, light transmission efficiency of on-pixels is too low–below 50%–and off-pixel transmission is ~1%, which is too large by two orders of magnitude (a factor 100 or so) to eliminate the extra night sky background contribution coming through the many off-pixels, at least for ground-based observations.

Another, at first highly promising, such component is the digital multimirror device (DMD), developed by Texas Instruments for overhead projectors (Figure 2.5). With millions of tiny plane mirrors that can be almost instantaneously flipped on/off to either reflect the light to the instrument or away from the instrument, this looked as a perfect solution, given the high reflectivity of the mirrors. An off-the-shelf TI DMD features 2048 × 1080 mirrors with a 13.7 μm pitch. Mirrors can be individually put in their on or off positions, respectively at ±12° tilt, in a small fraction of a second. When using, for example, a moderate

Figure 2.5 Picture of a TI. DMD used in digital projectors. (https://en.wikipedia.org/wiki/Digital_Light_Processing. Used Under Creative Commons License: https://creativecommons.org/licenses/by-sa/4.0/.)

size 3.5 m telescope and imaging the sky at F/4 on the DMD, each mirror corresponds to $0.2'' \times 0.4''$ on the sky, and the total field covered by the DMD is $3.4' \times 3.4'$, both quite respectable values. Note that DMDs are not buttable, making it impossible to pave a larger field. On the other hand, although not developed for working in a cryogenic environment, the TI DMD operates well in vacuum at $-40\,°C$: this makes it possible to use it at relatively long wavelengths, up to the full H-band, without a significant thermal emission contribution.

Alas, as with liquid crystal displays, DMD-based instruments suffer from diffused light coming from off mirrors and mirror to mirror interfaces, which are roughly at the 0.25% level with the array illuminated at F/4. This gives a similar problem as with slitless spectroscopy, if at a reduced level: a useful detector pixel gets illuminated by one spectral element from an object (1/5000 of the wavelength at a Nyquist spectral resolution of 2500), but also by the night sky background contribution integrated over the whole spectral bandwidth of the spectrograph, usually about one octave. This would be much improved by illuminating the DMD at a smaller aperture ratio, say F/40, as the diffused light component would be reduced by a factor of ~100, but the spatial field of the instrument would then become prohibitively small.

One ground-based MOS instrument of this type is in operation at the Kitt Peak Observatory 4-m Mayall telescope.[2] This is the IRMOS spectrograph, working

---

[2] For up to date information, see the NOAO IRMOS home page.

Figure 2.6 Two close-up views of the JWST micro-shutter waffle-like array. The entire array is made of four quadrants, each the size of a postage stamp. Credit NASA.

in the full 1–2.4 µm near-infrared region, even if performance in the K-band is substantially limited by the DMD thermal emission. In the optical region, the DMD-based RITMOS instrument [16] has been used on the Mees 24-in. telescope [16], and the BATMAN DMD demonstrator is being developed for the 3.5 m Galileo telescope [17].

A transmissive variant, the micro-shutter array (Figure 2.6), has been developed for the James Webb Space Telescope Near-Infrared Spectrograph cryogenic environment. It features an array of 248,000 shutter cells, each 100×200 µm in size. The cells have lids that open and close when a magnetic field is applied. Each cell can thus be controlled individually, allowing it to be opened or closed to view or block a portion of the sky: a magnet is moved over the array to activate the shutters, and electronic commands are sent to latch the shutters that are in open position through electrostatic forces. Multiplex is of the order of 100. The crucial gain of this expensive approach is a much better contrast of about 2000, which is fine in space, but would already add significant background light for ground-based use in the near-IR region.

Ground-based multislit systems are thus essentially based on mechanical systems. The simplest approach relies on a 1D stack of narrow slits that can be moved independently in the dispersion direction to address as many targets (e.g. MOS-FIRE on Keck, FORS1 on the VLT). While quite easy to implement, even if in a cryogenic environment, this approach provides small multiplexes, a few tens at most. Most systems are thus based on the more complicated approach of using interchangeable masks at the focal plane of the telescope, each custom drilled to address a specific science field. Masks are typically made of thin (to keep the slitlets in focus) Invar (to avoid slitlet–object mismatch due to thermal dilatation) foils, with hundreds of slitlets drilled by a high-accuracy commercial high power laser tool at the locations of the science targets.

Metrology accuracy is very good, close to 1 µm, with typically a few 100 µm wide slitlets. That might look as an overkill: yet, the reasonable requirement of getting a spectrophotometric accuracy of ~1% calls for slitlet widths that are uniform to within 1% or indeed a few microns. The masks are inserted in cassettes, holding enough masks (say 8–20) to provide good operation flexibility during the night. Note that cumbersome cryogenic mechanisms are required if one observes in the NIR region, longward of about 1.4 μm.

2.3.3 Multislit Systems

Getting this kind of accuracy over a typically 15–40 cm across mask with the drilling system itself is a good first step. However, one needs also equally accurate mapping of the on-sky astrophysical coordinates of the objects to the laser driller coordinates to get all slitlets at the corrected positions, as well as a way to actually point the telescope at the exact location to ensure that all targets are centered on the mask slitlets. Here is one typical way to perform that kind of metrological waltz:

- Design and build the masks and their holders with kinematic mountings to ensure that the masks are by design always inserted at the same position in their holders, to a few micrometers overall accuracy; the same is applicable for the mask holder's insertion inside the instrument.
- Insert at the instrument focal plane a laser-made "metrology" mask featuring a regular grid of small holes (plus one off to derive correctly the mask orientation) at accurately known locations. Put the instrument in its imaging mode (disperser out of the optical beam) and make an exposure with the mask illuminated by the on-board calibration lamp. Use the detector data to map the laser driller coordinates to the CCD coordinates and vice versa. This step is normally done only once, if the instrument and the laser driller are not modified.
- Get images of the astronomical fields with the multislit instrument itself (again removing the disperser), select the science targets, plus a few pointing stars, and measure their positions in detector coordinates.
- Apply the mapping conversion obtained in the second step to send mask slitlets/holes coordinates to the laser driller. This produces masks with slitlets/holes at the science target positions, plus a few bigger holes (say 4″ diameter) centered on the typically 2–3 pointing stars.
- Insert a mask in its holder, point the telescope on the associated field, image the field with a very short exposure and use the actual positions of the pointing stars inside their respective holes to correct field acquisition (moving slightly the telescope pointing, and if necessary turning its field rotator slightly) in order to get all objects dead centered on their slitlets.
- Insert a grating and (at last!) start the science exposure.

Similar to the slitless case (see previous section) and for the same reason, optical requirements for the associated spectrograph are quite demanding, easily doubling or tripling the cost compared to the equivalent optics for a long-slit spectrograph. The disperser subsystem is often a set of exchangeable zero-deviation grisms with different dispersion and/or wavelength range, which provides significant scientific flexibility, an easily implementable imaging mode (thanks to an extra empty position in the grism wheel or slide), and a simpler straight-on optomechanical system. Quite importantly, the grism exchange mechanism does not need to be extremely accurate, as a first-order rotation error in the repositioning of the grism leads to a second-order spectral shift only (see Exercise 5).

Table 2.1 List of optical multislit spectrographs on 8–10 m class telescopes with their main characteristics: host telescope (Tel.), field of view (FoV) in arc minutes, maximum multiplex M, spectral resolution $\mathfrak{R}$, and spectral range $\Delta\lambda$ in μm.

Name	Tel.	FoV	M	$\mathfrak{R}$	$\Delta\lambda$	References
LRIS-B	Keck1	7.8' × 6'	60	3000–5000	0.31–0.52	[18]
LRIS-R	Keck1	7.8' × 6'	60	3000–5000	0.52–1.00	[19]
DEIMOS	Keck2	16.3' × 8'	130	3000–6000	0.41–1.10	[20]
FORS2	VLT	6.8' × 6.8'	470	260–2600	0.33–1.10	[21]
VIMOS	VLT	4x7' × 8'	1000	200–2500	0.36–1.00	[22]
GMOS1	Gemini-N	5.5' × 5.5'	60	630–4400	0.36–0.94	[23]
GMOS2	Gemini-S	5.5' × 5.5'	60	630–4400	0.36–0.94	[23]
FOCAS	Subaru	6' dia.	30	250–2500	0.34–1.00	[24]
RSS	SALT	8' dia.	60	800–6000	0.32–0.90	[25]
MODS 1-2	LBT	6' × 6'	500	100–2000	0.32–1.00	[26]
OSIRIS	GTC	7.5' × 6'	250	360–2500	0.36–1.05	[27]

Table 2.2 List of NIR multislit spectrographs on 8-10 m class telescopes with their main characteristics: host telescope (Tel.), field of view (FoV) in arcminutes, maximum multiplex M, spectral resolution $\mathfrak{R}$, and spectral range $\Delta\lambda$ in μm.

Name	Tel.	FoV	M	$\mathfrak{R}$	$\Delta\lambda$	References
Flamingos 2	Gemini-S	2' × 6.1'	>100	1200–3000	0.95–2.40	[28]
MOIRCS	Subaru	4' × 7'	40	500–3000	0.90–2.50	[29]
MOSFIRE	Keck1	6.1' × 3'	46	3500	0.97–2.45	[30]
LUCI 1-2	LBT	4' × 4'	>100	1900–8500	0.90–2.50	[31]
EMIR	GTC	6' × 4'	>100	3000	0.90–2.50	[32]

2.3.4 Multislit Instruments

As of the mid 2010s, essentially all 8–10 m class telescopes feature at least one and often more multislit spectrographs in operation, as per Tables 2.1 and 2.2 for respectively the optical and NIR domains.

2.4 Fiber-Based Multiobject Spectroscopy

2.4.1 Multifiber Concept

The overall concept is to put individual optical fibers on as many astronomical targets and regroup the fibers at the entrance slit of a classical long-slit spectrograph (Figure 2.7). This approach was made possible by development in the 1970s of very low loss optical fibers for the telecommunication industry. First astronomical light with this mode was achieved by Hill *et al.* at Steward Observatory in 1980 [33].

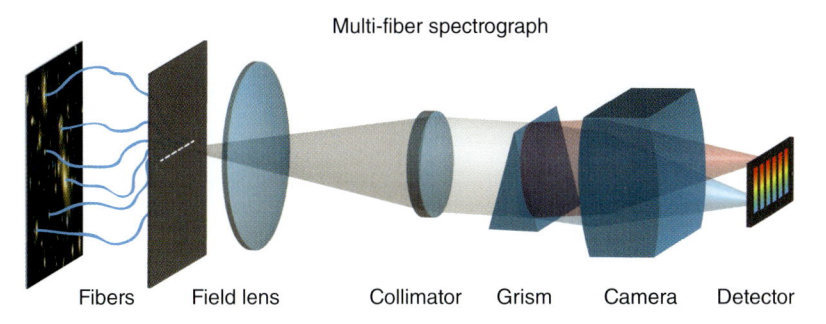

Figure 2.7 Principle of the multifiber spectrograph. Optical fibers are put on selected objects in the field of view and rearranged in a pseudo long-slit at the entrance of a classical long-slit spectrograph. The spectrograph gives a spectrum for each of the selected targets.

In principle, one could inject target light into the fibers with any beam angle up to the fiber acceptance angle, that is, at focal ratios of ~F/1.4 or slower. However, light entering a fiber at a slow aperture ratio, say F/6 to F/15, a typical value at a telescope Cassegrain or Nasmyth focus, would exit the fiber with a much wider beam, especially if the fiber is long (tens of meter) or/and bent, which is usually the case for multiobject spectrographic systems. This is a serious etendue loss, which can fortunately be mitigated by putting a microlens glued to the fiber entrance: light is then injected at, for example, ~F/1.8 in the fiber with very low insertion loss. This results in a moderate aperture ratio degradation, with typically a ~F/1.5 output beam. Microlenses are also generally used at each fiber output to adapt the output beam focal ratio to that of the spectrograph collimator. Note that this focal ratio degradation (an etendue loss) results in lesser brightness, but not less overall light, provided the spectrograph optics is oversized enough to fully accept the enlarged output beams.

Individual sky field injected in each fiber is usually of the order of the site median seeing (say 2″ on relatively poor quality sites–0.7″ on the very best ones), a tradeoff between collecting enough of the typically < 0.5″ diameter objects light and not too much of the sky background. This usually leads to multimode fibers with core diameters in the 50–150 μm range, which are readily commercially available. There is of course some related light loss with these fiber pick-up systems. This includes (i) internal absorption, quite significant (> 10%) when the fibers are long (tenths of meter) or are used below ~450 nm, and (ii) fiber insertion and extraction light losses, easily at a small percentage level each.

2.4.2 Positioning Systems

Multifiber positioning systems come in many variants. The simplest approach is to drill holes in plates at the location of the astronomical targets, manually glue the fibers in the holes, and finally insert the package at the focal plane of the telescope. This strategy is well adapted to extremely large individual surveys, such as the long-running SDSS survey (for more, see the SDSS homepage and/or its Wikipedia article). For more flexible needs aimed at a variety of surveys, the drawbacks of this approach are the lack of operation flexibility and the substantial manpower cost involved. Hence, a number of much more complicated automated "user-friendly" pick-up systems have been developed, as briefly presented below.

The Australian Astrophysical Observatory (AAO) has developed the pick and place 2dF (for 2-degree field) robotic system with the fibers inserted on magnetic buttons positioned sequentially on a steel plate. Since the robot takes a significant time, about 15 min, to put in place the 2dF 400-fiber complement, two focal plane plates mounted on a tumbler are located at the focal plane of the AAO 4-m telescope: while one plate fiber complement gets illuminated by the sky objects and sends the light to the spectrograph entrance slit, the pick and place robot configures the other plate for the next observational round. With typically half an hour to one hour integration on a given field, there is ample time for that step. Time lost between two successive exposures is then slashed down to the few minutes required anyway to slew the telescope to another sky direction and to center accurately the objects on the fibers: this is done by having three larger field imaging fibers that are put by the robot on relatively bright stars in the two-degree field. Note that the robot command software comprises inter alia a collision avoidance package to avoid any fiber entanglement. To learn more about this sophisticated system (Figure 2.8), you are encouraged to look at [34].

Other equally high-tech systems are based on separate remotely controlled mechanisms that are operated in parallel (or nearly in parallel when there can be potential collisions between the fibers). This includes fishermen pond type mechanisms, echidna-type spine motions and, most recently, autonomous fiber

Figure 2.8 Picture of the two-degree field multifiber positioning system installed at the prime focus of the 4-m diameter Anglo-Australian Telescope. The pick and place robotic arm is on top, with the currently addressed focal plane plate at the bottom. (Lewis *et al.* [34]. Reproduced with permission of Australian Astronomical Observatory.)

Figure 2.9 A highly schematic view of the fishermen pond multifiber positioner system. (Greg et al., 203.15.109.22/instsci/spie2004/5495-38-positioners.pdf. Reproduced with permission of Australian Astronomical Observatory.)

handling minirobots that walk on a glass window located on the focal plane of the telescope. Owing to their largely parallel operation, configuration setting time for these positioning flavors is usually of the order of a few minutes, and the tumbler approach sketched above is not required anymore. Each approach has its own set of trade-offs though.

Fishermen pond type systems (Figure 2.9) are the simplest ones, but true to their eponymous activity they are easily prone to fiber entanglement and cannot achieve multiplexes higher than a few tens.

The 2008 Echidna system [35] features hundreds of close by fiber holder spines that flex in two dimensions, using miniature piezoelectric motors, in order to pick the light from the selected targets (Figure 2.10). Multiplex is high, many hundreds, but can be actually used only when there is a very high density of potential astronomical targets, since each spine can move only inside its own individual small patrol field. Also, as fibers tilt to address their respective targets, a significant light fraction is lost: this is restricted to a manageable level (typically no more than 20%) by using Echidna systems only at high aperture ratios close to the fiber's acceptance angle. In practice, this means installing the positioning system at prime focus (around F/1.4–F/1.8 for 8–10 m class telescopes). The 2012 Cobra variant [36], produced by New Scale Technologies, gets away with spine tilts, using two rotary piezoelectric motors to move the fiber ends inside a 9.5 mm diameter patrol field owing to a planetary motion[3]: a record 2397 multiplex system is being developed for the Subaru 8.4-m telescope Prime Focus Spectrograph (PFS).

The 2012 versatile Starbug concept [37] is based on small autonomous walking robots that move in parallel to the inner surface of a window precisely located on the telescope focal plane, each carrying either a single fiber or a set of

---
3 That is, 360° rotation around two parallel but offset axes.

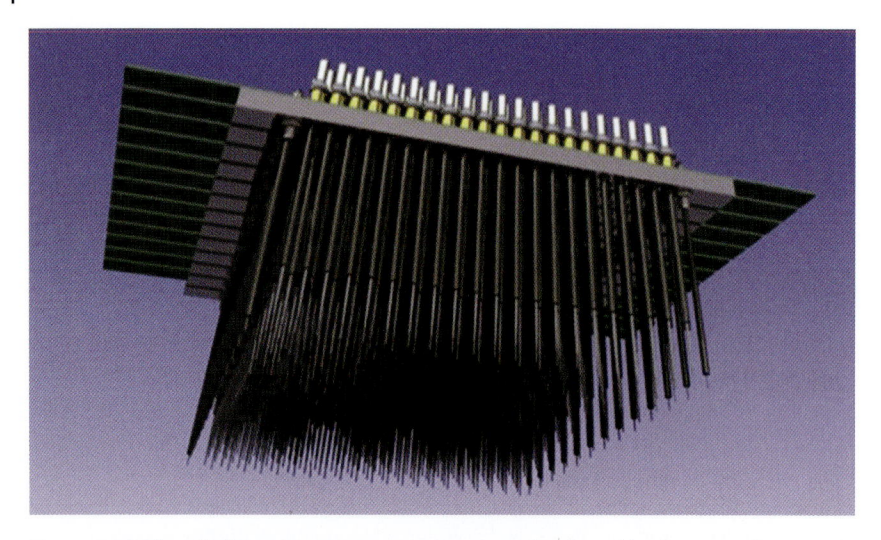

Figure 2.10 The Echidna positioning systems concept developed by the Australian Astrophysical Observatory. It uses piezo-activated fiber-carrying spines distributed evenly at the telescope focal plane. On-target pointing of each spine inside its own small 2D patrol field is achieved by spine flexing. (Akiyama *et al.* [35]. Reproduced with permission of Australian Astronomical Observatory.)

fibers (Figure 2.11). The walking gait of the robot is obtained with adequate voltages applied to two coaxial piezoelectric tubes. Ensuring that the bugs can first walk and then stay firmly at their working positions has been a major development headache. This has been eventually solved by applying a vacuum between the coaxial tubes. The bugs are then held in place by suction for any inclination of the telescope, even if positioned head down.

This approach offers considerable scientific flexibility, as the bugs may carry single fibers (classical multifiber system) or minifiber bundles (multi-deployable integral system; see Section 5.5), or more complex packages, all working in parallel on the same sky field. As of the mid 2010s, it has not yet been field-tested though. Note that it is possible in principle to move slightly the fibers during an exposure, for example, to compensate for atmospheric refraction effects as telescope elevation changes.

Whatever the positioning flavor chosen, there is a significant metrology problem to address. In particular, one must ensure (i) the ability to map the sky coordinates of the telescope focal plane with a minimum relative accuracy of typically 0.2″ on the sky, and (ii) a similar or even better fiber positioning accuracy. It is also important to miniaturize the fiber buttons in order to be able to address close by sky targets. Depending on the patrol field diameter and the focal plane scale, these requirements can easily translate in say a robotic system that must achieve a few microns positioning accuracy across a 1-m diameter field plate, a not so easy feat especially in telescope focal plane environments (variable temperature and variable inclination with respect to gravity). This is usually not possible with an open-loop process, and most fiber positioning systems work in closed loop, with, for example, the fibers back-illuminated with calibration lamps in order to

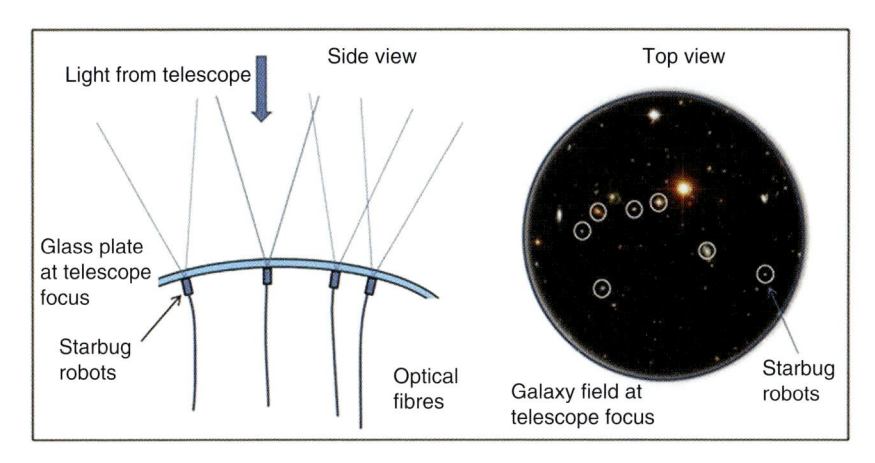

Figure 2.11 Highly schematic view of the Starbug concept, developed by the Australian Astrophysical Observatory. Piezo-driven semi-autonomous robots walk on the focal plane window to their designed target locations. They are prevented from falling by vacuum suction. (Gilbert et al. [37]. Reproduced with permission of Australian Astronomical Observatory.)

sense their actual positions with a vision camera during the positioning process and send real-time corrections to the positioning actuators.

2.4.3 Fiber-Based Spectrograph

If the deployment of fiber positioners and their associated fiber cables is a significant headache, on the other hand the fiber-based MOS approach leads to much simpler spectrographs than for the multislit approach. First, fiber outputs can be configured along a curved convex slit: this is an ideal input to a spherical mirror (essentially an inverted Schmidt telescope) acting as the collimator. Depending on the slit length and the fiber output light focal ratio, one might need to add a Maksutov[4] or Schmidt corrector or use a Mangin mirror (a lens with a reflective coating on its back surface), but that is still a rather simple solution. Cameras are also rather easy to design, since first order chromatic aberrations can be taken out simply by tilting the detector. In particular, direct Schmidt cameras (with field flatteners to match plane detectors) are often used, with the mechanical complexity of inserting a cryogenically cooled detector in the middle of the optical beams. Finally, at the cost of using tens of meters of fibers, the spectrographs can be put in a fixed controlled environment, which can easily cut overall cost and building timescale by a factor of 2. The negative side is an added significant light loss (say 20%) in the visible region, which turns into a dramatic one in the far blue to near UV region.

2.4.4 Fiber Systems Performance

All fiber systems suffer from a number of spectrophotometric shortcomings. Even with clever compensating systems, they are subject to variable stresses

---

4 A divergent thick lens with the two radii of curvature centered on the pupil, thus still an optical system enjoying full rotational invariance; see Figure 1.12.

as the telescope moves during the 30–60 min individual exposures to keep the targets on the fibers. This translates into different time-variable transmission efficiencies for each fiber, which are, in addition, wavelength dependent. Hence the need for extensive calibrations, and in particular, flat fields obtained by short integrations on the sky just after sunset or just before sunrise in order to get an almost even illumination over the whole patrol field and enough photon flux to get a high signal-to-noise ratio in a, by necessity, short exposure. This does not solve the time variable part, however, as the fibers move during the typically 1-hour long exposures: sophisticated data resolution packages have been developed to nevertheless get accurate subtraction of the sky background component, below the 1% level, and reasonably accurate spectrophotometry, within a few percent. Their gist is to use the huge 3D data sets obtained with this type of instrument to self-calibrate *a posteriori* the various observations. In any case, this is an essential part of the instrument system, without which even the best hardware would be useless.

2.4.5 Present Multifiber Facilities

A number of fiber-based MOS systems are in operation circa 2016 on 8–10 m class telescopes. The FLAMES optical spectrograph on one 8.2 m VLT Unit Telescope holds a 132-fiber system with an AAT-like pick and place robot; the optical-NIR FMOS spectrograph on Subaru uses a 400-fiber Echidna system. A huge spectrograph (PSF) with 2397 individual fibers and a Cobra-type positioner is in construction for the Subaru prime focus. The Cobra-type 100-fiber optical MEGARA spectrograph and the 1000-fiber near-infrared MOONS spectrograph are in development for respectively the GTC and the VLT.

Since the mid 1990s, wide-field MOS systems of the multifiber kind installed on smaller (2.5–4.2 m) telescopes have played a large astronomical role as the favored approach to build dedicated survey capabilities. Huge radial velocity surveys of up to a million galaxies at small spectral resolution ($\Re \sim 1000$) have revealed their "soap-bubble"-like distribution over the sky. Even larger radial velocity surveys of up to 10 million distant galaxies are now being prepared to constrain the nature of dark energy, a still mysterious component that dominates the present energy content of our Universe and gives telltale small distortions in the radial velocity 3-D distribution of galaxies at the ~1 billion light-year scale. Surveys at high spectral resolution ($\Re \sim 25{,}000$) are also in progress or in preparation to complement the one billion stars astrometric survey being currently done by the ESA Gaia satellite, to provide accurate abundance determinations of key nucleosynthesis elements on typically ~1 million selected Galactic stars. The list of present and planned future optical massive survey systems is given in Table 2.3. All surveys are extensively documented on the Web.

Patrol fields and multiplexes are more modest in the NIR region, especially for multiobject spectrographs working up to the thermal infrared, beyond ~1.75 μm, due to the very large size cryogenic vessel required. Even when the instrument is restricted to the non-thermal domain, they are still well below typical values in the optical domain, owing to the much higher cost (a factor ~10) of

Table 2.3 Wide-field optical MOS survey capabilities with their main characteristics: instrument name, host telescope (Tel.) name and diameter, spectral range $\Delta\lambda$ in μm, start date, patrol field area in square degree, multiplex M, spectral resolution $\mathfrak{R}$, and main current or planned surveys.

Name	Tel.	$\Delta\lambda$	Start	Field	M	$\mathfrak{R}$	Survey
2dF	3.9-m AAT	0.47–0.85	1996	3.1	400	1,300	5.10^5 galaxies
SDSS	2.5-m Apache	0.38–0.91	2000	3.1	640	1,000	1.510^6 galaxies
SDSS	2.5-m Apache	1.51–1.70	2011	3.1	300	22,500	10^5 stars
SDSS	2.-m Apache	0.36–1.04	2011	3.1	1000	2,000	1.510^6 galaxies
LAMOST	4-m LAMOST	0.37–0.91	2014	20	4,000	1,800	10^6 galaxies
HERMES	3.9-m AAT	0.47–0.79	2014	3.1	400	28,000	10^6 stars
PFS	8.2-m Subaru	0.37–0.85	2017	1.5	2,400	3,000	Legacy Survey
BigBoss	4-m Mayall	0.4–0.8	2018	7.1	5,000	5,000	All Sky (z 0.2-3.5)
Weave	4.2-m WHT	0.37–0.95	2018	3.1	1,000	5,000	2.10^7 galaxies
Weave	4.2-m WHT	0.37–0.95	2018	3.1	1,000	28,000	10^6 stars
DESpec	4-m Blanco	0.55–0.95	2020	3.8	4,000	3,000	8.10^6 galaxies
4MOST	4.1-m VISTA	0.37–0.95	2021	7	2,400	5,000	2.10^7 galaxies
4MOST	4.1-m VISTA	0.37–0.95	2021	7	2,400	20,000	10^6 stars

Table 2.4 Wide-field NIR MOS survey capabilities with their main characteristics: instrument name, host telescope (Tel.) name and diameter, spectral range $\Delta\lambda$ in μm, start date, patrol field area in square degree, multiplex M, spectral resolution $\mathfrak{R}$.

Name	Tel.	$\Delta\lambda$	Start	Field	M	$\mathfrak{R}$
Hydra	3.5-m WIYN	0.40–1.80	2005	0.8	93	1000–10,000
FMOS	8.2-m Subaru	0.90–1.80	2012	0.2	400	600–2200
PFS	8.2-m Subaru	0.90–1.45	2017	1.5	2400	3000
MOONS	8.2-m VLT	0.80–1.70	2019	0.14	1000	8000–20,000

infrared pixels. Table 2.4 shows the relatively few wide-field NIR survey facilities in operation or development by 2016.

2.4.6 Conclusion

Among Multiobject spectrographic instruments, the multifiber flavor is the system of choice when a huge patrol field, say at least 30′ for an 8–10 m class telescope, is required to achieve the scientific goals. There is a penalty to pay in the complicated/expensive positioning systems, and in significant extra light loss compared to the other multislit flavors. On the other hand, the stationary spectrograph (or spectrographs) fed by the fibers are easier/cheaper to produce, easily by a factor of 2. Fiber-based MOS is therefore essentially the province of huge spectrographic surveys over a significant fraction of the whole sky.

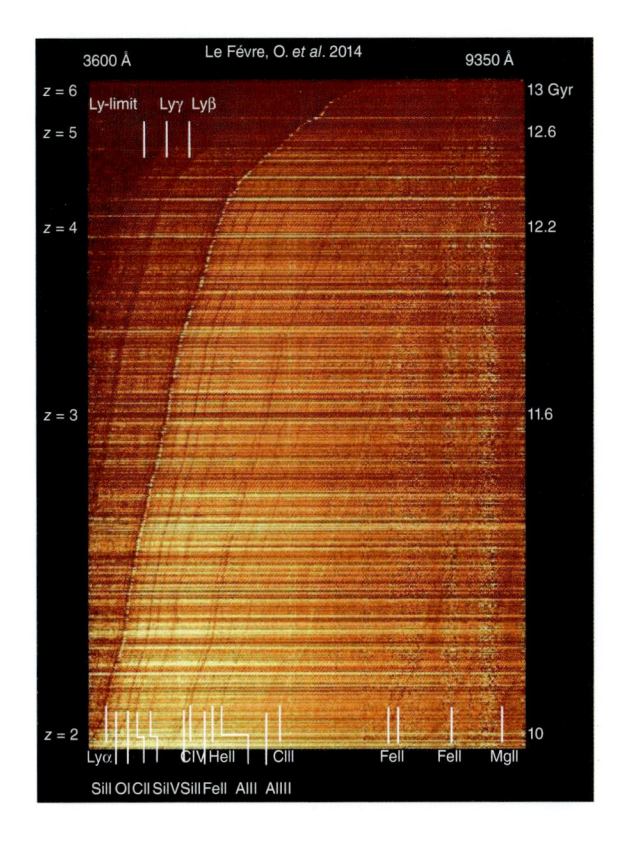

Overview of ~10,000 spectra of the VIMOS Ultra Deep Survey [122]. The spectroscopic redshift survey covers 1 square degree with targets selection based on an inclusive combination of photometric redshifts and color properties. Spectra covering 3650 < λ < 9350 Å have been obtained with the slit-based MOS VIMOS on the ESO-VLT with integration times of 14 h. The survey has a completeness in redshift measurement of 91%, or 74% for the most reliable measurements, down to $i_{AB} = 25$, and measurements are performed all the way down to $i_{AB} = 27$. The redshift distribution of the main sample peaks at $z = 3$–4 and extends over a large redshift range mainly in 2 < z < 6. At 3 < z < 5, the galaxies cover a large range of luminosities $-23 < MU < -20.5$, stellar mass $10^9 M_\odot < M_\star < 10^{11} M_\odot$, and star formation rates $1 M_\odot yr^{-1} < SFR < 10^3 M_\odot yr^{-1}$. ([122]. Reproduced with permission of Olivier Le Fevre.)

3

Scanning Imaging Spectroscopy

3.1 Introduction

Scanning over time offers a direct way to build a 3D data cube out of a 2D detector, with time providing the lacking third dimension. With the scanning long-slit spectrograph (SLSS), the 2D detector data have one spatial axis and one wavelength axis and the object is moved in discrete steps across the slit to provide the second orthogonal spatial axis. With the scanning Fabry–Pérot spectrograph, each exposure covers the full sky field but only one wavelength bin; the etalon spacing is moved in discrete steps to provide a (small) number of wavelength channels on each spatial pixel. In the case of Fourier transform spectrography, each exposure covers the full sky field too, but with only one frequency bin (hence the Fourier transform bit); the Michelson interferometer spacing is moved in discrete steps to cover a number of frequency channels. Note that, since almost any interesting astronomical object is quite faint, the number of steps for any scanning 3D spectrograph flavor cannot be large, typically a few tenths only. This contrasts with the huge pixel numbers provided on the other two axes of the "data cubes," typically a few thousands each.

3.2 Scanning Long-Slit Spectroscopy

3.2.1 The Scanning Long-Slit Spectroscopy Concept

The SLSS approach is arguably the most elegant, easiest, and cheapest way to get a 3D spectroscopic format: one has only to use a standard long-slit spectrograph, take an exposure on an object (typically a galaxy), move the object by typically one slit width, start a second exposure, and repeat this procedure until the full object is covered.

This approach is well adapted for day-time Earth observation from space, using an orbiting satellite. Spatial scanning is then automatically done by the satellite motion, with the Earth bright enough to give good enough signal-to-noise ratio within the very short integration time for each individual spatial channel (a few milliseconds only).

Optical 3D-Spectroscopy for Astronomy, First Edition. Roland Bacon and Guy Monnet.
© 2017 Wiley-VCH Verlag GmbH & Co. KGaA. Published 2017 by Wiley-VCH Verlag GmbH & Co. KGaA.

3.2.2 Astronomical Use

The concept has been developed by Wilkinson *et al.* [38]. Their first (and last) 1983 observation gave a huge data "slab" on the bright nearby galaxy Centaurus A, featuring 1650 spectral pixels and 1650 × 71 spatial pixels (Figure 3.1). The 4-h total integration time was divided into 71 subintegrations of about 20 s each. Despite this good result, the technique has never been reused for astrophysical purposes! Apparently, this a priori odd occurrence came out from three basic difficulties, or curses, deeply ingrained in that approach:

1) *The Atmospheric Curse.* Atmospheric transmission is rarely constant over the many hours needed to build the data cube, and atmospheric turbulence (which directly impacts image quality) is never constant. Both affect the different spatial channels in a different way, which is hard to correct a posteriori since the data cube itself does not give any information on either effect.
2) *The Sensitivity Curse.* Except for a very few bright objects such as Centaurus A, the total integration time required to get enough signal-to-noise ratio for accurate radial velocity fields and line ratios lies in the hundreds of hours regime, which would be next to impossible to get from any telescope time allocation committee.
3) *The Shape Curse.* Scanning spectrography is a priori an optimized tool to study highly elongated large individual objects, the perfect one being a nearby galaxy disk viewed close along the line of sight. Unfortunately, this is but a vanishing sample of all potential astronomical targets. Most such targets are actually

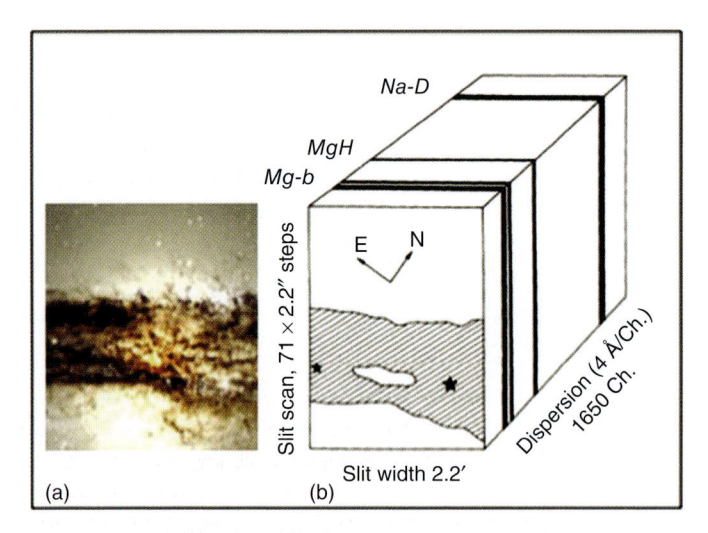

Figure 3.1 First scanning long-slit data cube on the galaxy NGC 5128 (Centaurus A) at the 4-m Anglo Australian Telescope by Wilkinson *et al.* 1986. (a) Optical image of the galaxy field, credit the European Southern Observatory, PR image eso0005b. (b) Data cube covering the blue to yellow optical region with ~10,000 spectral resolution. Strong magnesium (Mg) and sodium (Na) lines are indicated. The slit length is horizontal. The vertical direction has been filled by 71 successive sky integrations, moving the slit on the galaxy each time by about one full 2.2″ slit width. (Wilkinson *et al.* [38]. Reproduced with permission of the Royal Astronomical Society.)

more or less roundish and/or smallish galaxies, for which, as will be seen in Chapter 4, integral field spectroscopy offers a much efficient match between the object and the spatial shapes of the spectrographic data.

Note that for space-based astronomical observations, while the atmospheric curse does not apply, the scanning long-slit technique status is so far the same, with no instrument of this type being flown despite its great simplicity. One telling example concerns the near infrared spectrograph for the James Webb Telescope. Given the stringent limitations in volume, weight, energy dissipation, and operation complexity inherent in that project, the SLSS approach was indeed suggested early on by one of this book's authors (GM). Yet a multislit system has been ultimately selected, in spite of a much more complicated operation involving physical motion of a heavy magnet, a no small feat to achieve reliably over a decade at 1.5 million km from the Earth. Again, this is the shape curse that doomed the scanning long-slit approach: multislit data are well suited for identifying the relatively rare extremely high redshift galaxies ($z > 8$), which constitute the crucial JWST science driver, while a scanning long-slit approach with the same detector area would give much less object spectra in the same observation time, by a huge factor of ~ 30.

3.3 Scanning Fabry–Pérot Spectroscopy

3.3.1 Introduction

A multifilter imager is arguably the simplest 3D scanning instrument. Its modern version features a large 2D digital detector and a set of exchangeable interference filters, in order to image the sky field successively in perforce a small number of spectral channels. On large telescopes, the instrument is ideally put at their high aperture ratio (typically $\sim$ F/1.8) prime focus to better match the sky turbulence light smearing to detector pixel size. The detector 2D format can currently feature up to 1 billion optical (CCD) pixels, or up to 100 million infrared (IR array) pixels, and imagers typically hold a dozen or so large interference filters. While such an instrument is conceptually simple and can cover a large field times telescope diameter etendue (a record $34' \times 27'$ at the 8.2 m Subaru telescope), building the large detector mosaics and the huge filter exchange mechanism (based on a wheel or a sliding cassette) is anything but easy. Imagers also require large cryogenic systems to cool their huge detectors, plus the huge filter holders as well when working in the infrared region above ~ 1.65 µm.

Spectral resolution is however always very low, typically less than 100, which is not enough to qualify as a bona fide spectrographic technique. Much higher spectral resolutions, typically from a few thousands to a few ten thousands, can be obtained by adding a Fabry–Pérot interferometer, as presented below.

3.3.2 Fixed Fabry–Pérot Concept

Fixed Fabry–Pérot interferometers have been used for 2D astronomical spectrography since the pioneering observation by Buisson *et al.* [2] of emission lines in the Orion nebula. A colored – later interference – filter is put in the focal plane of the telescope in order to isolate a single order of the etalon. A collimator gives an intermediate image of the pupil (telescope mirror) on the etalon with the sky image located at infinity. A camera re-images the sky on the detector. If the field is filled with purely monochromatic light inside the filter bandpass, one gets a set of rings, covering $1/N$ (N being the etalon finesse, typically in the 8–24 range) of the detector area, the radii of which give the wavelength of the light. More realistically, when filled from variable wavelength nebula emission lines, one gets pieces of rings at the radii where by chance light happens to be at the right angular position to be transmitted, instead of reflected, by the etalon. Note that this happens at every field position with an $1/N$ a priori probability, and hence one statistically gets wavelength data over a fraction $1/N$ of the field only.

This technique has thus been used essentially to observe single ionized gas regions in our galaxy, or a full field of ionized gas regions in an external galaxy (e.g., see Figure 3.2), from a single emission line (e.g., $H\alpha$ 656 nm, [OII] 373 nm, [OIII] 501 nm) selected with an interference filter. Gas clouds that fortuitously happen to have a line emission that is just at the right location to be transmitted by the etalon appear on the detector, with measurement of their position giving their wavelength and hence their relative radial velocity.

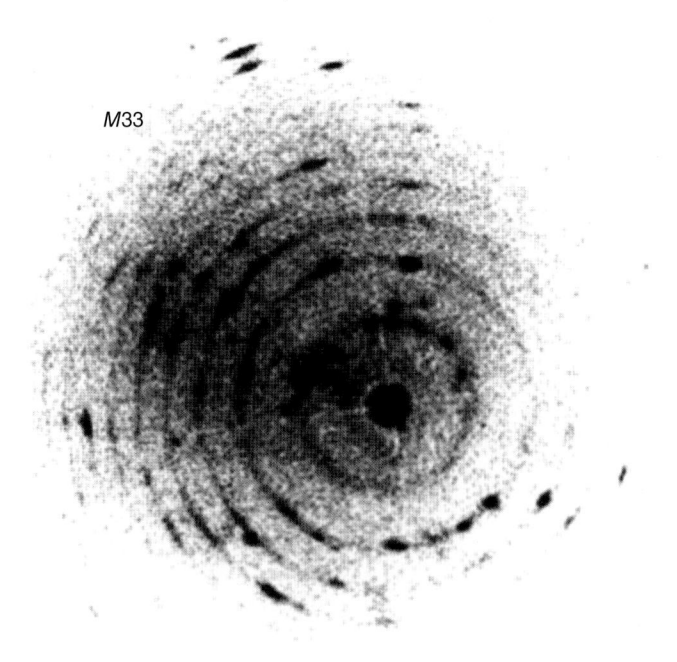

Figure 3.2 First 2D Fabry–Pérot data on an external galaxy by Carranza et al. [39]. One sees etalon ring fragments in the 656.3 nm line of ionized hydrogen, fed by the extended ionized gas regions in the Messier 33 galaxy that happens to have the right wavelength–radius combination to be transmitted by the etalon. 1024 different radial velocities were extracted in the 6 arcmin diameter field, giving the first global ionized gas radial velocity field in a galaxy. (Credit Observatoire de Marseille, France.)

This innate $1/N$ incompleteness is obviously a major limitation. Also, extra caution was needed to avoid possibly severe wavelength errors in the case of an ionized gas region with highly nonuniform brightness. Even worse, while this is a good technique to get accurate velocity fields from isolated emission lines, obtaining spectrophotometric data, for example, accurate line brightness ratios, was essentially impossible.

3.3.3 Scanning Fabry–Pérot

The next step toward a true 3D instrument is the brainchild of Tully 1974 (Figure 3.3). The concept is to make N successive field exposures, changing the etalon cavity optical depth by $\lambda/2N$ at a time. Each of the, for example, 16 millions detector (spatial) pixels then gives a short spectrum – typically 15–40 spectral pixels long – centered on a single emission line wavelength. In that way, full 3D coverage of the velocity field (a few 100 km s^{-1} range) can be achieved.

Developing a scanning Fabry–Pérot interferometer adapted to astronomical use has not been a small feat though. Historically, this was long done by putting a fixed etalon in a gas-filled pressure chamber, equipped with two plane parallel windows. Scanning the gas pressure over a few atmospheres modifies the gas index of refraction n, and hence the light optical path $2ne$ in the cavity. Two suitable gases are propane (which can easily explode) and freon (which is now banned because it efficiently destroys the Earth ozone layer). This was a very bulky approach, now superseded by the piezo-driven technique. Applying high voltages

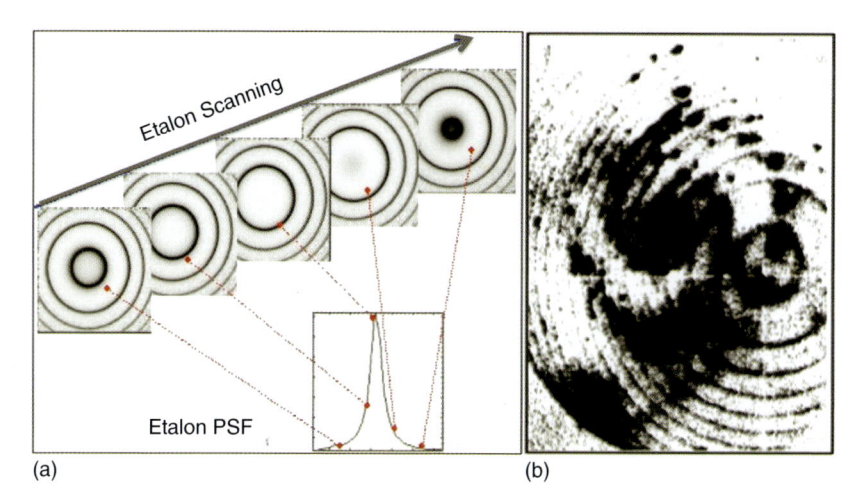

Figure 3.3 1974 first 3D scanning Fabry–Pérot data [4] at the KPNO 84 in. telescope. (b) One of the 16 successive exposures in the 6562.3 nm ionized hydrogen line in the Messier 51 galaxy, showing ring fragments fed by the galaxy ionized gas regions. (a) Illustration of the scanning concept. The etalon cavity is changed by 1/16 of an order for each of the 16 successive exposures and the light intensity curve on each spatial pixel is extracted. This gives a data "slab" made of a small spectrum for each of the ~10^5 spatial points in the ~6′ field. (Reproduced with permission of B. Tully.)

to three piezoelectric stacks, optically contacted at 120° on the two sides of the cavity, changes the thickness e of the cavity, hence its optical path. Since piezos suffer from strong nonlinearities and even worse hysteresis effects, this is done in closed loop with the cavity depth near the three piezos continuously monitored from the measurement of three capacitors deposited on the cavity sides. In this way, the daunting precision required, in the 2–5 nm range, can be achieved even with a moving instrument. There are very few vendors for this expansive component and the (small) astronomical market has been dominated since the 1970s by Queensgate Instruments Ltd, now IC Optical Systems.

Development of this concept from Tully's early prototype (unwieldy pressure scanning Fabry–Pérot and relatively low quantum efficiency image intensifier coupled to an analog photographic plate) to the modern scanning Fabry–Pérot interferometer benefitted from two major technological upgrades, namely, piezoelectrically scanned etalons and 2D digital detectors (first photon-counting detectors, then CCDs). Early examples were the 1980 Royal Greenwich Observatory TAURUS [40], the 1984 Observatoire de Marseille CIGALE [41], and the 1989 University of Hawaii HIFI [42]. As for every 3D spectrography avatar, a key ingredient has been the development – and maintenance – of huge data extraction and reduction pipelines, using quite sophisticated algorithms, with the accurate subtraction of Earth's upper atmosphere emission lines being arguably the largest headache at the time.

Figure 3.4 shows the canonical scanning Fabry–Pérot spectrograph optical design, with the etalon put on an image of the pupil and the rings projected on the sky. Note that an interference filter is put either near the telescope focus or near the pupil in order to select one etalon interference order p.

One version – dubbed the tuneable filter concept – uses a low-order etalon, typically $p \sim 100 - 250$, with a high finesse, typically $N \sim 25 - 40$. In this way, a significant spectral range (25–40 times the spectral resolution) can be scanned, addressing in particular wide emission lines from active galaxies with up to $1000 \, \text{km s}^{-1}$ radial motions. The classical Fabry–Pérot spectrograph, with $p \sim 500 - 1500$ and $N \sim 10 - 20$ is on the other hand well adapted to

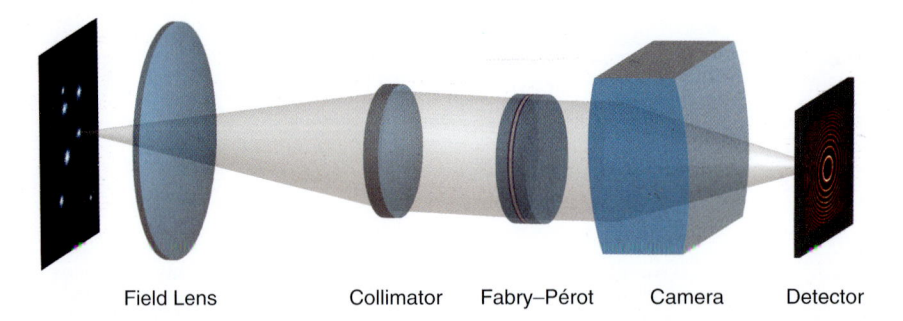

Field Lens   Collimator   Fabry–Pérot   Camera   Detector

Figure 3.4 Scanning Fabry–Pérot spectrograph concept. The 2D square or circular field is collimated, with the pupil image put on a Pérot–Fabry etalon. One etalon order is selected by an interference filter (not shown here) and the field is re-imaged by the camera on the 2D detector. Scanning the etalon builds the same small-size spectrum at each field/detector pixel.

get emission line radial velocity fields in quieter galaxies with only a couple of 100 km s^{-1} radial velocity range.

As for every scanning instrument, this approach suffers from the atmospheric curse, in particular change in transmission over the usually many hours of total integration time. This was elegantly solved when using a photon-counting detector: scanning over the, say, 20–50 steps is done in a time short enough to eliminate atmospheric transmission variations (30 s is a fair value at good astronomical sites) and repeated until one gets good enough signal-to-noise ratios by summing up all subexposures on each spatial pixel. Similarly, atmospheric seeing changes over the very long total integration time equally affect all wavelength channels and are thus effectively averaged out.

Contrary to photon-counting systems, CCDs offer strong practical advantages: they have a higher quantum efficiency (at least by a factor of 3), and are extremely rugged, essentially fool-proof (meaning astronomer-proof), and thus loved by large telescope staffs, who symmetrically positively hated photon counting when it was around. This is a real problem for the scanning FP technique, since the significant read-out noise of CCDs (typically 1–2 electron/pixel/readout) would dominate the photon flux of faint sources for such short individual exposures. Besides, even if CCD readout noise could be zeroed down, it just takes too long to read the detector.

In practice, CCD-based scanning FP individual integrations usually last a few minutes, with one full scanning taking an hour or so. There are some ways to correct from sky variable transmission and seeing: one elegant scheme when there is a relatively bright star in the field (normally the case given the large 2D spatial field covered) is to take its spectrum with a long-slit or better integral field spectrograph with at least twice the spectral resolution of the FP data. This reference spectrum can then be used as a spectrophotometric "ruler" to correct the successive exposures for atmospheric transmission variations. Less elegantly, one can in addition correct for seeing variations by recording stellar shapes in the field to get the spatial point spread function for each channel (scan) and degrade all data to fit the worst seeing. Not terribly appealing though!

These are however only mere trifles. That Fabry–Pérot scanning interferometers are now rarely offered as general-use instruments stems from two fundamental shortcomings. The first is the inherently small number of spectral pixels, while most (at least current) astronomical studies revolve around getting thousands of spectral pixels, for example, to get line ratios in ionized gas region, stellar component of galaxies, and, eminently, for spectrographic surveys of the distant Universe. The second, also strong, shortcoming comes from the near impossibility of measuring absorption lines: not that etalons have any problem per se with absorption lines, but their spectral output is modulated by the *fixed* spectral transmission curve of the interference filter used as an order sorter, a problem already identified by Charles Fabry in his seminal 1914 paper! This is a non-problem for a high contrast narrow emission line, but nearly impossible to disentangle for the low-contrast wide absorption lines coming from the integrated stellar component of galaxies. There is theoretically a way out, namely, putting two etalons in series, one high order to get the high spectral resolution, and one low order to act as the order sorter, and scanning them synchronously.

This should in principle solve both problems at once, that is, for astrophysical problems that would warrant hundreds of hours of total integration time.

There are presently only two general-use scanning Fabry–Pérot cum CCD detectors on a large telescope. One is mounted on the OSIRIS focal reducer [43] at GTC, featuring two tunable filters, one for the blue-green region and one for the yellow to far-red. The second one is mounted on the RSS multimode spectrograph on the SALT telescope [44]. It can hold three different interferometers, namely, a low-resolution (tuneable filter) etalon, a medium-resolution one, and a high-resolution one. Interestingly, two of the etalons can be put in tandem, one being the low-resolution one, which as indicated above should in principle open the huge absorption line observing domain. Initial implementation of this mode on SALT did not work however, owing to strong ghosts and other parasitic light resulting from light bouncing back between the two etalons (remember, an etalon of finesse N is mostly a mirror, reflecting back a fraction $(N-1)/N$ of the incident light), and by 2016 the jury is still out on that whole issue.

3.4 Scanning Fourier Transform Spectroscopy

3.4.1 Fourier Transform Spectrometer

The Fourier transform spectrometer (FTS) is a rather straightforward optical system, even if this apparent simplicity is largely deceptive: A beam slitter divides an incoming light plane wavefront in two equal flux wavefronts, the two beams are then reflected back by two plane mirrors with a path difference $2e$, and recombine/interfere on the beam splitter. With this classical Michelson interferometer mounting (Figure 3.5), output light is actually the intensity (I) versus frequency ($v = 1/\lambda$) Fourier transform of the incident light, hence the technique moniker. Specifically, for an input light frequency spectrum I_v, the modulated part of the output light at a $2e$ path difference is

$$I_e = \int_{v_1}^{v_2} I_v \cos(2\pi v e)\, dv \qquad (3.1)$$

With one of the two plane mirrors moving by n individual $\lambda_c/2$ steps, an inverse Fourier transform then gives back the classical intensity versus wavelength spectral format in the full spectral range covered by the instrument, with a spectral resolution $\mathfrak{R} = n$.

FTSs were first applied to IR astronomical observations, typically in the 1–5 μm range, in the early 1960s. At this time, there were no IR detector arrays, only noisy single pixels detectors. This made the FTS a highly attractive IR spectroscopic device: its etendue is large, the same that for a Fabry–Pérot of equivalent spectral resolution (acceptance cone half-angle $\theta = 1/\sqrt{\mathfrak{R}}$). Furthermore, the packing of hundreds of orders in a Michelson interferometer output gave a spectral multiplex advantage when, as it was then the case, signal-to-noise ratios (S/N) of observations are fully dominated by detector noise, not by photon noise from the object. In this case, the multiplex gain is the square root of the number of

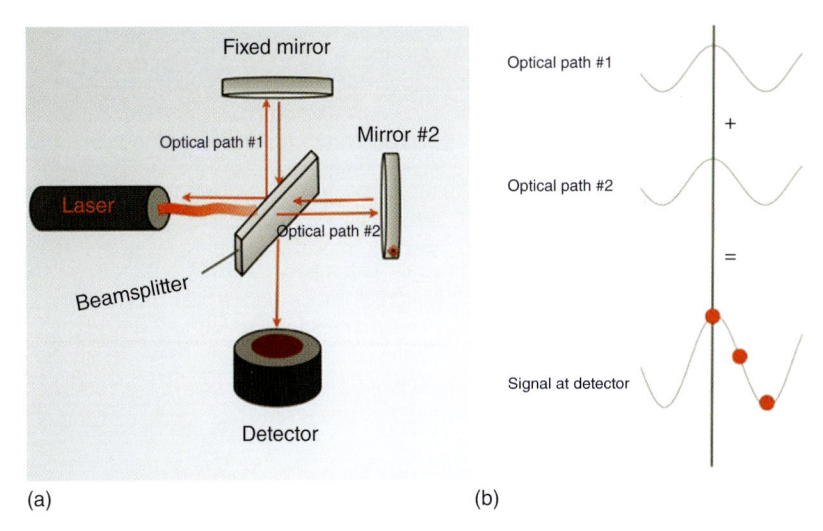

Figure 3.5 (a) Schematics of the simplest incarnation of a Fourier transform spectrometer. For astrophysical use, a second detector is added on the horizontal return beam to avoid losing 50% of the light. (b) Output signals (three red dots) for three moving mirror positions, respectively with fully constructive, half constructive, and fully destructive interference. (Credit L. Drissen, Université Laval, Québec.)

orders. Their main use has been making IR spectra of solar system giant planets and bright galactic stars at very high spectral resolution, usually greater than 10^5.

Note that this was an essentially 1D technique, giving a highly detailed spectrum of the integrated light of a single object, at a spectral resolution equal to the total number of $\lambda/2$ mirror steps. In order not to lose any light, a dual output system was used, with two InSb detectors, one on each beam out of the interferometer. Like for all scanning systems, the data quality was severely impaired by varying atmospheric transmission; this was solved by using a dual input, one on the astronomical target, and another on a blank sky position, and switching very rapidly between the two. Spectral quality in terms of diffused light, wavelength absolute precision, and spectrophotometric accuracy was outstanding, both for absorption and emission lines.

There were a number of rather harsh technical problems, with two main offenders. Firstly, for proper scanning, one of the two interferometer mirrors has to move in identical $\lambda/2$ steps with a few Å accuracy over say a 1 m length in total, generally at a rate of tens of steps per second. This challenge is solved using cutting-edge closed-loop control systems (a real feat in the early 1960s, before the era of even single transistors!), using a reference single-mode laser beam. Secondly, the daunting task of moving the mirror parallel to itself with the same few Å accuracy is neatly sidestepped by using instead a corner-cube, for example, a set of three rigidly attached orthogonal plane mirrors. This Cain's eye[1] contraption returns any light ray striking the three mirrors on turn at exactly the same angle, the output light ray being symmetrical from the input ones with respect

---

[1] Since looking at it from any angle results in only seeing one's eye, reenacting its namesake biblical curse.

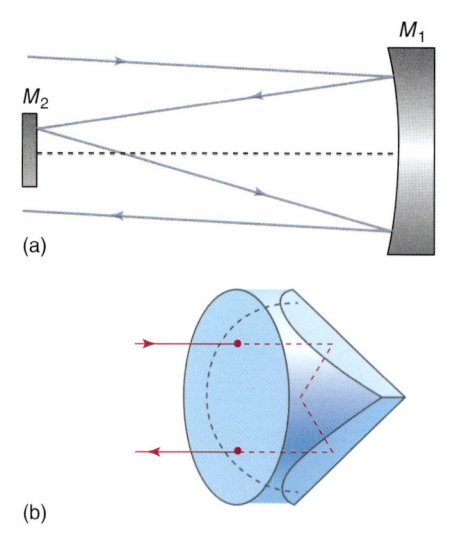

Figure 3.6 Schematics of retroreflectors used for Fourier transform spectroscopy. (b) Corner cube prism with three orthogonal faces, which reflects back any incoming ray parallel to itself. (a) Cat's eye using a parabolic main mirror M_1 and a small plane mirror M_2 at its geometrical/optical focus, with the incoming light beam first reflected by M_1, then by M_2, and finally sent back to the beamsplitter by M_1. Any moderately tilted incoming ray is reflected back in an almost parallel direction. (Reproduced with permission of Laurent Drissen, University of Laval.)

to the corner cube vertex. This arrangement works as a single plane mirror for the interferometric performance, but with parallelism requirements relaxed by many orders of magnitude. Actually, since building an accurate enough set of three orthogonal mirrors is tough, a further refinement uses a so-called cat's-eye mounting.[2] See Figure 3.6 for a schematic look at these two solutions.

By the 1980s, the availability of, at first hundreds of thousand pixels, soon millions of pixels, IR arrays led to the replacement of the handful of astronomical FTSs in operation by high-resolution grating spectrographs. One prime reason was the huge gains in detector noise exhibited by these IR arrays: with modern IR detectors, the observed S/N is no longer limited by the detector noise but by the object photon noise. In that case, the FTS observing efficiency suffers from a spectral multiplex disadvantage with respect to a conventional grating spectrometer.

3.4.2 Fourier Transform Spectrograph

In 1992, Maillard and Simons [5] transmogrified the 1–5 μm CFHT FTS into a bona fide 3D instrument, by replacing the two InSb detectors by two InSb arrays. The arrays are located on images of the entrance field. In that way, one gets an interferogram on each spatial pixel. This so-called BEAR instrument was used until 2001, essentially to provide low to high spectral resolution, 3-D spectroscopy on extended objects with a few IR emission lines selected with a wide-band interference filter: this limits the multiplex disadvantage to a few units. The final output, after inverse Fourier transforms on each detector pixel to get back a classical spectral format, is very similar to that of a scanning Fabry–Pérot, except that a much smaller set of wider interference filters is needed. Also, high-precision data cubes in typically a handful of close by emission lines are obtained simultaneously, instead of only one at a time.

---

2 A low-cost catadioptric variant mimicking cats' glowing eyes is used for the retroreflectors of vehicles.

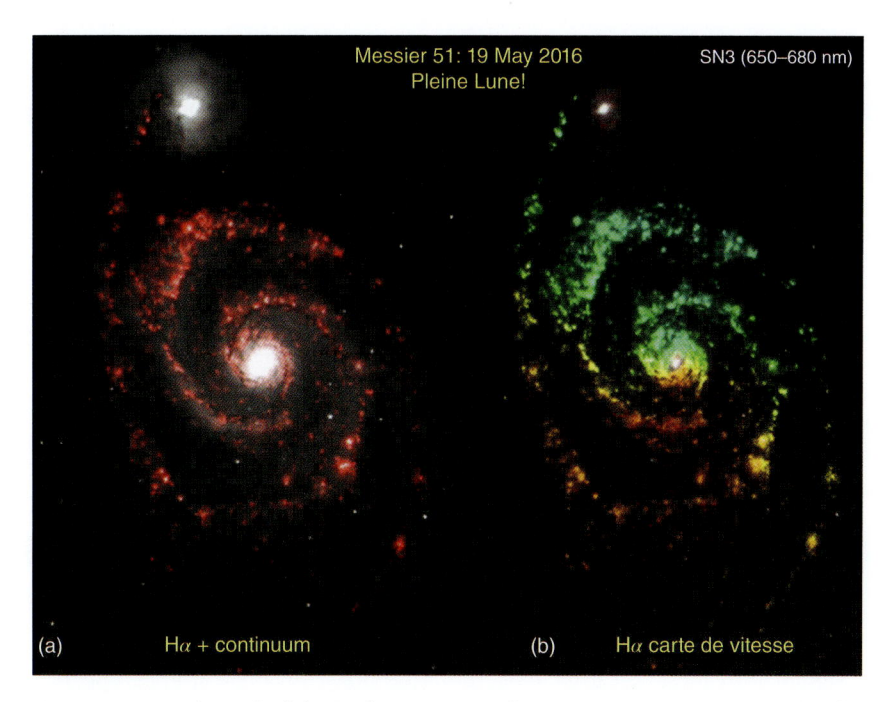

Figure 3.7 An early result of the Sitelle Fourier Transform Spectrograph at CFHT on the 9′ diameter nearby galaxy Messier 51. (a) Reconstructed image in the ionized hydrogen 656.3 nm H_α line. (b) Corresponding color-coded radial velocity field, mainly due to rotation of the ionized gas clouds around the galaxy center. (Reproduced with permission of Laurent Drissen, University of Laval.)

A similar imaging FTS is working in the (much tougher) optical domain (350–850 nm) at the 1.6 m Mount Megantic telescope. This SpIOMM instrument, developed by Drissen *et al.* [45] is a dual input dual output system with a 12′ field. It has mainly been used on extended emission line regions in a moderate wavelength range (free spectral range $R_c \sim 7$) to limit the multiplex disadvantage impact. SITELLE, an enlarged version, has been built for the 3.6 m CFH Telescope [46] and put in operation in 2015 (see Figure 3.7). It covers a very large 11′ × 11′ diameter field, with any spectral resolution up to 10^5 in the 350–950 nm range. Typical spatial resolution is ~0.7″. This instrument has the potential to reach unique pastures in ultra wide-field 3D astronomical spectroscopy, for example, high spectral resolution (~4.10^4) 3D coverage of a whole globular cluster at once [47].

3.5 Conclusion: Comparing the Different Scanning Flavors

A large variety of 3D scanning techniques appeared during the golden twenties (1975–1995) of astronomical instrumentation, in particular SLSS, Fourier transform spectrography, and scanning Fabry–Pérot spectrography. They represent a direct way to build huge 3D data files, even when using the relatively modest size 2D detectors available at the time. Yet, they have been and are still poorly represented among general-use instrumentation on large telescopes, especially compared to the current ubiquity of non-scanning 3D techniques, such as multiobject and integral field spectroscopy, both on the ground and in space. This suggests major genetic deficiencies, as briefly discussed below.

One common weakness of scanning techniques comes from unavoidable changes of observing conditions from scan to scan, in particular with respect to image quality, light transmission, and Earth upper atmosphere background emission. This used to be largely circumscribed at the time of photon-counting detectors by the simple expedient of scanning with enough speed to freeze atmospheric conditions, that is, with typically no more than 20–40 s per full scan. However, with the demise of photon-counting detectors and their replacement by integrating CCDs at the end of the 1980s, this hurdle has been back with a vengeance. This cannot be the only killing factor however, as shown by the equally poor portfolio of these scanning techniques for deep-sky astronomical observations from space, despite the ideally stable observing conditions there.

The Achilles' heel of scanning Fabry–Pérot technique is that it is far from optimal for the study of the 3D shapes of most current astronomical objects, that is, relatively small roundish spatial extents (say, a hundred or so pixels across) and significantly larger (at least a few thousand pixels for most studies) spectral extents. These are poorly covered with its much larger spatial coverage and much smaller spectral coverage, a few tens spectral pixels only. This can be in principle extended, but at the cost of a bank of tens of interference filters and total observation times of many weeks for a single target. This is equally true, if in a different way, for the SLSS, with adequate spectral coverage, but spatial coverage of a narrow rectangular shape, unless one is prepared for weeks of integration on a single faint object.

The scanning Fourier transform spectrograph could in principle cover any required 3D shape, but often at the cost of significant sensitivity losses. With quasi-continuum sources (in particular galaxies stellar emission), the small signal variation coming from the shallow absorption lines is easily swamped by the photon noise coming from the continuum light integrated over the whole spectral range: this is the so-called multiplex disadvantage, which grows as the square root of the number of spectral pixels for such continuum sources. The Fabry–Pérot technique is similarly hampered, if for a different reason, for continuum objects, as the shallow absorption lines can easily be lost owing to the spectrograph scanning of the non-scanning interference filter spectral shape. This is why both techniques have been so far mostly used for the study of large (arc minutes across) ionized gas regions for which most of the total light flux is concentrated in a few emission lines. With the mid-2010s revival of Fourier transform spectrography in the visible region on a reasonably large telescope, this might change though and it will be instructive to look at its scientific output in the next few years.

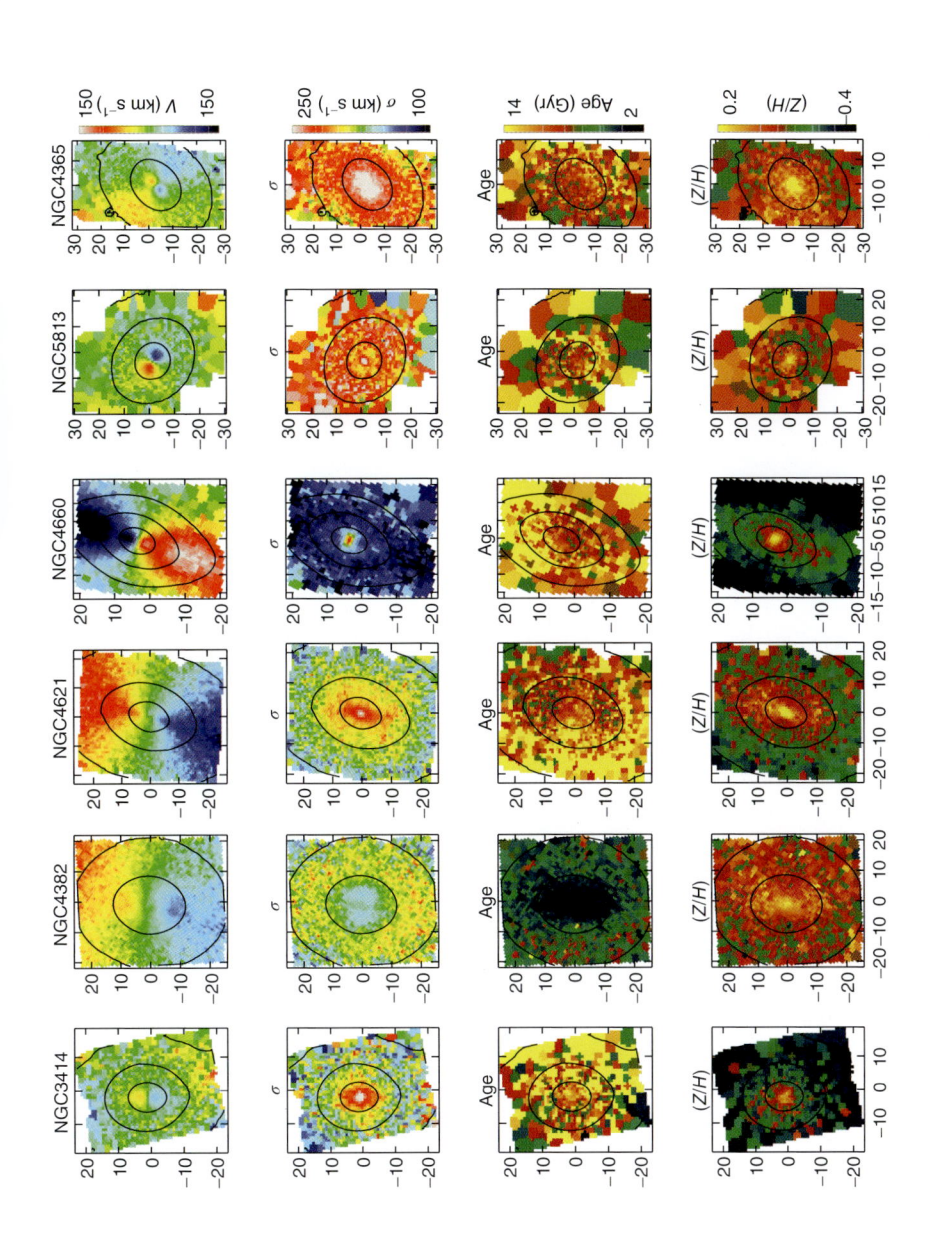

The SAURON survey [123] was one of the first surveys ever performed with an integral field spectrograph. A representative sample of 72 early type galaxies were observed with the lenslet-based IFS SAURON [53] at the William Hershell Telescope. A few examples are shown in this figure. In contrast to the smooth light distribution (overlaid as contour) of the observed galaxies, the kinematic maps (first two rows), the age map (third row), and the metalicity map (last row) exhibit a large variety of structure representatives of their formation and evolution history. ([123]. Reproduced with permission of Davor Krajnovic.)

4
Integral Field Spectroscopy

4.1 Introduction

Full spectroscopic study of an individual astronomical object requires getting a spectrum for each of the object 2D spatial pixels, a 3D format generally called hyperspectral imaging outside of the astronomical field, for example, for surveillance or environmental purposes. The so-called integral field spectrographs (IFSs), first developed in the 1980s, are able to get such data cubes in a single exposure. This approach offers much better sensitivity as well as more accurate photometric data than scanning techniques. The obvious tradeoff is that since a large number N of detector pixels – usually thousands – are needed to give a full spectrum for each object spatial pixel, an IFS field coverage area is smaller by at least the factor N compared to the equivalent scanning instrument: a typical field for a scanning interferometer might be $\sim 7'$ in diameter versus say $\sim 7''$ across with an equivalent IFS.

To give meaningful data, any IFS needs to separate the 2D spatial pixels from each other in order to get non-overlapping spectra on the detector. The three current IFS flavors stem from three different techniques, using an array of minilenses (or lenslets), a multifiber-based slicer or a multimirror-based slicer. They are presented and discussed in the following sections.

4.2 Lenslet-Based Integral Field Spectrometer

Lenslet-based IFSs use a microlens array to perform the bidimensional spatial sampling of the field. Each lens produces a micropupil, which is then dispersed, giving a many-pixel spectrum on the detector. The instrumental concept is illustrated in Figure 4.1.

Micropupil diameters are much smaller than the microlens diameter (typically by a factor 15–30). This leaves enough free space on the detector (a factor 225–900) to insert a many pixel spectrum for each micropupil. Micropupils light can be dispersed by a prism (for very low spectral resolutions), a grating,

Optical 3D-Spectroscopy for Astronomy, First Edition. Roland Bacon and Guy Monnet.
© 2017 Wiley-VCH Verlag GmbH & Co. KGaA. Published 2017 by Wiley-VCH Verlag GmbH & Co. KGaA.

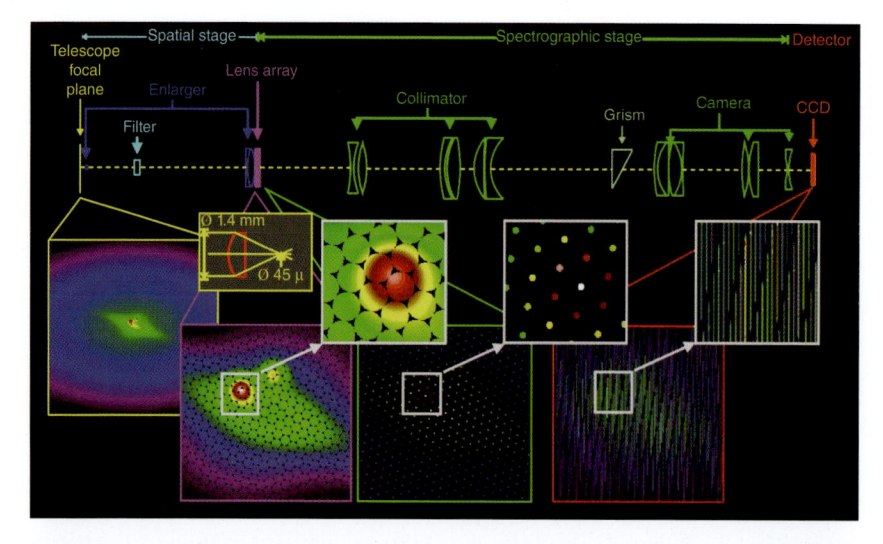

Figure 4.1 The lenslet-based integral field spectrograph instrumental concept. *Left to right*: An enlarger plus field lens combination produces a highly magnified telecentric image of the small field of view at the entrance of a 2D microlens array. Each lens samples a very small spatial field and produces a small circular exit pupil, called a micropupil. These micropupils act as the equivalent of the slitlets at the entrance of a classical multislit spectrograph, which gives a spectrum stack (one per each micropupil) on the detector. The (single) exit pupil at the level of the disperser is actually filled by highly enlarged stacked images of each small fraction of the field of view sampled by the microlens array.

or a grism combination. To avoid spectra overlaps, their length is limited by an interference filter and the microlens array axis is rotated by a few degrees with respect to the dispersion direction, so that each spectrum just "misses" the adjacent ones. Note that the spectrum stack is precisely aligned along the detector lines in order to improve data extraction.

Being images of the telescope pupil, the micropupils are uniformly illuminated to first order with the same geometrical shape[1] (Figure 4.2); they only differ by their global illumination. This ensures that the spectral PSF is independent of the astronomical target. As we will see later (Chapter 11), this strong a priori information is of much help in extracting the spatiospectral data cube from the detector output. Also note that field spatial sampling is done once for all at the entrance of the microlens array; moderate optical aberrations further down the optical path (i.e., collimator + disperser + camera) of course degrade the spectral resolution, but leave the spatial resolution untouched, that is, as long as adjacent spectra do not significantly overlap on the detector.

---

1 The exact spatial PSF shape is the convolution of the telescope pupil by the spectrograph spatial PSF.

Figure 4.2 Illustration of the uniformly illuminated spatial point spread function of one lenslet micropupil. One sees the telescope central obscuration by its secondary mirror.

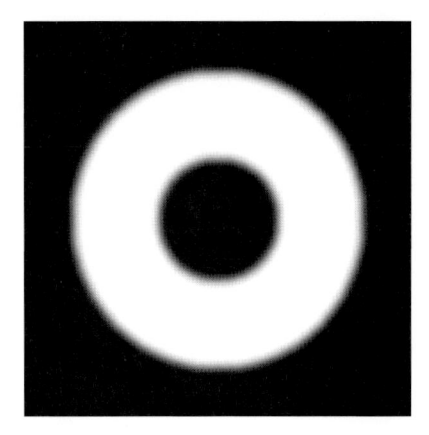

> Multipupil imaging is a key characteristic of this concept. In classical slit spectrography, the field of view is imaged on the detector. While spatial information along the slit length is left untouched, the one across the slit width is inherently mixed with the spectral information, producing the so-called slit effect. For an object with a strong light intensity gradient across the slit width (e.g., when the slit skims the central nucleus of a galaxy), measured radial velocities V_R from the spectra could be in error by as much as $c/2\mathfrak{R}$ (150 km s^{-1} for $\mathfrak{R} = 1000$). This is very significant, as radial velocity gradients of gas clouds around galaxy nuclei are typically of the same order of magnitude. This effect is essentially zeroed with multipupil imaging.

On the other hand, a drawback of the concept comes from the geometrical arrangement of its spectrum stack. One can see from Figure 4.1 that adjacent spectra on the detector do come from adjacent field areas, which is great, but are significantly shifted in wavelength, which spells trouble. This has indeed two negative impacts that must be mitigated by design:

1) The measured intensities of two neighboring spectra, taken along the same detector column x, do not originate from the same wavelength and are difficult to disentangle precisely at the data reduction phase. It is thus necessary to keep enough free space (δ_x) between each spectra to reduce this spectrum to spectrum contamination to a low enough level. This however lowers the number of spectra that can be put on a given size detector, further reducing the small spatial field inherent for any integral field instrument. There is then a definite tradeoff between spectra separation and the total number of information data points (spaxels) on the detector. For example, an instrument designed to achieving a good spectrophotometric accuracy [2] should have

---

2 The spectrophotometric accuracy measures the precision to which the flux can be measured as a function of wavelength. For example, to achieve a 5% accuracy we must be able to limit to less than 5% the total contamination due to neighboring spectra. This leads to $\delta_x \geq 1.9\ W_x$ for a Gaussian PSF.

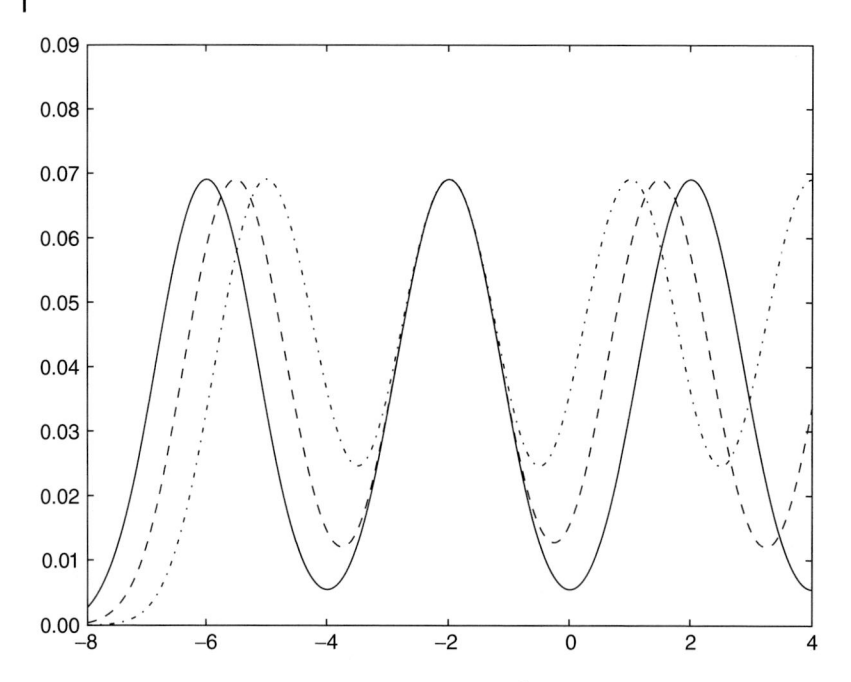

Figure 4.3 Examples of cross-dispersion profiles for $\frac{\delta_x}{W_x}$ values of 2 (solid line), 1.75 (dashed line), and 1.5 (solid-dashed line) in the case of constant intensity spectra. Increasing cross-talk as spectra are more closely packed on the detector can be clearly seen.

a large δ_x with respect to the instrument cross-dispersion PSF. Usually, the value of $\frac{\delta_x}{W_x}$, W_x being the full width half maximum (FWHM) of the instrument PSF along the cross-dispersion, will be between 1.5 and 2. Examples of typical cross-dispersion profiles are given in Figure 4.3.

2) For an overwhelming majority of astrophysical spectral investigations, observers need to get all spectra over a common spectral range. For a long-slit spectrograph, this is achieved by design; on the other hand, for this particular IFS flavor, this spectral shift does lead to truncated spectra. For a detector of a given size (n_x, n_y, y being along the dispersion direction), the fraction of spectra (q_S) that are truncated at the edge of the detector is related to the spectral length (l_S) in pixels. It is given by

$$q_S = \frac{l_S}{n_y} \qquad (4.1)$$

For example, with a detector with $n_y = 2000$ pixels, we shall limit the spectral length to $l_S < 400$ pixels to get 80% of non-truncated spectra.

Finally, the total number of spaxels n_L is given by

$$n_L = \frac{n_x n_y}{\delta_x l_S} \qquad (4.2)$$

The packing efficiency P_e is given by

$$P_e = \frac{n_L(1-q_S)l_S}{n_x n_y} = \frac{1 - \frac{l_S}{n_y}}{\delta_x} \quad (4.3)$$

In addition, as with any grating-based (as opposed to prism-based) spectrograph, one must take care that the spectra resulting from the other diffraction orders do not overlap with the observed spectra. In most cases, gratings are used in the first diffraction order to maximize efficiency over a sizeable wavelength range; we thus have to take care of unwanted light from the zero and second diffraction orders. The zero order integrated flux, while rather weak for such an optimized grating (a few %), is focused on a few pixels. This results in a spot bright enough to strongly impact the first order spectra. To avoid this overlap we should have the distance between the first and zero order larger than n_y. This is achieved when the resolving power $R = \frac{\lambda}{W_y}$ is

$$R > \frac{n_y}{W_y} \quad (4.4)$$

where W_y is the FWHM of the instrument PSF along the spectral direction. The second order diffraction integrated flux is about the same, but now distributed over many pixels, actually twice that for the first order spectra. Its impact is thus quite negligible. In addition, the recipe above to get rid of the zero orders by making them fall (just) out of the detector edge also guarantees the same for the second order spectra, now falling out of the opposite edge. Contrary to the fiber bundle integral field spectrographs discussed in Section 4.3, which are optimized for long spectra and smaller number of spaxels, lenslet IFSs are optimized for a large number of spaxels with relatively small spectral length. For low spectral resolutions (say well below a thousand), prisms are often preferred to gratings, as they fully eliminate by nature any risk of overlap with zero diffraction order spots (and with any higher order spectra as well).

Detailed presentation of the lenslet-based integral field spectrograph concept can be found in the original paper [48].

The making of a lenslet integral field spectrograph involves designing, manufacturing, and assembling four major subsystems: (i) the fore-optics the function of which is to enlarge the field of view in order to match the scale of the lenslet array, (ii) the lenslet array including an exit micropupil mask to minimize any diffused light sent to the spectrograph, (iii) the spectrograph, composed of a collimator, a grating (or a prism), and a camera, and (iv) the detector and its cryogenic and control systems.

The lenslet array is the most innovative element. It must fulfill the following requirements: high fill factor, accurate lens to lens pitch, uniform focal length, good image quality, low surface roughness (to avoid diffused light), and high throughput. In addition, it is designed to deliver telecentric micropupils (i.e., with all small subfields put at infinity) in order to collect all the light with the spectrograph collimator.

The first lenslet array (TIGER) was built from a set of a few hundred circular lenses of 1.4 mm diameter. After being polished separately, they were glued together on a parallel plate. This process is not only highly time consuming and then expensive, but also limited to circular lenses that provide only 91% fill factor. Today, various techniques to manufacture lenslet arrays are used: monolithic fused silica lenslet arrays manufactured by reactive ion etching, crossed cylindrical arrays, photoresist lenslet array manufactured by surface reflow, epoxy lenses on a glass substrate, monolithic fused silica lenslet arrays (see Figure 4.4), and monolithic polymethylmethacrylate lenses [49].

Lens shape can be hexagonal or square leading to almost 100% fill factor. In that case, however, the spectrograph pupil must be enlarged to accept all the beams. All optical surfaces (2 or 4 depending on the approach) are whenever feasible AR coated to avoid energy losses. Cheap off-the-shelf lenslet arrays can be procured, but generally with smaller diameters than the 0.5–1 mm size typically needed for the integral field application. It is nevertheless possible to find a few vendors

Figure 4.4 The picture shows the microlens array of the OASIS integral field spectrograph, made of 1500 identical hexagonal lenses. With a number of exchangeable enlargers before the lens array, sky samplings ranged from 0.04″ to 0.3″. Adaptive optics corrections (Section 9.4) were applied when working with every sampling but the coarser ones.

who can produce good quality lenslet arrays at a much higher, but still affordable, price.

The remainder of the instrument involves only classical optics and there is no feasibility concern for the IFS main optical train. There are a number of quirks however: (i) the enlarger must deliver a telecentric image (i.e., with the exit pupil at infinity) in order to get a regular 2D array of micropupils from the microlenses array; (ii) as is usual for astronomical spectrographs, the exit pupil delivered by the collimator must be located about 50–75 mm after the surface of the collimator last lens to minimize the grating and camera sizes, and the final image on the detector must be preferably close to telecentric; (iii) the collimator/camera combination must deliver a very good optical quality over the whole spectrum length and for the whole spatial field: in the optical domain, this demanding requirement mandates using custom lenses with a number of expensive and fragile fluoride glasses or crystals, and (iv) a fast system (typically $\sim$ F/2 camera focal ratio) is usually chosen; this empirically points toward a collimator/camera minimum pupil diameter size at least equal to the diagonal of the detector, for example, 73 mm for a circa 2016 state-of-the-art 4k × 4k, 12.5 µm pixel CCD. Such a high focal ratio design usually requires at least one expensive aspheric lens surface in the camera. As a result of the above, while there are no inordinate manufacturing challenges, building the optics of a modern lenslet-based IFS is not cheap, with the total optics cost dominated by the collimator–camera unit, with a usual price tag around 250 kEuro/kUSD.

Lenslet-based IFSs have been introduced in 1987 with the TIGER instrument, and operated at CFHT during 12 years [50]. Table 4.1 shows a list of similar instruments with their main characteristics. A typical IFS is shown in Figure 4.5.

In addition, two very small spectral resolution Tiger-type spectrographs, GPI and SPHERE, have been commissioned in 2015, respectively at Gemini-S and at

Table 4.1 List of lenslet integral field spectrographs with their main characteristics: name, host telescope, start and when applicable end year, N_L number of spatial elements, N_S, number of spectral elements and S_L, spectral length in pixels, $\Re$, spectral resolution, range, spectral range in µm, and Ref., bibliographic reference.

Name	Tel.	Start-End	N_L	N_S	S_L	$\Re$	Range	References
TIGER	CFHT	1987–1999	350	300	0.4–0.6	400 1800	0.4–0.8	[48]
MPFS	BTO	1990–2010	99	350	0.3–1.6	1000 500	0.4–0.8	[51]
OASIS	CFHT	1997–2002	1100	400	0.04–0.3	1000 3500	0.4–1.0	[52]
SAURON	WHT	1999–2015	1577	540	0.3–1.0	1400	0.5–0.7	[53]
OASIS	WHT	2003 –	1100	400	0.1–0.3	1000 4000	0.4–1.0	[54]
SNIFS	UH 2.2	2004 –	225	2500	0.4	1300 3400	0.3–1.0	[55]
OSIRIS	Keck2	2005 –	1000	1600	0.02–0.1	3500	1.0–2.5	[56]

Figure 4.5 The SAURON integral field spectrograph at the Cassegrain focus of the 4.2 m William Herschel Telescope (La Palma, 2000).

the VLT. Coupled to extreme adaptive optics systems, their common goal is to boost by about two orders of magnitude the direct detection of giant gaseous exoplanets (planets orbiting stars other than our Sun) via spectral signatures of wide absorption bands in their atmospheres.

4.3 Fiber-Based Integral Field Spectrometer

4.3.1 The Fiber-Based IFS Concept

The first ever IFS prototype was developed and tested by Vanderriest in 1980. It was based on an image dissector made of 205 optical fibers of 100 μm diameter. This industrial component was used to transform the small hexagonal field (about 10″ across on the UH Mauna Kea 2.2 m telescope) into a narrow rectangle, constituting the entrance slit of a classical spectrograph (see Figure 4.6). The first sky test of this new concept on the nebulosity surrounding the quasar 3C 120 was successful.

Yet, it took until 1988 to see the SILFID dedicated visitor instrument in operation, first at the Haute Provence Observatory 193 cm, then at the Mauna Kea 3.6 m CFHT telescope. The key changes were a reasonable size CCD detector and a spectrograph collimator collecting most of the light delivered by the dissector,

Figure 4.6 Fiber-based integral field concept (Courtesy C. Vanderriest). The figure shows its key ingredient, a multifiber image dissector that changes the input squared field format into a narrow slit-like output format, well-suited to directly fill a classical long-slit spectrograph. Credit Observatoire de Paris, France.

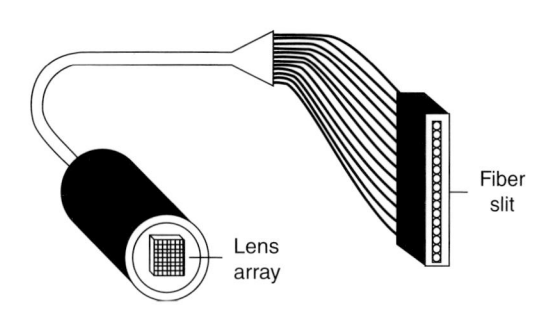

both ingredients lacking in the prototype (actually for the CCD, not available at all) almost a decade earlier.

4.3.2 The Fiber-Based IFS Development

Following this pioneering work, fiber-based IFSs have been rather quickly deployed, at first on many 4 m-class telescopes, as, compared to the lenslet flavor, they are conceptually simpler, can be easily retrofitted to existing classical long-slit spectrographs, and give data cubes relatively easy to reduce.

The first such systems used bare step-index fibers fed at a high aperture ratio ($\sim$ F/2.5 – 3.0) to minimize beam enlargement within the fiber slicer. One drawback was the significant field loss between the fibers due to both their circular shape and their outer cladding. This not only degraded the spatial resolution but, even more importantly, led to significant spectrophotometric uncertainties. This was solved by tessellating the field with square or hexagonal lenslets, each illuminating a single fiber, a neat lenslet/fiber hybrid approach first implemented in 1997 at the MPFS instrument on the Russian 6-m BTO telescope and now present in all such instruments. Similar microlenses are used at the output of the fiber slicer to adapt the light beam focal ratio to that of the spectrograph.

By the mid-2010s, a number of fiber-based IFSs are in operation or in development on 4-m class and 8-m class telescopes. Table 4.2 shows their main characteristics.

4.3.3 Conclusion

This fiber-based slicer approach to integral field spectrography offers better detector coverage than the lenslet-based approach, at least by a factor two. This translates into a larger spectral coverage and/or a larger field. On the other hand, light efficiency is significantly lower, with a typical 35% light loss due to the fiber slicer, and spectrophotometric accuracy is lower. It is interesting that the relative astrophysical use of the two approaches reflects this tradeoff quite well. Most fiber-based IFS science output lies in the study of relatively bright emission lines in active galaxies, while the main ecological niche of lenslet-based IFS is the study of faint absorption lines in quiet galaxies, for which accurate spectrophotometry is a must.

Table 4.2 List of fiber-based integral field spectrographs with their main characteristics: host telescope, start and end date, N_L, spaxels number, N_S, spectral pixel number, S_L, spatial sampling ("), $\mathfrak{R}$, spectral resolution, and maximum spectral range (μm).

Name/Tel.	Start-End	N_L	N_S	S_L	$\mathfrak{R}$	Range	References
ARGUS/CFHT	1993–1998	397	600	0.40	$(0.6)\,10^3$	0.39–1.0	[57]
MPFS/BTO	1990–2010	256	1024	1.0	$(2.5)\,10^3$	0.46–0.72	[51]
INTEGRAL/WHT	1997 –	195	2000	0.5–2.0	$(0.5)\,10^3$	0.48–1.0	[131]
SPIRAL/AAT	1998 –	512	1024	0.70	$(1–9)\,10^3$	0.48–1.0	[132]
PMAS/Calar Alto	2003 –	256	1024	0.50	$(3–5)\,10^3$	0.35–0.9	[104]
GMOS/Gemini	2004 –	1500	6000	0.20	$(8)\,10^3$	0.36–0.94	[23]
VIMOS/VLT	2002 –	6400	2048	0.3–0.7	$(0.2–2)\,10^3$	0.36–1.0	[22]
IMACS/Magellan	2005 –	2000	1600	0.20	$(3.5)\,10^3$	0.36–1.0	[58]
LRS2-B/HET	2015 –	300	940	0.60	$(1–2)\,10^3$	0.37–0.7	[59]
MEGARA/GTC	2018 –	400	1100	0.62	$(6–19)\,10^3$	0.37–0.98	[60]

4.4 Slicer-Based Integral Field Spectrograph

4.4.1 Introduction

Mirror-based image slicers are a priori ideal for transforming a classical long-slit spectrograph into an integral field one, with a better detector filling than the fiber-based slicer (no more etendue loss) and a much better one than with a lenslet-based system (no more need for a lot of free space between adjacent spectra on the detector). This advantage is especially important in the infrared region, with IR detector pixels 10 times more expensive than visible ones. If we add that optical fibers do not work well above 1.6 μm and that the tough rugosity specifications (level of mirrors surface smoothness) of narrow (typically a few 100 μm width) slicer mirrors are less demanding in the IR region, it is no great surprise that the first ever mirror-slicer-based IFS, the MPE-3D/SPIFFI instrument, worked in the NIR 1 − 2.4 μm region.

The instrument slicer concept is based on a circa 1930 German patent intended for optomechanically based TV sets. As seen in Figure 4.7, sets of two small plane mirrors are used to slice a squarish field into thin slices and stitch them along a single long slit. Since the field slices are actually on top of each other, the various slitlets are slightly displaced from each other, giving a small wavelength range mismatch on the IFS detector. Crosstalk between spectra is small, essentially due to residual surface rugosity and to the small defocus of the field on the staircase field mirrors. Besides, this all-mirror system is (relatively) easy to adapt for the cryogenic environment inevitably associated with IR operation beyond 1.5 μm. Light efficiency is also excellent, easily above 90%. This two-flat-mirror design is remarkably totally aberration free. In practice, the slicer is fed by a very narrow beam, for example, at F/40 or more: even then, the primary mirrors are extremely

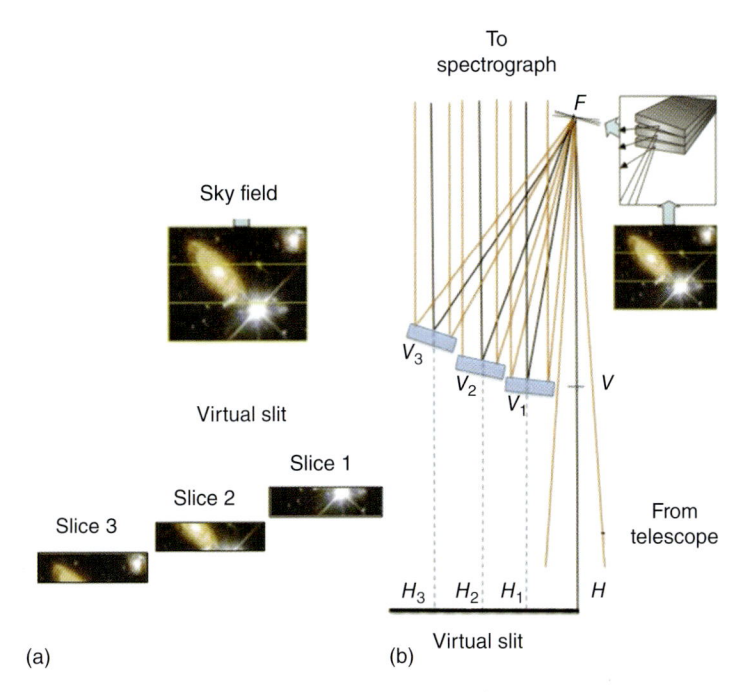

Figure 4.7 MPE 3-D/SPIFFI mirror slicer concept, shown here with 3 of its original 16 slices. (b) Telecentric beams from the telescope fall on a "staircase" made of a stack of N thin (0.1–0.2 mm) tilted flat primary mirrors at the telescope focal plane F. The stack slices the small 2D field and sends the light back to N flat secondary mirrors that redirect all light beams parallel to their common initial optical axis. With the secondary mirror vertexes lying on a paraboloid of focus F, Fermat's principle (constancy of the optical paths $FV_n - V_n H_n$, here all nulls) ensures that all output light beams originate from the virtual slit shown here. (a) Illustration of field slicing and its reorganization into a staggered (virtual) pseudo-slit.

thin, $\sim 0.1 - 0.2$ mm and difficult to make and to align with the secondary mirrors.

The one significant drawback of the approach is the large physical size of the slicer, 30 cm across for SPIFFI. This has been much improved by Content [61] by replacing the first set of plane mirrors with curved ones that re-image the telescope pupil on the second matching mirror set. A real slit image is formed at the output of the slicer and mirrors/lenses are added to send a single common pupil to the spectrograph (Figure 4.8). A number of even more compact variants with large slit demagnification have since been developed, for example, for the MUSE instrument (Section 5.4.2).

Optical fabrication and alignment of the advance slicer versions are quite difficult too. For the NIR, this is generally solved by fabricating the whole optomechanical slicer with the diamond-turn technique. More complicated approaches are needed in the optical region; see Section 5.4.2 for one such development. This is from now on the clearly preferred approach for most integral field instruments,

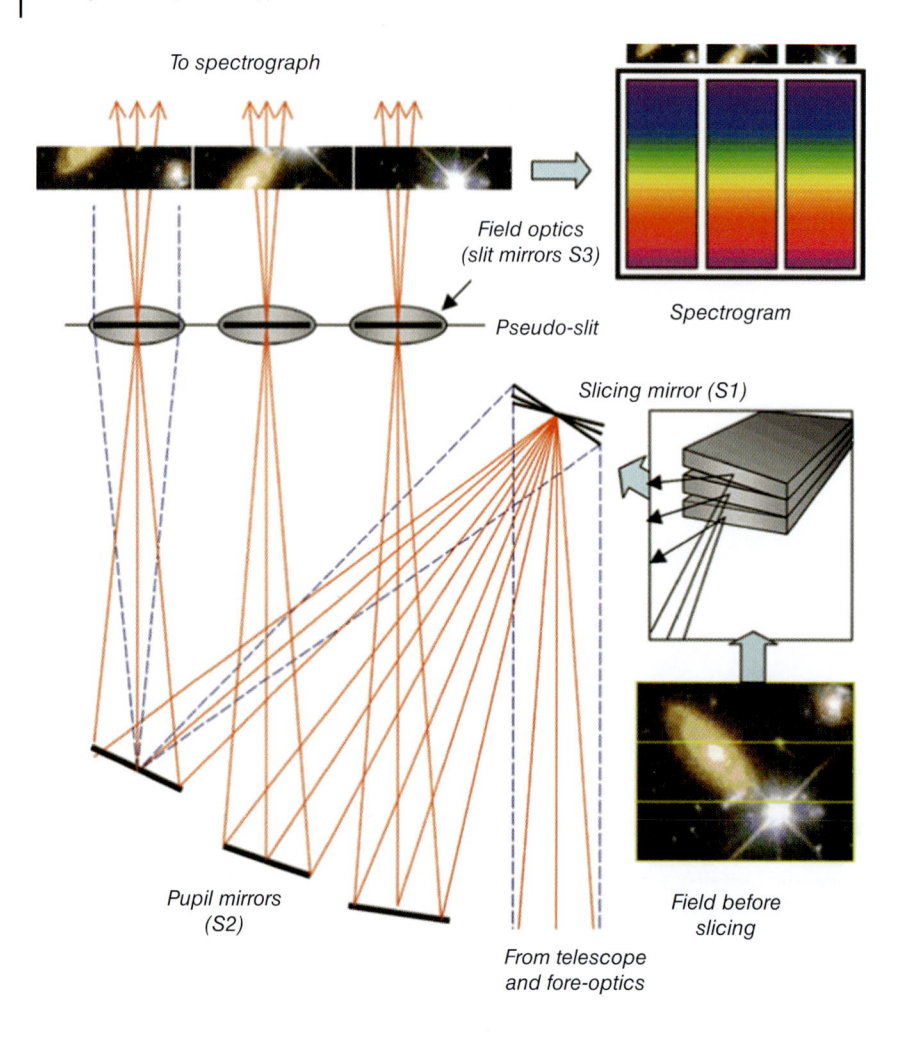

Figure 4.8 Advanced image slicer concept. Telecentric input beams (bottom right) from the telescope fall on a staircase stack, made of *N* thin *curved* primary mirrors S_1, which slices the small sky field. *N* small *curved* secondary mirrors S_2 are put on the respective pupil images given by the primary mirrors. They re-image the *N* field slices on a *real* pseudo-slit. Field lenses or mirrors S_3 are added near this pseudo-slit to give *N* telecentric output beams (top left) that input the associated spectrograph. Figure from the Gemini Telescope web site (http://www.gemini.edu/node/21). (Reproduced with permission of Gemini Observatory.)

with the best light efficiency (shared with the lenslet kind), the best spectral packing on the detector (better than with fibers and much better than with lenslets), and minimal crosstalk. Table 4.3 shows the main such instruments currently in operation on large telescopes.

All these instruments feature a zoom-like capability: the sky field can be imaged at different magnifications on the slicer, giving a large range in spatial sampling. Actual choice by the observer depends on the target characteristics: one major tradeoff is the observable field, which of course decreases in parallel

Table 4.3 List of mirror slicer based integral field spectrographs on large telescopes with their main characteristics: host telescope, start date, N_L, spaxels number, N_S, spectral pixel number, S_L, spatial sampling (″), Range, spectral domain (μm), $\mathfrak{R}$, spectral resolution, Ref., bibliographic reference.

Name/Tel.	Start	N_L	N_S	S_L	Range	$\mathfrak{R}$	References
SINFONI/VLT	2005	1024	1024	0.025–0.25	1.10–2.45	1500–4000	[130]
NIFS/Gemini	2005	1024	1024	0.04–0.1	0.95–2.40	5300	[133]
SWIFT/Palomar	2008	3916	2680	0.08–0.24	0.65–1.00	3825	[134]
MUSE/VLT	2014	9.10^4	4096	0.025–0.2	0.46–0.93	4000	[62]

with finer spatial sampling, the second being sensitivity to extended regions, which goes as the square of spatial sampling. Coarse sampling is thus used for relatively large objects (seconds of arc across) with large structures, finer one for smaller objects – typically 1″ diameter or less – with high-contrast small substructures. To efficiently exploit their small spatial sampling capability, all instruments listed in Table 4.3 use the much improved image quality delivered by adaptive optics systems: this technique, presented in Section 9.4, beats atmospheric turbulence above the telescope and its 0.3″–0.4″ image quality barrier, otherwise unavoidable even at the best terrestrial sites.

4.4.2 Integral Field Spectroscopy from Space

Mirror slicer-based IFS with its compact design (for advanced versions), no moving parts, excellent light transmission, and superb detector pixel filling factor, looks a priori well-suited for space-based observations. This has been quite slow to come though, and the first use has been with the far-infrared PACS instrument on board the far-infrared Herschel Space Telescope launched in 2009. It provides integral field spectroscopy in the 51 − 220 μm range over a 47″ × 47″ field, with spectral resolutions ranging from 4000 to 1000.

The 6.5 m diameter JWST space telescope, scheduled for a 2018 launch, features near-infrared and mid-infrared integral field spectrographic modes on respectively its NIRSPEC and MIRI instruments.

The NIRSPEC IFU slicer covers a 3″ × 3″ field of view, built up from 30, 0.1″ × 3″, slices. Spectral resolutions of 100, 1000, and 2700 are provided by a set of three grisms. To avoid overlapping grating orders, the full 1 − 5 μm instrument spectral range is divided into three adjacent wavelength intervals, each requiring a separate exposure with a different order-sorting filter.

The MIRI integral field medium resolution spectrograph uses four different image slicers to split its huge 5 − 27 μm spectral range in four adjacent spectral windows. The four fields of view are centered on each other and range from 3.6″ × 3.6″ for the 'bluest' window to 7.6″ × 7.6″ for the 'reddest' one; spatial resolution similarly ranges from 0.18″ to 0.64″. Spectral resolution is ∼ 3000.

4.5 Conclusion: Comparing the Different IFS Flavors

As seen in this chapter, there are presently three main flavors of integral field spectrography, respectively lenslet-based, fiber-based, and mirror-based. Of these three, the last and the most recent one looks a priori the winning solution, with the best detector filling and the best light efficiency. This is especially true in the infrared (as opposed to the optical) domain, for which the higher cost of detectors makes the near-perfect detector filling factor especially worthwhile. It has also a much better cryogenic compatibility than the fiber-based flavor, the spectral window of which is currently limited to below ~ 1.65 μm.

Fiber-based or lenslet-based systems are nevertheless still working options, mainly for practical reasons. For example, both offer a compact "no frizzle" way to transmogrify long-slit spectrographs for the former, and multislit spectrographs for the latter, in an integral field version. Also, as indicated in Section 5.2, some lenslet IFS systems have been found particularly effective in the diffraction-limited regime, that is, when adaptive optics techniques are used to get image quality at the ultimate limit set by light finite wavelength.

IFS is still a vibrant instrumental field, and the next chapter looks closely at recent trends in this domain.

** Exercise 7   Designing a Tiger-type instrument

You have access to a diameter D telescope ($D = 3.6$ m) at its F/ω_1 Cassegrain focus ($\omega_1 = 8$). You already hold some Tiger subsystems: a collimator–camera unit, with a d_2 diameter pupil ($d_2 = 45$ mm) and respective focal ratios $\omega_a = 4.2$ and $\omega_c = 3.5$; and a n_x, n_y pixel (size $p \times p$) CCD, with $n_x = 640$, $n_y = 1024$, $p = 15$ μm.

Your mission, should you decide to accept it, is to derive to first order the basic parameters for a lenslet integral field spectrograph that includes in addition an enlarger with magnification γ, a microlens array made of $N \times N$ circular F/ω_A microlenses of diameter d_1 and a l lines/mm zero-deviation grism (prism angle A, refractive index $n = 1.5$). We set the sampling of the micropupil images by the detector at $g = 2$ pixel, a minimum spectrum separation of $g' = 10$ pixels, and non-truncated spectra length l_s of 534 pixel along y on the detector. Good luck James/Jane.

1. Express and compute the sky sampling α in arcsec ($1'' = 4.85 \times 10^{-6}$ rad). Hint: use linear etendue conservation between the sky and detector planes.
2. Express and compute the microlenses array format ($N_x \times N_y$ lenses), then the instrument on-sky field f, the (small) angle θ between the microlenses array and the detector axis, and finally the microlenses diameter d_1. For simplicity, neglect the second-order terms in θ.
3. Find and compute the enlarger magnification γ. Hint: use linear etendue conservation between the microlenses array and detector planes.
4. Assuming a spectral PSF $g'' = 2.5$ pixel wide with an associated spectral resolution $\mathfrak{R}_s = 1000$ at a central wavelength $\lambda_c = 525$ nm, find and compute

the grism line frequency l, the prism angle A, and the wavelength range $\Delta\lambda$ covered by the spectra.

Answer of exercise 7

1. Linear etendue conservation between the sky and detector planes gives $D\alpha = gp/\omega_2$. Numerical value: $\alpha = 0.74''$.
2. The uniform pitch $N_x \times N_y$ micropupil array covers $(n_x) \times (n_y - l_s)$ pixels on the detector. Hence, $N_x/N_y = n_x/(n_y - l_s)$. The $N_x \times N_y$ micropupils also cover the detector n_x pixels along the x axis, with g' pixel between adjacent ones. Hence $N_x N_y = n_x/g'$. Numerical values: 63 lenses with $N_x = 9$; $N_y = 7$. With $\alpha = 0.74''$ per lens, field of view is $6.7'' \times 5.2''$ on the sky.
Spectra geometry on the detector gives $\tan\theta = g'N_y/(n_y - l_s)$. Numerical value: $\tan\theta = 0.143$, or $\theta = 8.1°$.
On the detector, micropupils pitch is $g' \times N_y$ pixel. On the microlenses array, it is multiplied by the camera to collimator magnification factor ω_a/ω_c. This is also by design the microlenses diameter d_1. Hence $d_1 = g'pN_y\omega_A/\omega_C$. Numerical value: $d_1 = 1.26$ mm.
3. The F-ratio at the entrance plane of the microlenses array is $\gamma\omega_1$. Linear etendue conservation between the microlenses array and detector planes gives $d_1/\gamma\omega_1 = gp/\omega_2$ or $\gamma = d_1\omega_2/gp\,\omega_1$. Numerical value: $\gamma = 34.7$.
4. A resolved spectral element on the detector subtends an angle $\delta i'$ at infinity from the grism plane, with $\delta i' = g''p/d_2\,\omega_c$. We have also classically $\delta i' = (n-1)\tan A/\mathcal{R}_s$. This gives $\tan A = g''p\,\mathcal{R}_s/(n-1)d_2\,\omega_c$. Numerical value: $\tan A = 0.48$, or $A = 25.6°$. The zero deviation condition gives $(n-1)\sin A = l\lambda_c$. Numerical value: $l = 440$ l mm^{-1}. To first approximation, resolved spectral element width is $\lambda_c/\mathcal{R}_s$ spread over g'' pixels; common spectral coverage $\Delta\lambda$ is $l_s\lambda_c/g''\mathcal{R}_s$. Numerical value: $\Delta\lambda = 112$ nm.
Congratulations, you have just repeated the original computations by the authors that led to the development of the first "TIGER" lenslet-based IFS in the early 1990s. Note the small field and the small spectral format, due to the small detector formats available at the time.

** Exercise 8   Building a cheap IFS

For less demanding applications that install an IFS on a giant telescope for cutting-edge astronomical observations, a much cheaper instrument can be built, albeit with significant transmission losses. You are to design such a device, using only off-the-shelf components found on the Web. Hint: you might use the "astrophotography" keyword to select a fair performance, yet not horrendously expensive, detector system. Most other key components can be found in optical catalogs, for example, from Edmund or Newport.

Iterating from available components, present a coherent design, with a rough cost estimate.

Answer of exercise 8

There is no single solution as (i) you can choose more or less professional components (usually a tradeoff between better or cheaper) and (ii) components availability, performance, and price may vary widely over the years, not necessarily in the right direction. Here is, as an example, one mid-2010s vintage semi-professional package:

Detector: Canon ID Mark 4; 3888 × 2592 px; 7.2 × 7.2 µm pixel. Camera: Canon EF 50 $mmF/1.8$. Microlens array: 8 Suss-microoptics 18 – 00035; 25 × 3 mm; 250 µm pitch; NA 0.18. Volume phase holographic grating: Edmund NT 48-587; 1" dia.; 600 l mm^{-1}.

** Exercise 9   Scanning Fabry–Pérot / Lenslet IFS Hybrid

Taking the lenslet integral field spectrograph (IFS) defined in Exercise 7, you add a scanning Fabry–Pérot located just before the lenslet array (telecentric beam aperture F/278). The goal is to keep the IFS spectral range (112 nm wide, centered at $\lambda_c = 525$ nm), but boost its central spectral resolution from the original $\mathfrak{R}_1 = 1000$ (for a $g'' = 2.5$ pixels spectral PSF) to $\mathfrak{R}_2 = 8000$. The Fabry–Pérot order is $p = 800$ at λ_c and its defect finesse is $N_d = 24$. The full data cube is obtained by making 10 successive exposures (say each 12-min long), scanning the etalon one step at the time. This recreates the original Pytheas concept LeCoarer et al. [63], once successfully tested on the sky, but only to be left aside for lack of a sufficient customer base.

1. Find the etalon reflective finesse N_r for which the combination spectral resolution is indeed equal to 8000 at λ_c. Hint: Use the Fabry–Pérot cooking book.
2. Describe the data output on the detector. What happens when scanning the etalon? Comment on the behavior at the lowest and the highest wavelengths.
3. The etalon is loosely held in position and can move as a whole a couple of millimeters in any direction, plus tilt by up to 0.5° around its normal position: would it cause any harm to system performance? The IFS part also suffers from sizeable flexures, giving as much as a 1-pixel spectra shift per hour on the detector: is there any harm?

Answer of exercise 9

1. The beam half-angle θ on the etalon is 1/556, giving a huge beam finesse $N_\theta = 2 \times (556)^2/800 = 773$. With $N_d = 24$, the etalon overall finesse N is equal to 10 for $1/N_r^2 = 1/10^2 - 1/24^2 - 1/773^2$, or $N_r = 11.2$.
2. Each IFS spectrum is transformed by the etalon multi-bandpasses into a "string of pearls," separated by a small but manageable $g''\mathfrak{R}_1/p = 3.1$ pixels, corresponding to the etalon free spectral range. As the etalon is scanned, the pearls move by up to this interval around the central wavelength. At the blue edge of the IFS bandpass, the etalon order is $p = 895$, and total scan is of 1.12 order. Conversely, at the red edge, $p = 723$ with a total scan of 0.90 order. To get the full data cube, one must then overscan at all wavelengths but the red

edge ones. Spectral resolution might be kept constant with custom-made etalon coatings giving a reflective finesse proportional to wavelength, a trick indeed used on the Pytheas device.

3. Moving the etalon in any direction – including along the optical axis – does not change anything, provided the optical beams are still fully accepted by the etalon. Tilting the etalon by $0.5° = 1/114.6$ rad lowers the beam finesse, which is now $N_\theta = (114.6 \times 278)/800 = 39.8$. This lowers the overall spectral resolution and efficiency, but at negligible levels. IFS flexures move the "pearls" by a fully manageable 0.2 pixel during an etalon integration time. This concept has thus the intriguing feature of giving accurate spectrographic data even with loose handling of its key components.

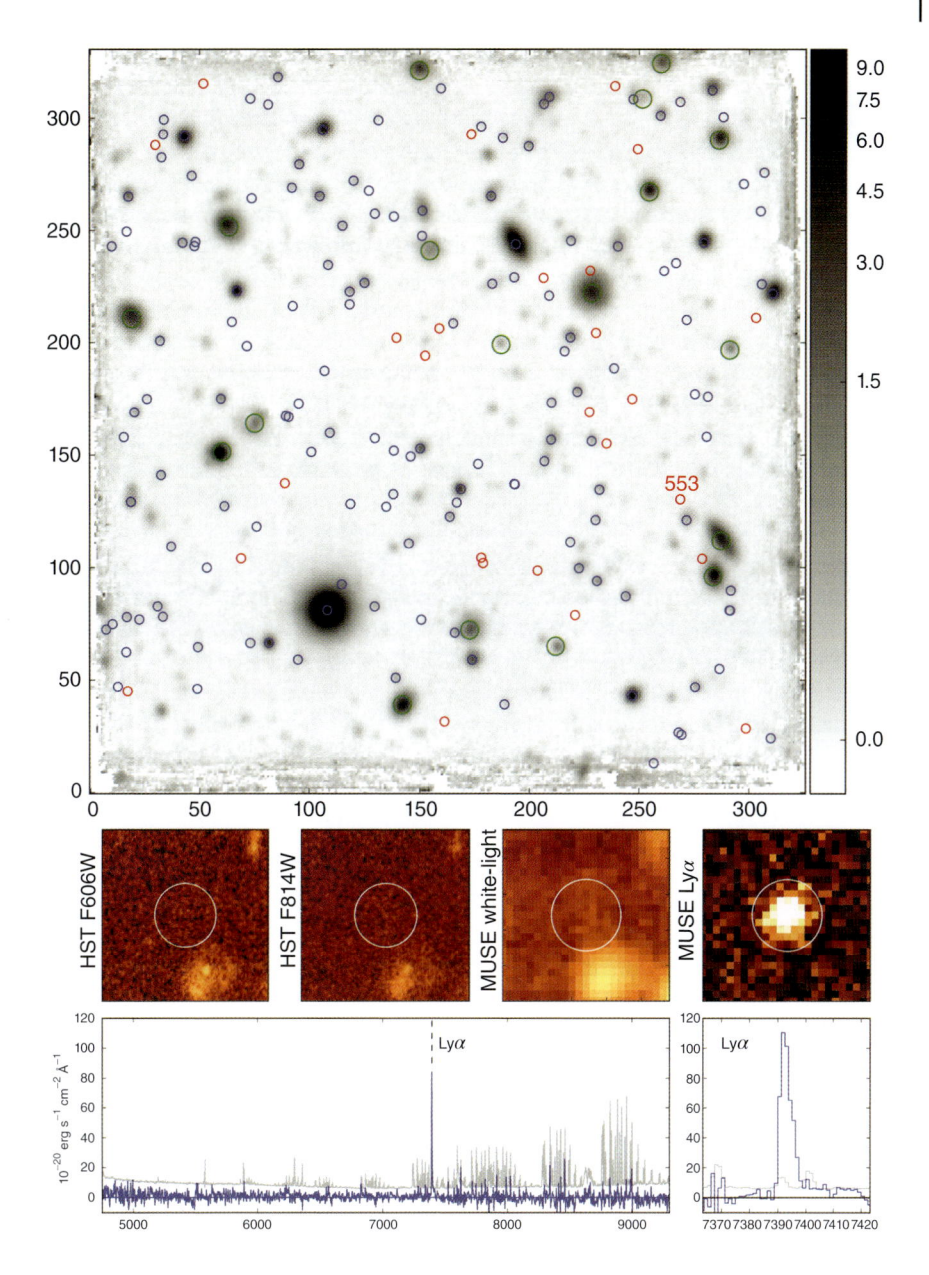

Discovery of a population of faint Lyα emitters in the Hubble Deep Field South with MUSE [101]. A field of 1 arcmin² in the HDFS was observed for 27 h with the slicer-based IFS MUSE at the ESO Very Large Telescope. The upper figure displays the reconstructed white light image in 10^{-20} erg s^{-1} cm^{-2} flux units. The overlaid symbols show the location of (i) the 18 previously known spectroscopic redshift (in green), (ii) the additional 144 measured redshifts with MUSE (in blue), and (iii) the 26 new Lyα emitters detected by MUSE without HST counterpart (in red). The lower figure shows one of the new Lyα emitter (ID 553), a $z = 5.08$ galaxy without HST counterpart. The HST images in F606W and F814W filters are shown at the top left, and the MUSE reconstructed white-light and Lyα narrow band images at the top right. The one arcsec radius red circles show the emission line location. The spectrum is displayed in the bottom figures, including a zoom at the emission line. The sky spectrum is shown in grey (more at muse-vlt.eu/science).

5

Recent Trends in Integral Field Spectroscopy

5.1 Introduction

Integral field spectroscopy (IFS) is a much alive field with new facilities constantly under deployment. We will cover here three current trends, namely, (i) its use for the detection of ultra-faint companions of bright stars, most notably exoplanets; (ii) the development of wide-field versions, covering up to a hundred thousand spatial pixels, and (iii) a hybrid of multiobject and integral field spectrographs with dozens of very small field Integral Field Units (IFU) deployed in a large patrol field.

5.2 High-Contrast Integral Field Spectrometer

5.2.1 Exoplanet Detection

Direct detection of exoplanets, in term of instrumental challenge, is an exercise in extremely high contrast detection. For instance, should we image in the optical/NIR region a clone of our solar system located at 150 light-years, Jupiter would appear as a reasonably bright point of light, but only 1″ away from a 500 million times brighter point-like central sun, a hopeless task with present technology. Detecting a young – a few 10 million year old–giant planet requires a more manageable, but still stiff contrast of about 5 million, and should be achieved by 2020. The whole problem is to extract such a small light smudge from the surrounding huge parasitic light spilling out from the central star image.

For ground-based observations, the first step toward this goal is to apply adaptive optics corrections (see Section 9.4.2) to a small – say 2″ diameter–field around the bright parent star. With such a bright central star right in the middle of a very small field, conditions are ideal to minimize light spilled out from the diffraction-limited central core of the star image, possibly by up to a factor of ∼5 in the H and K bands compared to a direct seeing-limited image. The largest gain, however, comes from putting, by the same token, nearly 80% of the putative planet light in a diffraction-limited core. An initial H-band

Optical 3D-Spectroscopy for Astronomy, First Edition. Roland Bacon and Guy Monnet.
© 2017 Wiley-VCH Verlag GmbH & Co. KGaA. Published 2017 by Wiley-VCH Verlag GmbH & Co. KGaA.

turbulence-limited 0.4″ diameter planet image then shrinks to 0.04″ diameter with an 8-m telescope, achieving an additional contrast gain by a factor of ~100. Occulting in addition the central star with a suitable coronagraphic system gives an additional gain of a few times only, but also results in a more stable image, a key property for the subsequent steps.

5.2.2 High-Contrast Integral Field Spectrometer

To go even further, one must apply some differential technique to further disentangle the planet point-like image from the remaining stellar light. This can be done with say two filters, one centered on a key spectral feature of the putative planet (e.g., on a strong IR methane absorption band common in Jupiter-like giant planets), and another, a reference filter just outside the absorption band. Extreme care in the overall optical system is needed to get an extraction gain in the hundreds. An example of this technique is shown in Figure 5.1.

An even more powerful approach uses instead a small field integral field spectrometer (IFS). Simultaneous information at many different wavelengths provides further discrimination between the signal from a real companion planet and the remaining artifacts. The expected resulting gain with respect to the use of differential filters should be ~10. Note that a modest spectral resolution of 20–30 is sufficient to resolve the strong absorption bands from molecules (CH_4, NH_3) in a giant planet atmosphere.

A further bonus to using an IFS is to give direct physical information on the companion. This mode has been implemented on two exoplanet imagers recently put into operation, GPI at Gemini South in 2014 and SPHERE on the VLT in 2015. They should be joined soon by CHARIS [64] on Subaru. Contrast is expected to reach 10^7 at a few arcseconds from the central star. All three systems work in the near-infrared region. It is crucial to get the best possible spectrophotometry of the small 2D field (a couple of arcseconds across) to extract such a small contrast object. Given, in addition, the relatively small data cube format (~200 × 200 spaxels and 50 spectral pixels), the three instruments use lenslet-based IFS.

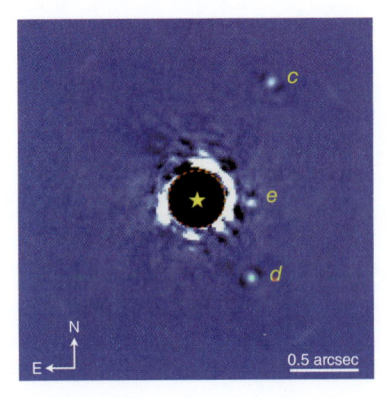

Figure 5.1 November 2013 direct detection in the K band of three giant (~10 Jupiter mass) exoplanets around the young (60 million-year old) star HR 8799 by the Gemini-South GPI instrument. Image credit: Christian Marois (NRC Canada), Patrick Ingraham (Stanford University) and the GPI Team. From the Gemini Telescope website (www.gemini.edu/node/12314). (Reproduced with permission of Gemini Observatory.)

5.3 Wide-Field Integral Field Spectroscopy

5.3.1 The Rationale for Wide-Field Integral Field Spectroscopy

One hallmark of the first generation of integral field spectrographs on large telescopes has been their very small sky fields, seldom more than 10″ across. This came directly from the unavoidable sharing of say the 16 million pixels of a single standard 4k × 4k detector between the instrument 2D spatial field and its 1D spectral range. This has so far essentially driven the scientific use of IFSs toward the study of small individual objects, admittedly already a whole zoo ranging from small bodies in our Solar System vicinity to faint galaxies at the far end of the Universe.

Enlarging the spatial field to, say, at least 1′ across on an 8–10 m diameter telescope opens new potentialities for IFS, in particular for conducting either massive surveys of far-away galaxies or blind discoveries of ultra-faint objects. These are indeed the respective main scientific goals of two major such projects launched around 2005, HETDEX/VIRUS on the HET for the former and MUSE at the VLT (Figure 5.2) for the latter. MUSE was completed and put in operation in 2014. By 2016, VIRUS is undergoing full deployment, with the start of the massive HETDEX survey expected soon thereafter.

5.3.2 Current Wide-Field Projects

Technically speaking, achieving such a wide-field integral field capability meant building spectroscopic instruments with hundreds of millions pixels, a huge number reserved until then for the much simpler direct imagers. The brute force approach would have been to build a huge (say a few hundred cubic meters) single spectrograph with a large detector array: there is, however, no way that this could have been fitted on the relatively small instrument platforms of 8–10 m telescopes, and, besides, their putative cost would have been much too high, in the many tens of millions of USD/Euros range. The approach chosen by both teams has been instead to deploy a large number of small-sized identical spectrographs, each addressing a smaller subfield, 150 spectrographic units, each featuring a 2k × 2k detector for VIRUS, and 24 units, each with a 4k × 4k detector for MUSE.

In terms of instrument development, this has been a paradigm change, as classical astronomical instruments are once-built prototypes, with at most four identical optical systems. With the new VIRUS/MUSE approach and in order to fully reap the gains provided by serial production,[1] much emphasis has been laid on the careful design of cheap/easily replicated spectrographic units and automated test benches for the acceptance, tuning, and testing of all major subsystems, as well as for the full instrument. All these early developments,

---

[1] Actually, the VIRUS team, with its 150 units replication scheme, went past serial production to nearly industrial production.

Figure 5.2 This picture shows the MUSE instrument installed at one Nasmyth focus of the ESO VLT 8.2 m diameter Unit Telescope #4. The instrument is largely hidden behind the small forest of cables needed to operate its main body and the 24 subunits. ©The European Southern Observatory.

including that of optimized spectrographic unit prototypes, made for slow progress at the onset, but have been more than repaid both in time and costing in the following phases of serial/industrial production, integration, and testing.

Optimization of the individual units is project dependent and actually the MUSE and VIRUS slicing/spectrographic units cannot be more different from each other. To get maximum efficiency and very good spatial resolution, as mandated by its main science goal, MUSE relies on two-mirror image slicers and all-dioptric spectrograph optics (see the next section for a detailed presentation). To get the largest possible field of view with an affordable overall cost, VIRUS relies instead on fiber-slicers and higher aperture ratio (F/1.32 instead of F/1.92)[2] catadioptric spectrograph optics: this allowed particularly cheap camera optomechanical systems using replicated optics and cast mechanical bodies.

In terms of global size, the two instruments are about as large across as the equivalent field classical ones would have been, as this is pretty much set by field size, but are smaller in length, a critical issue to fit the instruments on their telescope platforms. Theoretically, length gains should go as $1/N$, where N is the

---

2 At first sight, that might look like a mild difference; however, with technical difficulty scaling as the inverse cube of the aperture ratio, that means a factor of 3 impact.

number of units; in reality and due to the need for transferring the N subfields to the N slicer/spectrograph units, length gain is about 4 times smaller.

5.3.3 Wide-Field Systems 3D Format

Both MUSE and VIRUS work by reformatting their square spatial field on the individual spectrograph's long slits (24 for MUSE, 150 for VIRUS), using mirror-based slicers for MUSE and fiber-based slicers for VIRUS. With this approach, the full individual detector's length (4096 pixels for MUSE, 2048 for VIRUS) is used for the extensive spectral coverage of both instruments. On the other hand, this limits the spatial coverage to large but not huge numbers, 300 × 300 spaxels for MUSE and 374 × 374 for VIRUS.

Sharples *et al.* [65] proposed a new microslicer concept aimed at getting an even larger 2D spatial field (Figure 5.3). The basic approach is to use a highly anamorphic variant of the basic lenslet-based IFS to get a high packing ratio, almost as good as with its more complex mirror-slicer cousin. Feeding 27 spectrographs, each with a 6k × 6k detector, would provide a record 1.8 million spaxels (5′ × 5′ field on an 8-m). Compared, for example, to MUSE, the tradeoff is a much smaller spectral resolution in the visible domain ($\mathfrak{R} \sim 500$ vs ~2400), and a significantly smaller one in the near IR region ($\mathfrak{R}$ 1500 vs 3600 at 940 nm).

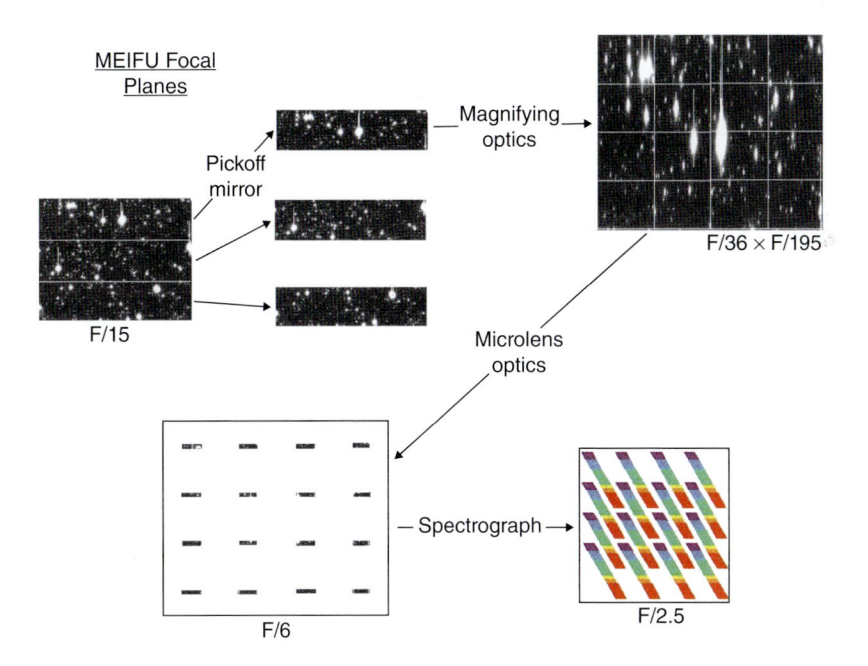

Figure 5.3 MEIFU Concept. Spectra are dispersed at a 13° angle to avoid spectral overlap. Each spectrum covers 200 × 12 pixels in the blue and 675 × 12 pixels in the red region on the detector (0.15″ per pixel). Interspectra gaps are 26 pixels (spectral) and 3 pixels (spatial). (Reproduced with permission of Simon Morris.)

5.4 An Example: Autopsy of the MUSE Wide-Field Instrument

5.4.1 MUSE Concept

The MUSE instrument has been developed over more than a decade, with first light on one of the four 8.2 m diameter telescopes at the ESO VLT Observatory at the beginning of 2014. Its main scientific goal is to perform ultra-deep 3D coverage on a relatively large field of 1 arcminute squared in order to detect early phases of galaxy formation in the Universe. Technically speaking, this translates into getting a highly efficient instrument ($\geq 50\%$ efficiency from photon input on the instrument to photoelectron output from the detector), delivering a huge 3D data set, covering one full octave in wavelength (465–930 nm) at a relatively high spectral resolution (about 4000) for each $0.2'' \times 0.2''$ spatial pixel in the $1' \times 1'$ total field of view. This corresponds to 90,000 simultaneous spectra, each about 4000 spectral pixels long, requiring a huge 360 million pixels total detector format.

In order to achieve this rather daunting manifest, the instrument holds 24 identical spectrographs (Figure 5.4) cum identical image slicers (Figure 5.5), each addressing an 1/24th section of the field through a field splitter and recording the resulting spectra on a 4k × 4k (16.8 million pixel) CCD. Given the extremely ambitious goal, the underlying approach was to save in project risk, cost, and timeline through serial production of the many identical units and subcomponents.

5.4.2 MUSE Approach

To get full advantage of the serial production approach, it proved essential to

- design a very cheap spectrographic unit, complete with alignment tools, which could be easily (cheaply) integrated and replicated;
- design an image slicer with the fewest number of components, again with easy to use alignment tools and procedures;
- build early on prototypes for both of the above, test, and upgrade them carefully before embarking on the serial production of the 24 units;

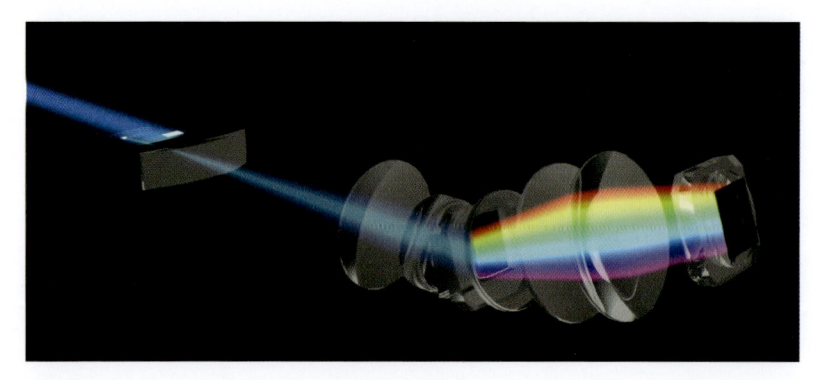

Figure 5.4 MUSE image slicer and spectrograph unit. This drawing shows light propagation inside one of the 24 MUSE spectrographic Units. Note that the drawing is to scale.

5.4 An Example: Autopsy of the MUSE Wide-Field Instrument | 121

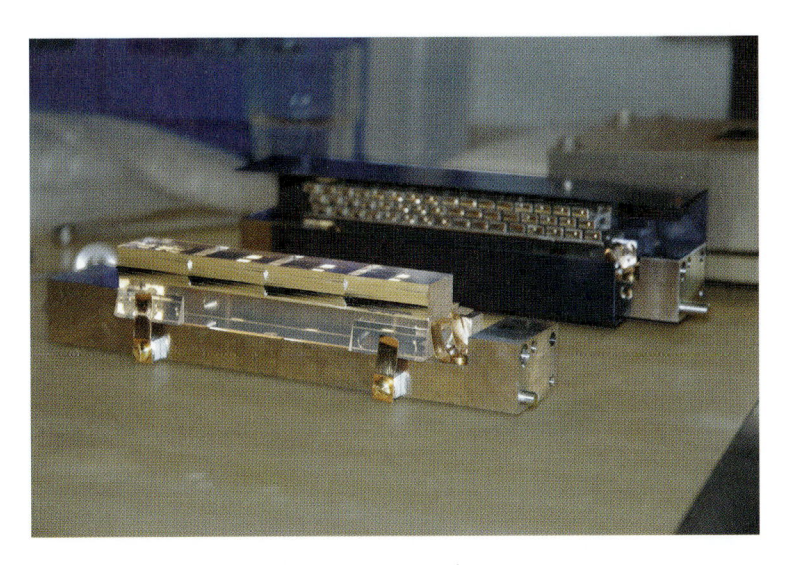

Figure 5.5 Picture of one of the MUSE image slicers.

- design and develop in parallel automated test benches, one typical example being one for acceptance[3] and setting up the 24, 4k × 4k, CCD detectors.

Specifically for MUSE:

- The collimator/camera main optical train, including the single volume phase holographic grating and the CCD window, encompasses a relatively meager (for its F/1.92 output) 21 optical surfaces with only one aspheric, made of only three different cheap glasses. This has been possible because each spectrograph is illuminated from a single slit, alleviating the need for apochromatic corrections via fragile/expensive fluoride glasses. The excellent image quality has thus been obtained by tilting and/or decentering most camera lenses: this has required developing more complex tools for the cameras' optical integration, but the extra-cost is shared between the 24 cameras, with a small overall impact. For the same reason, the camera exit focal plane is also tilted.
- For cost and efficiency purposes, advanced mirror-based image slicers featuring a field slicer block and a slit block, both with 48 individual mirrors, have been selected and improved. Overall length has been shortened from about 2 to 1 m and, for cost saving purposes, inside each block all mirrors are spherical and identical. Developing a working production line has been a protracted process. Two fast/cheap techniques have been tested at first, namely, fly-cutting and diamond-turning of metallic mirrors. The first ones alleviates most alignment problems by directly providing monolithic mirror blocks, but ultimately could not produce the field slice mirrors with their desired spherical shape. Instead, cylindrical shapes with zero curvature along the very small slice width have been systematically obtained. The second technique gave the required optical shapes, but alignment of mirrors inside the field slicer block with the

---

3 This was an essential procedure as the manufacturer agreed to meet tough technical specifications at relatively little cost increase, provided the significant extra testing effort was borne by the customer.

necessary few microns accuracy proved impossible. After these two painful failures, a more classical approach has been used successfully with glass mirrors polished and then assembled by the optical manufacturer.

- With every input photon going through a total of 27 optical interfaces (25 transmissions and 2 reflections) before impacting on the detector, getting long-lived (decades) efficient wide-band antireflection coatings and super-reflective coatings is essential, and less than 0.4% mean loss per surface with multilayer all-dielectric coatings has ultimately been achieved. Modern deposition systems can (relatively) easily fit the bill, but the devil is as usual in the details: in practice, after difficulties with prototype mirrors coating, handling and cleaning of the ∼2400 optical surfaces have been coordinated by the optical manufacturer at the coating manufacturer's premises, in order to achieve the stringent specifications for every single surface.
- Since photon wavelength linearly increases from one CCD detector edge to the other, a graded antireflection coating has been deposited by the manufacturer on each CCD front silicon surface; this gives a ∼20% efficiency gain at the blue (465 nm) and red (930 nm) edges, compared to a standard uniform AR coating: that could have been done long ago for any classical long-slit spectrograph, but here, the extra cost for that very nonstandard procedure has been shared between the 24 identical detectors with a small cost impact on each.
- The 24 slicer/spectrograph units are fully passive systems, with the few MUSE motorized functions incorporated upfront in the common fore-optics part. This is quite important since the consolidated cost (including manpower) of any motorized function, complete with its software interface, is usually in the 40–80 kEuro range. In particular, the spectrograph optical design has been well athermalized over the 5–20° temperature working range of the instrument, eliminating the need for 24 motorized systems in order to refocus the spectra on the 24 detectors whenever there are significant temperature changes.
- The 24, 139 g mm^{-1} identical volume phase holographic gratings are used at their Bragg angle for simplicity and to reduce light ghosts. Following prototype fabrication and testing, the grating fabrication process has been optimized by the manufacturer to reduce efficiency variations within the grating area from initial ∼20% peak to valley excursions to 5% at most.

5.4.3 MUSE Conclusions

Overall, the validity of the multi-instrument approach for MUSE has been largely validated in terms of performance, cost, reliability and timeline. In particular, unitary costs for the spectrograph and the image slicer are each in the 40 kEuros range, a very good value for such high-tech units. While on many aspects this approach indeed led to more manageable subsystems than with the classical monolithic approach, one fly in the ointment has been the need for delivering very high reliability and vibration-free electric power distribution, vacuum feeds, cryogenic coolant, dry nitrogen flux, and data bus for each of the 24 detector systems. This has been ultimately achieved, but relies on an extensive and somewhat unwieldy "plumbing" system, as well apparent in Figure 5.2. Also, the mirror-based fore-optics that split the field in the 24 subfields feeding the slicer/spectrograph unit has proved difficult to align and prone to internal flexures. It would have been great to find a more compact and stable way to get this essential function.

5.4.4 Validity of the Multi-instrument Approach

As shown above with the VIRUS and MUSE successful developments, the multi-instrument scheme offers substantial advantages over the classical "monolithic" ones, with generally a single – at most four – very large optical system. There are a number of provisos though:

- This approach is particularly well suited for dedicated instruments delivering a single rigid data format. On the contrary and at the opposite end of the instrumental spectrum, it would poorly fit a Pokemon-style (Gotta Catch 'Em All) flexible spectrographic facility, featuring, say, a dozen different gratings with associated interference filters, in order to offer a large variety in spectral resolution and wavelength range on each of the 24 units. Additional cost and manpower to design, fabricate, control, document, and commission these remotely controlled subassemblies would have been untenable. It might even be impossible to install them physically.
- Detailed project planning is even more essential than for standard monolithic instruments. In particular early-on development of alignment tools and automated test benches for components/subsystems acceptance, integration, and testing is a must. Working closely with the relevant industrial manufacturers is invaluable for any kind of astronomical instrument development; this is even more true when under the multiunits scheme.
- Optomechanical design optimization of the multiunits must be fully achieved through a lengthy prototype development, with much testing and refining before at last embarking into serial/industrial production. That takes time, but any shortcut in that process would have been courting disaster.

5.5 Deployable Multiobject Integral Field Spectroscopy

5.5.1 Concept

As soon as the first integral field spectrographs started operation, the idea of combining the multiobject and integral field concepts by deploying a number of small integral field units (IFUs) over a large patrol field, each centered on a promising astrophysical target, emerged quite naturally.[4] Its obvious tradeoff, at equal detector size, is a smaller field of view on each target than with the equivalent single integral field unit. The approach is thus quite optimum for rough characterization of roughly 1.5″ – 3″ diameter targets densely distributed over the sky, with moderately redshifted galaxies being the prime example.

---

[4] In 1989, while waiting in the dark on top of the 4200-m high Mauna Kea volcano for the first astrophysical data set coming out of the TIGER IFS, these two authors dreamt about this mythological-like chimera.

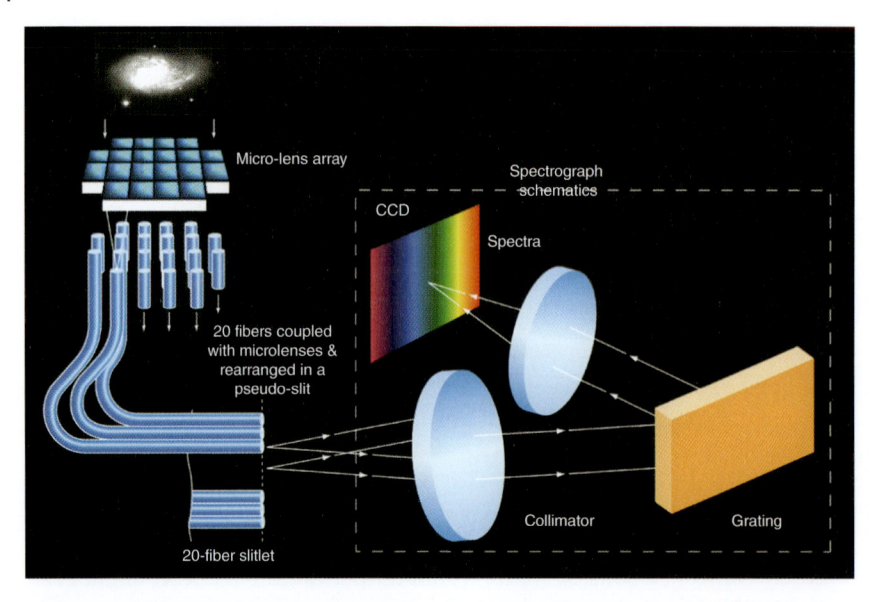

Figure 5.6 Schematics of the FLAMES deployable integral field units (IFU) system. (Used Under Creative Commons License: https://creativecommons.org/licenses/by/4.0/.)

5.5.2 The First Deployable Integral Field Units System

In 2002, this interspecies cross-breed mode, developed by Hammer (Observatoire de Paris) and integrated on the multiobject FLAMES instrument, had the first ever light on the ESO Very Large Telescope (see Figure 5.6).

The FLAMES deployable IFUs system (see the eso0203 press release on the ESO web site), developed by a European Consortium led by Hammer (Observatoire de Paris), works in the optical range. It features 15 probes. Each probe holds 20 fibers fed by 20 minilenses, covering *in toto* $2'' \times 2''$ on the sky. Mean light efficiency of the probes is a reasonably good 68%. The probes are deployed in the $30'$ diameter patrol field by a positioning robot, developed by the Australian Astrophysical Observatory (AAO).

This mode is especially suited to the survey of intermediate distance galaxies (say with a moderately high redshift z in the 0.3–0.8 range). A number of strong absorption features are located in the optical domain. The spatial multiplex is small but adequate given the small galaxy population sky density, the small probes field is well adapted to the median galaxy size, and the data cubes, with 20 spatial pixels on each galaxy, are just detailed enough to derive the basic properties of the galaxies, namely, central mass (from absorption line widths), gas angular momentum (from emission line wavelength shifts), and a few metalloid/metal abundance ratios (from absorption line intensity ratios).

5.5.3 Near Infra-Red Deployable Integral Field Units

With no strong emission line below 121.6 nm in the rest frame, going to highly distant ($z > 7.2$) galaxies requires shifting the observations to the near-infrared domain, longward of 1 μm and preferably up to at least 2 μm. Technically

Figure 5.7 Photographic image ann12071a of the KMOS focal plane, showing its pickoff arms and image slicers. (Used Under Creative Commons License: https://creativecommons.org/licenses/by/4.0/.)

speaking, this means that the deployable IFUs must be able to work in a fully cryogenic environment. In practice, this means that the whole instrument is operating in a vacuum at low temperatures, typically in the 70 – 140 °K range.

The KMOS instrument [66] at the ESO VLT has been designed along these lines. Twenty-four robotic arms position small mirrors at selected locations in a 7.2′ diameter patrol field (Figure 5.7). Each arm selects a 2.8″ × 2.8″ subfield, which is imaged with an anamorphic magnification (different magnification along the dispersion axis and the orthogonal axis) onto one of 24 mirror-based image slicers. Each subfield is cut into 14 slices, with 14 spatial pixels along each slice. Light output from the slicers is dispersed by three cryogenic grating spectrometers, with a maximum spectral resolution of 4200. This generates 14 × 14 spectra in the 0.8–2.5 μm range, each with ~1000 independent spectral resolution elements, for each of the 24 subfields, or a 6.10^6 element global data cube. Closest on-sky field separation between two field centers is 6″, and there can be at most three arms within 1′ × 1′.

Two subassemblies have been particularly difficult to produce and implement within the stringent technical requirements. One is the stepper motors driven robotic arms complement, which must achieve 50 μm position accuracy in quasi-open loops. The other one is the image slicer complement, for which all mirrors and their mechanical holders have been diamond-machined to ensure the extreme position and stability requirements. A combination of diamond-turning and fly-cutting has been necessary to achieve very small surface roughness, below 10 nm r.m.s.

A similar instrument (MIRADAS), operating in the 1–2.5 μm range, is under development for the GTC [67] by an international consortium led by Eikenberry *et al.*, University of Florida. It will feature 20 arms in a 5′ diameter patrol field, each covering a 3.7″ × 1.2″ subfield. With its higher spectral resolution

Table 5.1 List of deployable multi-integral field spectrographs with their main characteristics: start date, P_N, number of probes, P_S, number of spaxels per probe, PF, patrol field in square degrees, $\Re$, spectral resolution, and spectral range in micrometer.

Instrument	Start	P_N	P_S	PF	$\Re$	Range	References
FLAMES-VLT	2002	15	29	0.14	11,000	0.37–0.95	[68]
SAMI-AAT	2013	13	61	0.79	1,700	0.37–0.95	[69]
KMOS-VLT	2014	24	196	0.011	4,200	0.8–2.5	[66]
MANGA-SDSS	2015	17	19–127	7	2,000	0.36–1.03	[70]
MIRADAS-GTC	2019[a]	20	108	0.006	20,000	1–2.5	[67]

a) Provisional date.

($\Re = 2.10^4$) provided by a cross-dispersed echelle spectrograph, its main observing goal, in addition to intermediate redshift galaxies, is to study the chemical evolution of galactic stellar populations. The instrument will also hold a spectro-polarimetric capability. It is provisionally scheduled for first light in 2019.

5.5.4 Deployable Multi-Integral Field Systems: Conclusion

This recent addition to the 3D instrumentation zoo is now operating on the ground both at optical and NIR wavelengths. Its natural contender is the multiobject modes, which offer much higher multiplex, but scanty and potentially distorted physical information on observed galaxies. The multi-IFU approach on the other hand gives basic physical information of good quality, even in a crowded field environment, at the cost of a much smaller multiplex. In short, multislit or fiber systems are great for massive galaxy surveys aimed at getting very few global parameters (fast but dirty), and deployable IFUs for smallish surveys looking at getting a few reliable and accurate physical parameters (slow but clean).

A fast and clean combination would be technically feasible, at least in the optical domain, for example, by combining a huge fiber positioning system as in COBRA/SUBARU (Section 2.4) with multi-spectrographic units as in VIRUS (Section 5.3): that would require a huge technical and financial effort though. Note a first step in that direction with the SAMI prototype at the AAT, using the new cheaper "hexabundle" technology to produce the many imaging fiber bundles required for that kind of instrument (see Table 5.1).

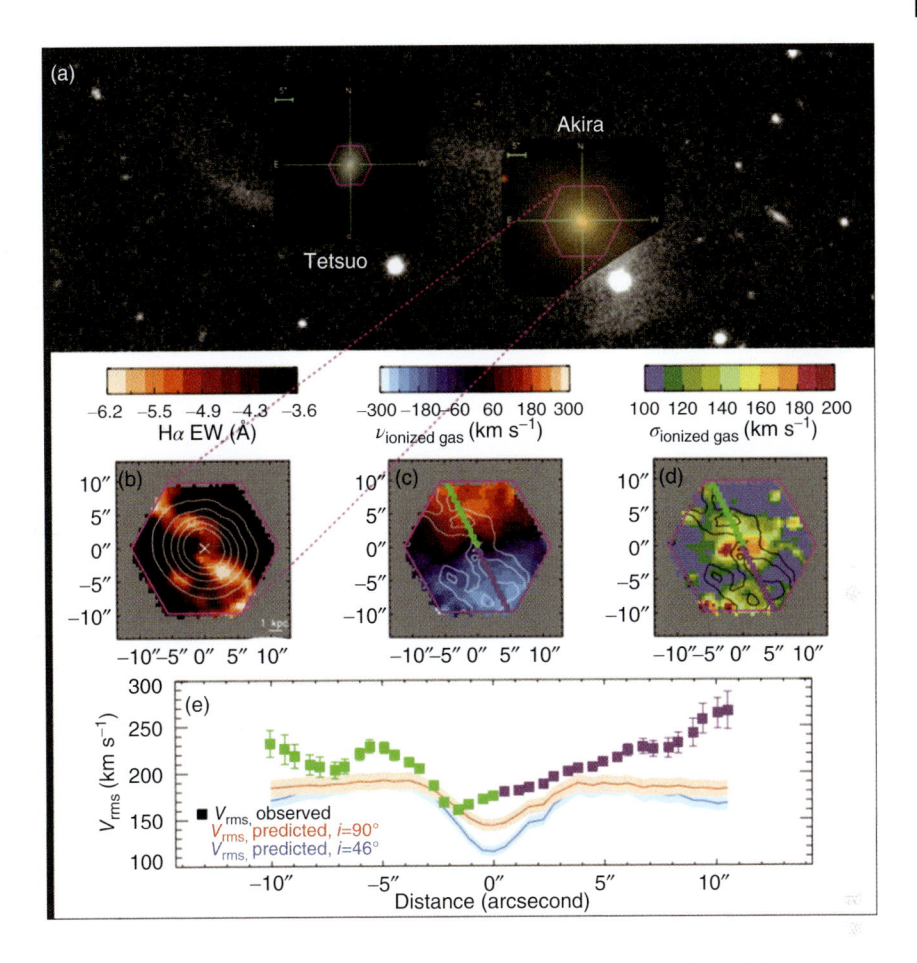

Highlights from the MaNGA survey showing a prototypical example (Akira) of the new red geysers class of quiescent galaxy [129]. (a) The SDSS gri color images of Akira (West) and Tetsuo (East) embedded in a larger SDSS r image, with the MaNGA footprint in pink. (b) The Hα equivalent width map of Akira, with contours tracing the stellar continuum. (c and d) The ionized gas velocity and ionized gas velocity dispersion maps of Akira, with the Hα EW contours overplotted. (e) The observed ionized gas second velocity moments from the highlighted spaxels exceed the predicted values from the gravitational potential, ruling out disk-like rotation. The MaNGA survey [70] is part of the fourth-generation Sloan Digital Sky Survey (SDSS-IV) and is based on a fiber-based multi-IFU instrument operated at the Apache Point Observatory. ([129]. Reproduced with permission of Nature Publishing Group.)

6

Comparing the Various 3D Techniques

6.1 Introduction

Comparing the various 3D techniques is a complicated, often largely subjective, affair. This chapter will thus just briefly touch upon a few promising avenues.

6.2 3D Spectroscopy Grasp Invariant Principle

As repeatedly stressed in this book, the various spectroscopic techniques are all ultimately limited by the number N of volume pixels (voxels) actually covered by the 3D data: note that N is by principle at best equal to the total detector real estate times the number of scans (in case of scanning techniques). In real life, it is substantially smaller when only part of the detector is actually covered by science photons, as for the lenslet integral field spectroscopy variant. In any case, the end result of an observation is a data cube (x, y spatial pixels and p spectral pixels) of volume $x \times y \times p$ equal to N. This is the so-called instrument grasp. For identical detector size and number of exposures (scans), it is essentially independent of the instrument flavor, narrow-band imaging, scanning interferometry, scanning long-slit spectroscopy, and integral field spectroscopy. For example, a data cube of $50 \times 50 \times 50$ voxels can be obtained with a scanning interferometer with 50 scans on a tiny 50×50 pixels detector, or alternatively a single exposure on an integral field spectrograph with a 50 times bigger 350×350 pixels detector.

This geometrical invariant does not however translate automatically into equal astronomical efficiency, which for any observation is driven by how well (or badly) the 3D volume shape covered by the instrument matches the corresponding shape of the astrophysical object. Integral field instruments are, for instance, at their best for moderate size galaxies, scanning interferometers for extended single emission-line clouds, scanning long-slit spectrometers for near edge-on galaxies, multiobject spectrometers for moderately dense distant galaxy fields, and so on. In addition, this general grasp invariance is in practice much modulated by the many quirks of actual instruments, as briefly covered in the next section.

6.3 3-D Techniques Practical Differences

A comprehensive analysis of the various instrumental flavor subtleties is well outside the portfolio of this book. Any instrument builder – and user – must nevertheless be aware that "details" such as sky subtraction limitations, poor rejection of bright objects in the field, and residual photometric inaccuracies might lead to rejection of some instrumental flavors or subflavors, depending on the observational purpose.

There are, however, two generic properties that can be quantified and thus objectively compared, namely, the packing and observational efficiencies.

6.3.1 Packing Efficiency

A first level quantitative assessment of an instrument's efficiency is the fraction of detector pixels that is actually used to get on-sky 3D information. This so-called packing efficiency much depends on the instrumental approach actually chosen.

For the scanning Fabry–Pérot or the Fourier transform interferometer, packing efficiency is unity, with a one-to-one mapping between sky spaxels and detector pixels. For the long slit scanning spectrograph, packing efficiency is some 15–30% less because of the need for some minimum field overlap between successive offset exposures.

For integral field spectrographs, the highest packing efficiency is obtained with the slicer-based concept: as contiguous spaxels on the sky are contiguous on the slices, they can be packed without any gap. The only lost pixels are then the ones that must be left between neighboring stacks when the slicer has more than one stack of slices. This is however a very small fraction of the detector pixels, and slicer-based integral field spectrographs can typically achieve packing efficiencies up to 98%.

For fiber-based integral field spectrographs, a similar perfect mapping between sky spaxels and individual fiber outputs can be achieved in principle. This however requires extremely fine adjustments of the fiber's outputs, and some free "elbow" space on the pseudo-slit is usually factored in, with typical packing efficiencies around 85%.

This contiguity is not true any more for the lenslet-based integral field spectrograph, for which adjacent spaxels give spectra that are considerably shifted in wavelength. This negatively affects the packing efficiency in two ways: (i) a significant fraction of wasted detector pixels to avoid mixing light at different wavelengths and (ii) because of this wavelength shift, a global common spectrum range smaller than the number of detector pixels along the wavelength axis. In practice lenslet integral field spectrographs' packing efficiencies are then at most $\sim 60\%$ and often less.[1]

A fiber-based multiobject spectrograph has by design a similar packing efficiency as its integral field version. Multislit MOS are much less efficient in this respect, given that the location of the sources is projected on the detector without any geometrical reorganization. Avoiding spectra overlap, spectra truncation

---

[1] This strong shortcoming is partly offset by better spectrophotometric performance than the other variants, with stable and uniform sky sampling performed at the level of the lens array.

and unwanted grating orders usually takes a significant hit in the actual multiplex achieved. In practice, even for the optimum case of highly dense fields with many more putative targets than can be accommodated by the instrument, multislit MOS packing efficiencies are rarely higher than $\sim 50\%$.

Given the high cost of infrared detector pixels, packing efficiency is a crucial parameter in this wavelength range. But even in the optical range for which a CCD pixel is cheaper by about a factor of 10, this is still an important factor, given that the cost of optics and mechanics usually scales with detector area.

The packing efficiency discussed so far describes how spatial and spectral pixels are packed on the detector. These geometrical properties are indeed important instrumental characteristics, but leave out the important issue of resolving powers. The true packing efficiency of the detector should also take into account how many detector pixels are needed to get a spatially and spectrally resolved information element. Indeed, even with a geometrical packing efficiency of 100%, from Shannon's theorem (Section 8.2), at least 8 detector pixels must be used for only one spatially and spectrally resolved element. Measuring this efficiency is less obvious because it involves the whole system from the atmosphere, telescope, and the instrument to the data reduction software. Frequently, the instrument PSF is larger than 2 detector pixels and the real packing efficiency is decreased in proportion. Also, detector pixel size (typically 15 μm) is usually much smaller than the seeing disk for large aperture telescopes. These effects are important and should be considered when designing an instrument.

6.3.2 Observational Efficiency

What ultimately matters is the telescope time needed to obtain a given signal to noise ratio for a spatially and spectrally resolved element of information (or for a number of sources in the case of multiobject spectroscopy). This is the key parameter that should be considered when comparing different instrument variants. This indicator is however complex to derive because it depends on many factors, including actual target properties. Although a comprehensive analysis of the many possible cases is outside the scope of this book, here are a few hints on the main factors into play.

For scanning instruments, the multiple exposures used to cover the third dimension have a direct impact on observational efficiency. This in practice limits their use to science cases with a mandatory large field of view, if at the expense of a small spectral range for a scanning Fabry–Pérot, moderate spectral resolution or moderately large spectral range for a Fourier transform spectrograph. For the Fourier transform spectrograph, there is an additional penalty due to the multiplexing scheme, which adds the background noise summed over the full studied spectral range to the monochromatic source noise on each spectral frequency channel.

Another critical parameter is instrument throughput, as overall efficiency scales at least linearly with its value. For faint sources, it can scale even more rapidly, as the system, because of lower throughput, moves from the photon noise to the detector noise dominated regime (see Section 8.3).

Because it is total telescope time used for a given observation that counts, overheads must be fully taken into account. Detector readout and instrument setup time shall be minimized. This last parameter is especially important in the case of multiobject spectrography for which target acquisition can have a major impact on overall efficiency. The various instrument calibrations can also create major overheads; this can be much minimized by building highly stable instruments, developing a comprehensive calibration strategy (calibrating the instrument per se rather than its observations), and whenever possible performing on-sky calibrations at twilight and dawn, that is, outside of precious astronomical time.

Finally, all sources of noise directly impact the efficiency. Detector readout noise and dark noise are obvious components, but they are often on top subtle instrument or detector related effects that left systematic uncertainties not fully zeroed out by the data reduction pipelines (see Section 11.2).

6.4 A Tentative Rating

Besides these objective parameters, there are others that are more difficult to quantify, for example, the availability of a robust data reduction system. These evaluations are difficult and partly subjective, and it is only practice that tells which instrument is the most appropriate for a specific application. In Table 6.1, we take the risk of hazarding comparative ratings based on our experience. It is neither exhaustive nor fully objective.

Table 6.1 Tentative ratings of the various 3D spectroscopic instrumental flavors.

Instrument	Pros	Cons	Science case
Scanning slit	Easy and cheap instrument	Poor final image quality	Not much used
	Large spectral range	Low observing efficiency	
Fabry–Pérot	High spectral resolution	Very small spectral range	Extended single emission line objects
	Large field of view	No absorption line	
FTS	Large spectral range	Photon noise penalty	Relatively bright objects
	Large field of view	Technically demanding	
Slitless spectrograph	Simple instrument	Sources overlap	Faint emission line objects (space based)
	High throughput	Low spectral resolution	
	Large field of view	Difficult data reduction	
Slit MOS	High throughput	Large overheads	Extragalactic spectroscopic surveys
	Large field of view	Average spectral resolution	
		Moderate multiplex	
Fiber MOS	High multiplex	Complex instrument	Wide-field spectroscopic surveys
	Very large field of view	Large overheads	
	Large spectral resolution	Moderate throughput	

Table 6.1 (Continued)

Instrument	Pros	Cons	Science case
Deployable fiber IFUs	Spatially resolved data Very large patrol field	Complex instrument Large overheads Moderate throughput Small multiplex	Galaxies detailed spectroscopic surveys
Deployable slicer IFUs	Spatially resolved data Large patrol field Good throughput	Complex instrument Large overheads Small multiplex	Galaxies detailed spectroscopic surveys
Lenslet IFS	No slit effect Good throughput	Moderate spectral range Low packing efficiency	Nearby galaxies High contrast objects
Fiber IFS	Flexible design Large spectral range	Low throughput	Any compact object
Slicer IFS	High throughput High packing efficiency Large spectral range	Complex subcomponent	Any compact object

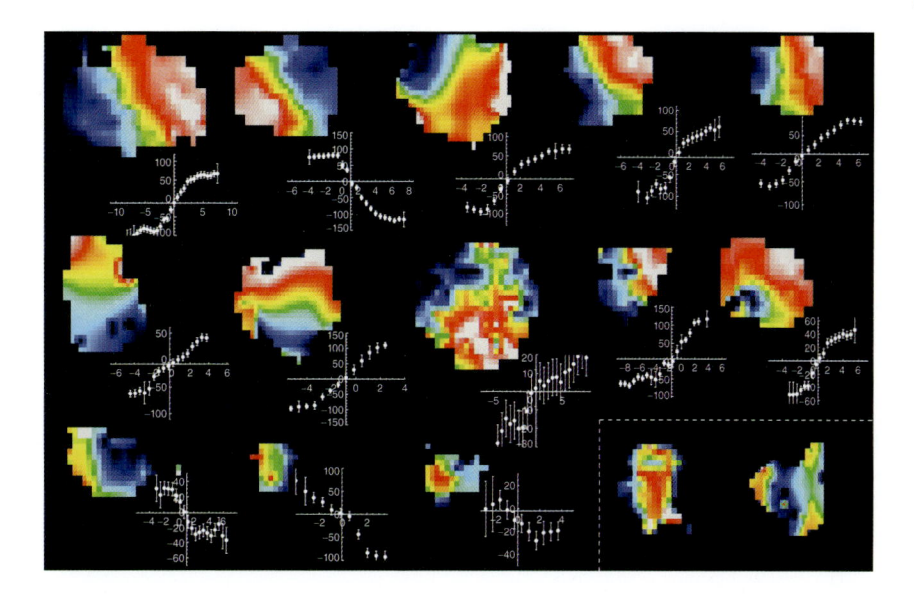

The KMOS kinematic survey (K2S). Examples of data from the KROSS survey of redshift one field galaxies [124]. This figure shows a small subset of the data from the survey illustrating the velocity fields that can be measured with the ESO slicer-based multi-IFU infrared KMOS instrument. The rotation curves of each galaxy are shown as an inset beside each galaxy. In total, the survey has so far obtained resolved Hα rotation curves for 624 galaxies. ([124]. Reproduced with permission of Richard Bower.)

7

Future Trends in 3D Spectroscopy

7.1 3D Instrumentation for the ELTs

3D instrumentation as a whole is now past its tumultuous youth, if with still steady technological advances. Overall performance, especially in terms of image quality and light efficiency, is good and adequately covers immediate observing needs. There is however a big challenge looming ahead with the upcoming 3D instruments[1] for the next generation of extremely large telescopes (ELTs), as discussed in this section.

Worldwide, three ELTs, namely, the 24-m diameter GMT, the 30-m TMT, and the 38-m E-ELT are currently being designed and approaching their construction phase (see their respective home pages for in-depth current information). In parallel, detailed instrumentation plans are being developed by the respective project teams. This is essential to ensure that major science questions can be addressed – and hopefully solved – as these new observing capabilities enter into operation, well in the 2020s. Two main ingredients, both directly related to etendue issues, conspire to make most instrumentation flavors on an ELT extremely challenging:

- Firstly, scientific requirements for the major science goals (such as the early evolution of our Universe and the characterization of exoplanets) led to similar, if not larger, field of views than for equivalent instruments presently on 8–10 m class telescopes: direct extrapolation then leads to the instruments scaling up in cost, size and weight, by the ratio of telescope areas, that is, multiplying factors between 9 and 25. This could easily push single ELT instruments in the unworkable (i.e., for ground-based astronomy) 100-ton plus and \$\$ 100-million plus regime.
- Secondly, there is a huge discrepancy between the atmospheric turbulence etendue and a 2-pixel detector etendue on an ELT. A typical linear value of the former is $\sim 90\,\mu m$ radian on a 30-m ELT (assuming somewhat optimistically that under median turbulence conditions 70% of the energy from a point source falls in a 0.6" diameter circle), while the latter is at most $20\,\mu m$ radian ($2 \times 15\,\mu m$ pixels and an F/1.5 camera) that is, 4.5 times smaller. That is an area imbalance of $(4.5)^2$ or a significant multiplying factor of 20. This mismatch can

---

1 And equally strong challenges for most of the non-3D ones.

Optical 3D-Spectroscopy for Astronomy, First Edition. Roland Bacon and Guy Monnet.
© 2017 Wiley-VCH Verlag GmbH & Co. KGaA. Published 2017 by Wiley-VCH Verlag GmbH & Co. KGaA.

be efficiently solved around 2 μm wavelength through full adaptive optics (AO) correction, but with rapidly decreasing performance for shorter wavelengths. In addition, an AO corrector for an ELT is a very expensive proposition, with overall size, actuator numbers, wavefront sensor detector format, etc., scaling again as primary mirror area.

Given their huge multiplex advantage, 3D instruments figure indeed prominently in the first generation instrumentation plans for GMT, TMT, and the E-ELT. Specifically,

- GMT plans cover four first generation instruments, including two 3D ones, GMACS [71], an optical multislit spectrograph, and GMTIFS [72], an NIR integral field spectrograph with AO correction. Also, MANIFEST [73], a facility fiber optics positioner to feed the instruments with hundreds of selected astronomical targets in the full 20′ diameter GMT patrol field.
- TMT plans cover three first light instruments, including two 3D ones, WFOS-MOBIE [74], an optical multislit spectrograph, and IRIS [75], an NIR integral field spectrograph with AO correction.
- The E-ELT plans cover four first generation instruments, including two 3D ones, HARMONI [76], an NIR integral field spectrograph with AO correction, and ELT-MOS/MOSAIC [77], a multiobject spectrograph combining high multiplex and high spatial resolution.

As telescope upper sizes inexorably increase, most associated instruments – including 3D ones – become bigger and bigger (many tons), expensive (many tens of millions USD/Euros), and highly complicated, requiring hundreds of person-years over a decade for study, building, and installation. There is a real concern that ELT instrument development might eventually lead to a dinosaur-like demise, collapsing under unmanageable size, weight, cost, time scale, and complexity.

The next two sections explore potential so-called transformational technologies that if/when successful would represent paradigm shifts in the conception and development of 3D instrumentation for small and large telescopes alike.

7.2 Photonics-Based Spectrograph

The developments presented in this section belong to the emerging field of "Astrophotonics," via the adaptation of (H-band) telecommunication photonics devices (also known as optical chips) for astrophysical instrumentation. The main pioneering centers involved by 2016 in this endeavor are located in Australia (University of Sydney/Macquarie University/Australian Astronomical Observatory) and in Europe (Potsdam-Germany, Grenoble-France, and Durham-the United Kingdom).

7.2.1 OH Suppression Filter

One of the first photonics applications to astrophysics, the so-called OH-suppression fiber, is not a bona fide spectrographic device, rather a

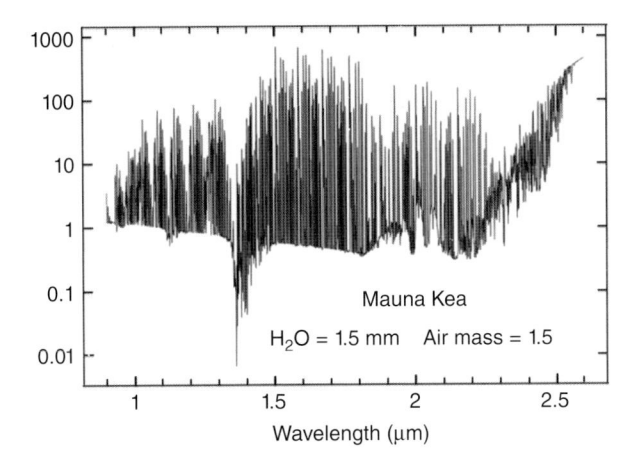

Figure 7.1 Night-sky emission spectrum from 1 to 2.5 µm at the 4200-m high Mauna Kea Observatory, in dry conditions. Given their huge range, OH line intensities are shown in a semi-logarithmic plot, with the flux unit in ph/sec/(arcsecond)^2/nm/m^2. Note the interlines low intensity level, at least 1000 times fainter than the brightest OH lines. The strong positive slope beyond 2.25 µm is due to atmosphere thermal emission. (Reproduced with permission of Gemini Observatory.)

multiline filter. Its goal is to boost detection of very faint extended objects in the near-IR domain, in particular for the study of faint high-z galaxies, by canceling the huge photon noise coming from the hundreds of bright emission lines in the H and J bands. These lines come mainly from the hydroxyl radical (OH) in a, roughly 90 km high, Earth upper atmosphere layer; see Figure 7.1 and Chapter 9.

For a detector pixel $g_{\mu m}$ illuminated with a half cone angle α over a wavelength interval $\delta\lambda_{nm}$ and a telescope + instrument efficiency τ, one unit on the diagram corresponds to a flux of $0.17\,\tau g^2(\delta\lambda)\sin^2\alpha$ photons per pixel per second. For typical values, $\tau = 0.35, g = 15\,\mu m, \alpha = 18.5°$ (F/1.5 camera), $\delta\lambda = 0.5$ nm ($\mathfrak{R} = 3000$ at 1.5 µm); this gives 0.67 ph/px/sec. For the brightest H-band line fluxes, around 8000 units, one gets 5400 ph/px/sec. This huge background emission has a number of negative effects:

- The lines are highly variable in time, and hence difficult to subtract accurately.
- Even if perfectly subtracted, it is physically impossible to subtract their photon noise. This noise usually dominates detector noise, at least at the location of the brightest OH lines.
- The usual strategy to limit their effect is to boost the spectral resolution of the spectrograph in order to get enough detector space between successive lines, the so-called OH software avoidance technique. This requires a minimum spectral resolution of 3000–4000, generally larger than the resolution strictly needed on the astrophysical side (e.g., ~1000 for the study of moderately distant galaxies). This usually results in much larger spectrographs that would otherwise be needed.
- With still relatively small detector space between two successive OH lines (say a 10-pixel mean value), light diffusion, especially by the spectrograph disperser,

fills the detector at a level greater than the very faint interline natural sky background. Since the OH airglow lines are exceedingly narrow, the optimum spectral resolution would actually be of the order of 2.10^5, but would result in a gigantic instrument with a very small spectral range or a very small spatial range or most likely both.

It would thus be highly attractive to cut this parasitic emission by rejecting all OH lines before they enter into the spectrograph. The technical solution is to use a laser to imprint suitable periodic refractive index variations along the core of a mono-mode fiber. The resulting Bragg grating strongly reflects back incoming light at the precise wavelength of hundreds of OH lines, while letting all other wavelengths through. Note that since upper atmosphere molecules have essentially zero mean relative radial velocity with respect to a telescope on the ground, there are no adverse Doppler–Fizeau shifts between the real OH line wavelengths and the imprinted ones, as the telescope slews from one direction on the sky to another. This OH suppression hardware rejection concept is shown in Figure 7.2.

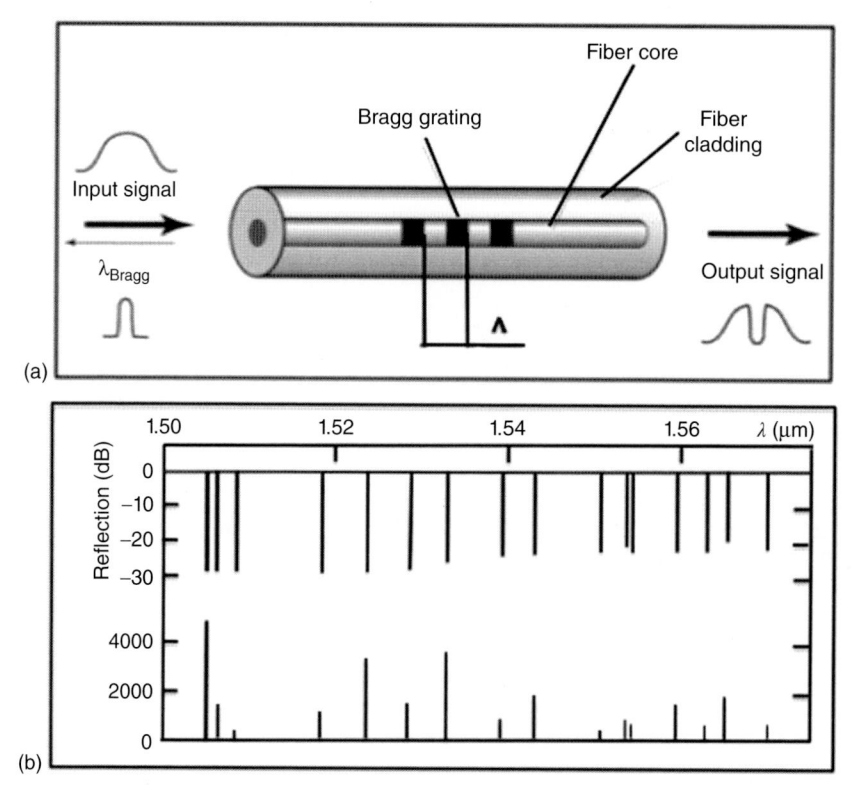

Figure 7.2 OH-suppression filter concept. (a) Schematic of an individual fiber with a laser-written internal Bragg grating. (b) Rejection of a large number of airglow lines in the H-band. Note that 10 dB corresponds to a rejection factor of 10, and 20 dB to 100. (Reproduced with permission of Prof. J. Bland-Hawthorn, University of Sydney.)

First demonstration of OH-suppressing fibers has been done in December 2008 on the Anglo-Australian Telescope with its IRIS2 spectrograph [78]. The fibers covered only about half of the H-band, but development is ongoing to extend their working range to the full J + H bands in one go.

7.2.2 Photonics Dispersers

Photonics dispersers neatly sidestep the extensive and expensive optical fabrication, integration, and maintenance processes required for the classical bulk optics-based spectrographic instruments presented so far in this book. For example, a photonics grating working at a spectral resolution $\mathfrak{R} = 10^3$ needs in principle only to be about 10^3 wavelengths long, that is, about 1.5 mm in size, and could cover one near-IR band in one go. Such arrayed waveguide gratings (AWG) devices have been developed by the telecommunication industry in the H-band (1550 nm central wavelength) in order to inject thousands of different wavelengths of laser light in a mono-mode fiber, and vice versa to extract thousands of communication channels at the fiber end. Coupled to OH-suppression fibers, this would constitute an ideal faint-object near-IR spectrograph, one fundamental staple of large telescopes instrumentation.

The first such device was demonstrated in 2009 at the AAO; The subsequent version 2 gives a spectral resolution of $R = 7000$, and covers a bandwidth $\Delta\lambda = 52$ nm (240 spectral pixels around 1550 nm); see Figure 7.3.

Direct photonics coupling of each such spectrographic device to its high-performance minidetector would be ideal, but this looks difficult to achieve and most probably expensive (see, however, the Fourier transform spectrometer below); the present photonics-based dispersive systems thus still need bulk optics to send the spectra emerging from the photonics elements onto a classical detector array.

7.2.3 Photonics Fourier Transform Spectrometer

A photonics version of the FTS has been invented and developed by LeCoarer at Observatoire de Grenoble (France). Christened SWIFTS for stationary wave integrated Fourier transform spectroscopy [79], it is based on intensity detection of a standing wave in an optical fiber as shown in the figure. Detection is done by optical nanoprobes located around the fiber and working in the evanescent field[2]; see Figure 7.4.

Quantum efficiency is, perhaps surprisingly, rather high, ~66%. Spectral resolution can easily be quite high, for example, in the $2.10^4 - 5.10^4$ range, and the spectral range can cover a full octave, for a small matchbox size device. Unfortunately, as for classical FTS, the approach suffers from the multiplex disadvantage, that is, observations signal to noise ratio (SNR) is divided by the square root of the number of spectral bins for a continuum source; the effect is considerably smaller if the object light is made of a few emission lines. This remarkably rugged

---

2 Within geometrical optics approximation, light trapped in a high index medium by internal reflection entirely stays inside the medium. In reality, there is a very small (wavelength-scale) outside leak of the electromagnetic field, called the evanescent field, which can be tapped.

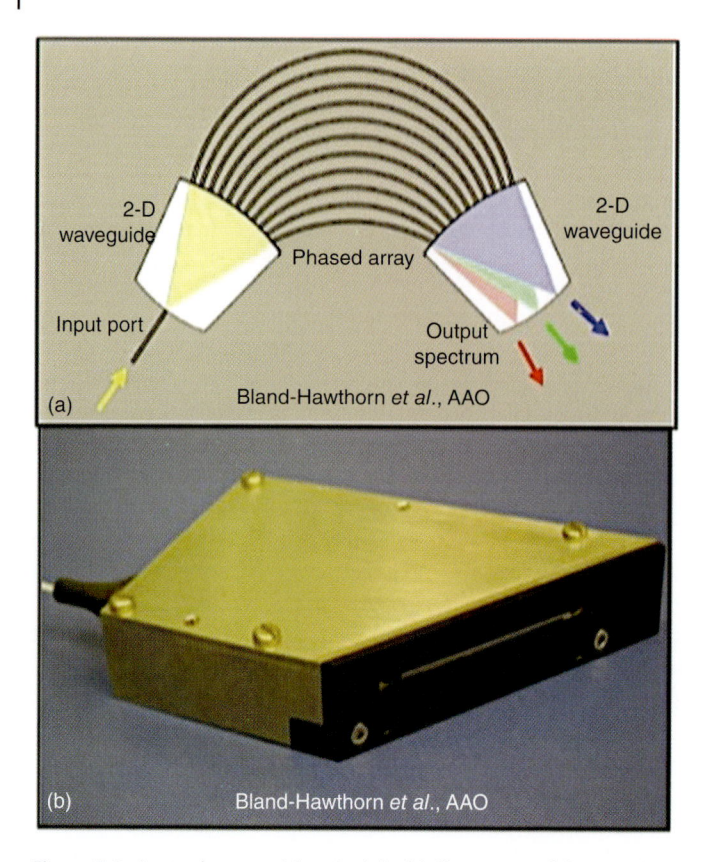

Figure 7.3 Arrayed waveguide principle. (a) Illustration of the wavelength dispersing effect of this component. (b) Picture of an actual AWG device. (Reproduced with permission of Prof. J. Bland-Hawthorn, University of Sydney.)

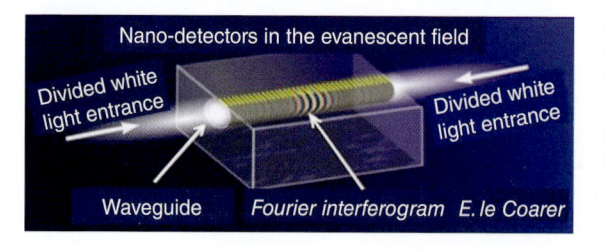

Figure 7.4 Photonics Fourier transform spectrometer SWIFTS concept. (Le Coarer [80]. Reproduced with permission of Nature publishing group.)

and powerful component is thus largely unsuited for the study of faint astronomical objects, but much useful, for example, for studies on the Earth's airglow from above.

7.2.4 Analysis

The photonics devices presented above look at first sight ideal. Unfortunately, they share a common fundamental limitation, namely, calling for mono-mode fibers that accept only the very small light input etendue of $\pi\lambda^2$ corresponding to one diffraction-limited core. For example, for a 25-m telescope working around 1 μm, an individual device covers only 10 mas on the sky. Almost all astrophysical applications call for much larger fields, and hence would require huge numbers of such devices.

One notable exception is when using a number of identical unit telescopes, all pointing the same field and with their light outputs combined coherently, the so-called multiaperture interferometry. Recording the fringe pattern given by light recombination permits reconstructing the objects in the field of view with a spatial resolution corresponding to the diffraction limit of a virtual telescope with a diameter equal to the largest distance between the unit telescopes (1 mas for a 250 m separation at 1 μm). The field of view is usually limited to the diffraction limit of any of the unit telescopes (100 mas, for a 2.5 m telescope at 1 μm), the so-called 'zero-field' mode. This particular observing technique would thus require only one OH-suppression fiber per unit telescope, and a few hundred waveguide arrays to measure the fringe pattern given by light recombination of all telescopes.

On the other hand, a diffraction-limited integral field spectrometer for a 25-m telescope with a small $0.6'' \times 0.6''$ field, would already require 4000 individual devices. The situation would even be harsher, for example, for wide-field high-z galaxy spectrographic surveys. In this case, about 4000 waveguide arrays would still be needed, but now for each of, say, at least 25 science probes deployed in a wide patrol field. One would also need a practical way to inject light from the telescope focal plane into the thousands of mono-mode fibers required. This has been made possible by the development by Sydney University of photonics lanterns with 19 mono-mode fibers coupled to a multimode. This offers a way to efficiently couple the coarsely sampled input light to mono-mode photonics components and vice versa to inject the light to the detector. However, many such lanterns would generally be needed, for example, ~ 2000 to cover the full height of a 4k × 4k detector.

The efficiency of photonics devices is also a crucial parameter. Present devices do offer light transmission of the order of 50–60%. This is below the corresponding bulk optics spectrographic efficiencies by typically a significant factor 1.5. Efficiency improvement would certainly increase their attractiveness.

In conclusion, photonics devices are already in use for zero-field interferometry. They are highly promising for diffraction-limited imagery and spectroscopy, but only if high-throughput waveguide arrays and OH-suppression fibers can be mass-produced at very low individual cost and mass-assembled at reasonable cost, while at the very least keeping present light efficiency. This looks problematic, as each of these three improvements is most likely possible, but unfortunately tend to be mutually exclusive.

Photonics devices might ultimately be applied to multiobject and integral field spectroscopy, but only if in addition efficient photonics lanterns can be mass-produced and mass-assembled. Another general prerequisite is that these devices should cover at least the near-IR range (J to K'), and ultimately would venture down to the optical domain. In short, photonics devices are already being used for interferometric devices and constitute a highly promising way, but the jury is still very much out for mainstream applications.

The photonics FTS has the same $\pi\lambda^2$ optical etendue limitation as the other photonics devices, that is, if one keeps the full SWIFTS spectral resolution. It can however accept a larger etendue at the cost of a lower spectral resolution, by a factor inversely proportional to the linear etendue gain.

7.3 Quest for the Grail: Toward 3D Detectors?

7.3.1 Introduction

While photonics devices offer a possible, if still highly uncertain, path to alleviate the shortcomings of bulk spectrography, it would be even better to get rid of spectrographic devices altogether by using a true 3D detector, that is, a 2D array of detector pixels, where each pixel is able not only to record each photon impact but also its individual energy, and hence its wavelength.[3]

Such 3D detectors are already in mainstream use in the X-ray domain, thanks to the high energy of each individual photon. For instance, the EPIC CCD camera on the XMM-Newton satellite [81] works as a multipixel photon-counting detector. It features 170,000 spaxels, and cover the 0.2–10 keV energy range (6.2 to 0.12 nm wavelength range) with a 10% energy resolution and an up to 90% quantum efficiency. The CCD array readout time is a relatively long 2.6 s; this is easily manageable since astrophysical X-ray photons usually impact each detector pixel at a glacial pace. Note on the other hand the detector's extremely low spectral resolution ($\mathfrak{R} \sim 10$) and consequently the small number of spectral pixels, ~ 40 only.

7.3.2 Photon-Counting 3D Detectors

Achieving similar photon-counting 3D detection in the optical domain is a lot more difficult, given the much lower energy of individual photons. This has nevertheless been achieved, based on incoming photons modifying the physical state of a supraconducting material. There are currently three variants, namely, (i) superconducting tunnel junctions (STJ), (ii) transition edge sensors (TES), and (iii) microwave kinetic inductance detectors (MKID). They all operate at frigid temperatures, below 0.1 °K. They can cover a huge spectral range, with, however, small spectral resolutions that vary as $\lambda^{-1/2}$.

Both STJ and TES devices had first light on the sky in 1999 with very small spatial and spectral formats (see respectively [82] and [83]). STJ development by the European Space Agency (ESA) led to successive S-CAM cameras at the 4.2 m diameter William Herschel Telescope in La Palma, up to a 10 × 12 pixel Ta/Al array covering 9″ × 11″ on the sky over the 339–750 nm range in 2006, with a spectral resolution of 10 at 500 nm. A 6 × 6 pixel TES array was installed in 2000 at the 2.7 m Harlan J. Smith telescope at McDonald Observatory. It covered a huge spectral domain from 350 to 1650 nm with a spectral resolution of 16 at

---

3 One could even further dream of a 5-D detector that would also record the angles of arrival of the photons, alleviating also the need for a telescope! And a 7-D detector could add the light polarization vector!

500 nm. Larger formats detectors were clearly needed, but turned out to be very difficult to obtain with these two techniques.

This is easier to achieve with the MKID approach, and by 2013 a substantial format (32×32 pixels, now upgraded to 46×44 pixels) MKID has been put in operation at the 200" Hale telescope. This so-called ARCONS camera operates at 0.085 °K and covers the full optical range (400–1100 nm) in one go. Like for the other two techniques, quantum efficiency is quite high, around 70%, and each incoming photon is time-tagged with a huge accuracy, to about 2 µs here. This latter capability makes these detectors perfectly adapted to rapid characterization of fast variable objects, usually stars in peculiar evolution states. As for the other optical (and X-rays) 3D detectors, energy resolution over the whole spectral range is only about 10%, resulting for ARCONS in a meager 14 spectral pixels over its wide spectral range. This is actually a fundamental limitation of all superconducting-based detectors, unless some miraculous new material is found.

The data format is thus quite similar to that obtained with a multifilter imager, albeit with the advantage of requiring only one exposure instead of a dozen separate ones with different filters. On the other hand, modern CCD mosaic imagers cum filters now feature a billion spatial pixels, if covering only one spectral pixel per integration. MKIDs cannot thus compete, by huge factors around a few 10^5, for extensive surveys aimed at finding rare interesting objects (from distant asteroids in our solar system to the most distant galaxies in the Universe) over huge – many square degrees – fields. Their relatively narrow ecological niche lies instead in the follow-up identification of selected intriguing objects, with the bonus of directly detecting if they are rapidly variable.

7.3.3 Integrating 3D Detector

The 3D detectors presented above are based on real-time photon counting. A totally different 3D detector is the Spectral Hole Burning Device (SHBD) developed around 1995 by Keller et al. [84], which can be viewed as a high-tech version of the color photographic plate. A dye-doped polymer film immersed in liquid helium gives a wide absorption band. Spectral hole burning happens as absorption of an incoming photon leads to localized wavelength-selective bleaching in the film, with as a result increased material transmission at the original photon location and wavelength (hence the spectral hole denomination). This is unfortunately less common than the absorbing dye molecule simply going back from its excited state to its ground state, but does happen a few times for some selected organic dyes. After integration, the 3D image can be read by illuminating the detector with a tunable laser and recording a sequence of monochromatic images covering the dye absorption band.

First – and sole so far – astronomical use was done in 1999 by Keller [85] who obtained a 15 nm wide solar spectrum centered at 634 nm at a huge spectral resolution of 6.10^5. The $2\,cm \times 2\,cm$ detector gave 5000×5000 spatial by 15,000 spectral elements, or a whopping 375 billion elements data cube. No further use was done, as the detector had a too narrow bandwidth and especially a much too low quantum efficiency (<0.1%) to be effective for nonsolar astronomical observations. A combination of dyes should be able to enlarge the bandwidth to a more reasonable value; unfortunately, it is unclear if quantum efficiency can be boosted to the absolute minimum requirement of a few percentage to render this whole approach competitive with respect to classical 3D instruments.

7.4 Conclusion

Photonics spectrographs and filters are already competitive for interferometric systems, where their intrinsically small etendue is just fine, and their robustness is invaluable. How much and how fast they will eventually migrate to replace conventional bulk-optics integral field spectrograph is uncertain and much depends on being able to integrate thousands (and more) of photonics building blocks in a single instrument and at an affordable cost, while keeping or even better improving present throughput.

3D detectors could in principle someday totally replace photonics and non-photonics 3D instruments alike. Within presently known technology, this looks highly problematic: photon-counting versions feature much too small spectral resolutions and still meager spatial formats; integrating ones have much too small quantum efficiency and wavelength range; the FTS photonics approach is not well adapted to astronomical faint object spectroscopy. Yet, 3D detectors would represent such a paradigm change in astronomical (and non-astronomical) instrumentation that keeping an eye on any emerging technology in that domain is a must.

7.5 For Further Reading

- Visible/Infrared Imaging Spectroscopy and Energy-Resolving Detectors, Eisenhauer F. and Raab W., Annual Review of Astronomy and Astrophysics, vol. 53, p. 155–197.

** Exercise 10   Designing a wide-field multislit instrument for an ELT

Basic design parameters: telescope diameter $D = 30$ m; slit sky width $\alpha = 0.6''$; slit height $0.9''$; patrol field $6' \times 6'$; central spectral resolution $\mathfrak{R} = 3.10^3$; spectral range 420–840 nm (1 octave); 15 µm detector pixel; F/1.5 camera.

1. Using the patrol field × telescope size linear etendue, find the field image size in detector pixels.
2. Using the slit width × telescope size linear etendue, find the slit width and height in detector pixels.
3. Find the overall detector format required to accommodate the full spectra originating from any location in the patrol field.
4. Find the grating format $d \tan \phi$ (d pupil diameter. ϕ blaze angle). Any suggestion on how to split it?

Answer of exercise 10

1. The 6′ patrol field linear etendue is $6 \times 2.91 \times 10^{-4} \times 30 \times 10^6$ μm rad., or 5.24×10^4 μm rad. A pixel linear etendue is $15/1.5$ μm rad., or 10 μm rad. The imaging field thus covers a respectable 5240×5240 pixels on the detector.
2. The slit width linear etendue is $0.6 \times 4.85 \times 10^{-6} \times 30 \times 10^6$ μm rad., or 87.3 μm rad. This corresponds to 8.73 detector pixels, a large but not too outrageous value. Slit height is $0.9/0.6$ times bigger, or 13 pixels.
3. Central wavelength λ_c is 630 nm. One slit width (8.73 pixels) corresponds to $\Delta\lambda = \lambda_c/\mathcal{R}$, or 0.21 nm. Spectral scale is $0.21/8.73$, or 2.405×10^{-2} nm per pixel. Spectral length is then about $(420/2.405) \times 10^2$ or 17,464 pixels. Total detector format is 22,704 (dispersion direction) $\times$ 5240 (orthogonal direction) pixels. The large detector size is 340×79 mm, with a 350 mm diagonal.
4. With $2d \tan\phi = D\alpha$, we get $d \tan\phi = 13.5$ cm. With a 35 cm diagonal field, the rule of thumb is to choose a $d = 35$ cm disperser/camera pupil. This gives $\tan\phi = 0.386$, or a reasonable 21° blaze angle.

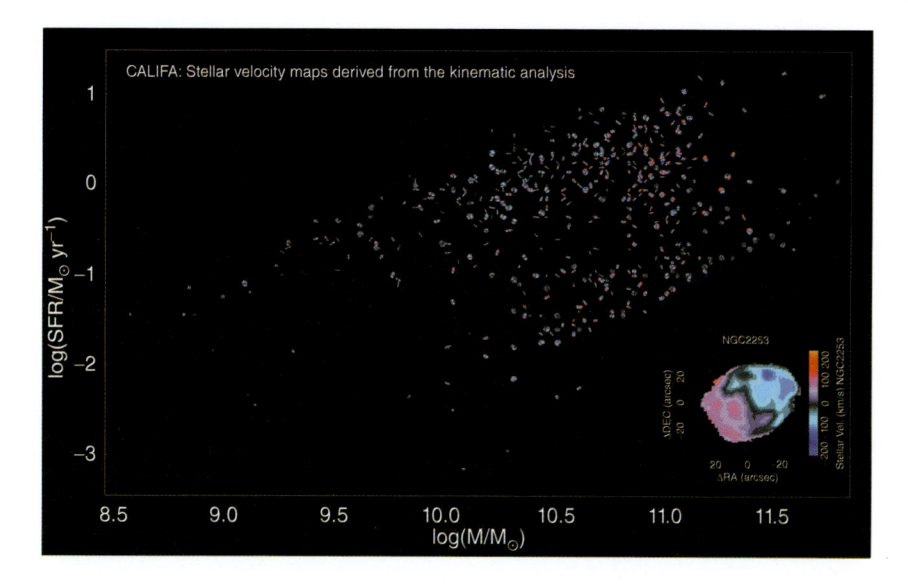

The CALIFA survey [125] was one of the first IFS surveys designed to characterize the full optical extension (up to 2.5 Re) of a representative sample of the galaxies of the Local Universe (excluding dwarf galaxies), covering all morphological types. Foreseen as a legacy survey, CALIFA delivered the reduced data in successive data releases. The last one (DR3) comprises 667 galaxies and 1576 data cubes [126], covering a wavelength range between 3700 and 7200 Å. An example of the information gathered for these galaxies is shown in this figure, where the stellar velocity maps for 476 galaxies extracted from DR3 are shown distributed along the star formation rate versus stellar mass diagram. The figure comprises the information included in the data cubes, showing results derived from stellar kinematics, emission line analysis (SFR was derived from dust corrected Halpha luminosities), and the study of the star-formation history (stellar mass). The survey was performed at Calar Alto with the fiber-based PPAK integral field unit. ([125]. Reproduced with permission of Sebastian Sanchez.)

Part II

Using 3D Spectroscopy

8

Data Properties

8.1 Introduction

Any instrument delivering analog or digital data has a finite data resolution, quantitatively expressing its ability to differentiate close (in time or space) signal features. For 3D instruments, there are two separate resolution types: the angular resolution, which is the smallest angular distance between two point-like objects that can be measured; and the spectral resolution, which measures the smallest spectral interval that can be distinguished between two spectral lines.

Resolution should not be mixed up with data sampling. Data sampling occurs when a device converts an input continuous signal into a series of discrete samples. This is the case for both CCDs and IR arrays, where the input signal is sampled by the detector in a series of pixels.

Data sampling and resolution play an important role in the instrument properties and performance. They can be very different from one instrument to another; for example, data sampling is radically different on a Fourier transform spectrograph, which samples the signal in the frequency domain, than in a classical spectrograph, which samples the signal in the wavelength domain.

Every signal comes along with its associated noise. There are many different sources of noises; some are intrinsic to the statistical nature of the signal (e.g., photon noise), and others are due to the instrument (e.g., readout noise). The signal-to-noise ratio (SNR) is one key parameter of an observation. Understanding the noise properties of observing data is very important to getting their confidence.

In this chapter, we discuss these two aspects for the whole set of instruments covered in the book.

8.2 Data Sampling and Resolution

The Nyquist–Shannon sampling theorem [86] gives the minimum sampling needed to fully reproduce a bandwidth-limited continuous signal, that is, with a given maximum spatial or temporal frequency. If the signal's highest frequency is N then it should be sampled at $2N$. For example, with a spatial sampling of

Optical 3D-Spectroscopy for Astronomy, First Edition. Roland Bacon and Guy Monnet.
© 2017 Wiley-VCH Verlag GmbH & Co. KGaA. Published 2017 by Wiley-VCH Verlag GmbH & Co. KGaA.

an observation 1 arcsec, we could only recover features in the signal with less than 0.5 arcsec^{-1} frequency. However, the theorem is valid only if the signal is perfectly bandwidth limited. But this is never the case in real observations and thus the recovered signal will suffer from aliasing. Aliasing creates artificial oscillation in the recovered signal. To minimize this effect one could sample the signal at a higher frequency, for example, $3N$, than what is required by the theorem.

8.2.1 Spatial Sampling and Resolution

Spatial sampling is the sampling performed in an image of the field of view. Each sample is called a *spaxel* (spatial pixel) to differentiate it from the generic name of pixel (picture element). The unit of spaxel is arcsec^2 and it comes in different shapes: for example, square, circular, hexagonal. The best shape is hexagonal because it is compact (i.e., it fills completely the surface without any loss) and more isotropic than the square shape for which the sampling is 1.4 times larger in the diagonal direction than in the grid direction. Circular shapes are fully isotropic but do not fully tile the field of view, resulting in some light loss (9% for a compact hexagonal arrangement). Spatial sampling location differs widely according to the instrument type:

For scanning Fabry–Pérot, scanning Fourier transform spectrographs, and slitless-based MOS, spatial sampling is performed at the detector level. In this case, the spaxel is identical to the detector pixel, which is always square shaped.
For scanning long-slit spectrographs and multislit-based MOS, the spaxel is defined by the detector pixel in the axis along the slit and by the width of the slit in the orthogonal axis. As a consequence, the spaxel shape is generally rectangular.
For lenslet-based IFS the spaxel is defined by the microlens and its shape can be circular, hexagonal, or square.
For purely fiber-based IFS or MOS the spaxel is circular with its size defined by the fiber core diameter. Note that because of the fiber cladding, the spaxels are not densely packed. In this case, the sampling size to take into account with respect to the Nyquist criteria must be the diameter between two neighboring fibers, which is larger than the core diameter of the fiber. However, in the usual case where a microlens array is used in front of the fiber bundle, the spaxel is defined by the microlens as in lenslet-based IFS.
For slicer-based IFS, the spaxel is defined by the width of the slice in one axis and by the detector pixel on the orthogonal axis.

Provided that the spatial sampling satisfies at least the Nyquist–Shannon sampling theorem, the spatial resolution is defined by the spatial point spread function (PSF) of the system. The PSF is the response output of the system to a point source input. The spatial resolution is often defined as the full-width at half maximum (FWHM) of the spatial PSF. For perfect (aka diffraction-limited) optical systems, that is, with image deterioration only due to diffraction effects, the spatial resolution is also defined as the size of the central Airy disk of the diffraction pattern. In the case of a circular aperture, the a-dimensional spatial angular resolution θ is given by the following formulae:

$$\sin \theta = 1.220 \frac{\lambda}{D} \tag{8.1}$$

where λ and D are respectively the wavelength and the diameter of the circular aperture.

There are many contributors to 3D observation PSFs. The main ones are listed below, in decreasing impact order:

1) *Atmospheric turbulence.* Atmospheric turbulence has a major impact on ground-based observations for any telescope roughly larger than 15 cm diameter in the optical or 30 cm in the NIR regions. See Chapter 9 for further details.
2) *Instrument PSF.* This is in general the second contributor to the angular resolution loss, at least for MOS and IFS instrument types.
3) *Guiding and focusing errors* and more generally any uncorrected instability taking place during an observation. Their effect increases with integration time. For scanning instruments this can be a major contributor, resulting in a different PSF for each scan.
4) *Atmospheric dispersion.* This is important only for high zenith distance observations and then only for instruments that do not incorporate an atmospheric dispersion compensator.
5) *Telescope PSF.* Most modern telescope optics are nearly diffraction limited and thus their contributions to the overall PSF budget is low.

PSFs are generally a function of wavelength, location of the objects with respect to the field of view, and time. Time dependence arises from system instabilities and, in the case of scanned techniques from the ground, to unavoidable time changes of observing conditions (in particular atmospheric transmission and turbulence). The wavelength dependence is due to atmospheric turbulence as well as chromatic aberration of the instrument. Variations within the field of view are generally due to the instrument optical aberrations. Evaluating the true spatial PSF of a given 3D ground-based observation is usually a difficult task because of variable atmospheric conditions. An example of the process involved in the case of the integral field spectrograph MUSE is given in Section 11.4.10 (Figure 8.1).

8.2.2 Spectral Sampling and Resolution

Spectral sampling occurs in the spectral dimension. The spectral sample is sometimes called a *spectel*, for spectral pixel, but this denomination is less used than spaxel. We will however use it in the rest of this book to differentiate it from the general pixel. Spectel unit is the typical wavelength unit (e.g., Å or nm). The exception is for Fourier transform spectroscopy, which operates in the frequency domain with a spectel unit in cm^{-1}.

For classical spectroscopy with dispersive elements, such as grating or prism-based spectrographs, spectral sampling is performed on the detector.

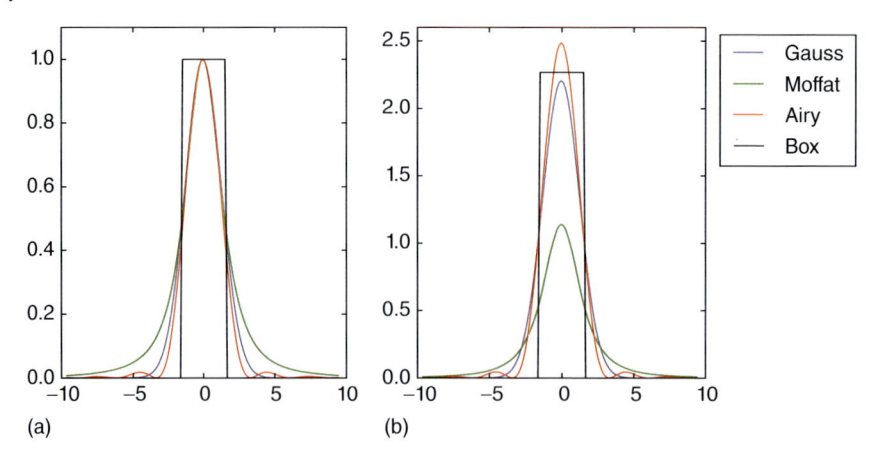

Figure 8.1 Examples of various analytical forms of 2D PSF. As shown in (a), all PSFs have been set to the same FWHM. However, their contribution can be very different when proper normalization by the total flux is taken into account (b). Note the impact of the very extended wings of the MOFFAT model with $\beta = 1.5$, which decreases the contrast of the central peak.

- This is the case for MOS, IFS, or scanning slit spectrographs. The spectel size is then given by the linear dispersion of the grating (or prism) times the size of the detector pixel.
- In scanning Fabry–Pérot, spectral sampling is performed by the scanning process, changing the etalon width, which translates into a shift of the central wavelength on each spaxel. The spectel size is then the wavelength shift between two successive steps. It is usually expressed in Å.
- For scanning Fourier transform spectrometer, the sampling is performed by changing the path length in uniform steps. The process results in an interferogram (the Fourier transform of the spectrum in wavelength unit) and the spectral sampling is the interferogram unit step. As a consequence, the recovered spectrum is not uniformly sampled in wavelength.

In astronomy, we are often interested in determining the radial velocity field of stars and/or gas inside a galaxy and thus would prefer to get a sampling that is uniform in km s^{-1} rather than in wavelength. This is also the case when studying distant objects because of the expansion of the Universe, also known as the Hubble flow. In that respect, the almost uniform linear dispersion given by a grating is not ideal, since it translates into a nonuniform velocity sampling ΔV:

$$\Delta V = c\frac{\Delta \lambda}{\lambda} \propto \frac{1}{\lambda} \tag{8.2}$$

Note that using a prism in the place of a grating does not help since a typical glass dispersion law roughly gives $\Delta \lambda \propto \lambda^3$ and thus $\Delta V \propto \lambda^2$.

For the FTS, this results into the again nonconstant law:

$$\Delta V = c\frac{\Delta \lambda}{\lambda} = c\lambda \Delta(\lambda^{-1}) \propto \lambda \tag{8.3}$$

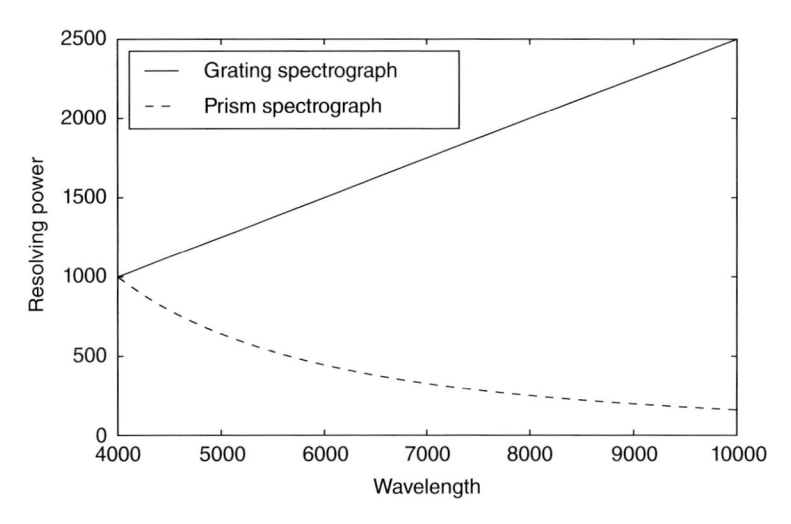

Figure 8.2 Example of resolving power as a function of wavelength for two types of dispersers: a grating and a prism.

The spectral resolution or resolving power is defined by

$$\mathfrak{R} = \frac{\text{FWHM}}{\lambda} \qquad (8.4)$$

where FWHM is the full-width at half maximum of the line spread function (LSF). The LSF is the response of the system to a monochromatic emission input. An example of the evolution of resolution with the wavelength is given in Figure 8.2.

As for the angular resolution, spectral sampling should satisfy at least the Nyquist–Shannon sampling theorem. While there is some possibility to overcome a spatial undersampling (see Section 10.4), this is generally not possible for the LSF and thus specific care should be taken to get the requested spectral sampling values.

For multislit MOS, scanning slit spectroscopy, or slicer-based IFS, the LSF is given by the convolution of the slit with the spectrograph PSF, provided the slit is uniformly illuminated. It is readily measured from a uniformly illuminated calibration with a monochromatic source. Unfortunately, this is not usually the case and the LSF is instead the convolution of the source brightness distribution inside the slit with the instrument PSF, and is at best poorly known. This results in quite serious effects, in particular systematic wavelength determination (radial velocity) errors plus uncertainties in the true spectral resolution. This phenomenon, called the slit effect, can be quite significant for multislit MOS and scanning slit spectroscopy, where the slit width is usually quite large (say 1–2 arcsecond on the sky) in order not to lose too much object light for the first flavor, or minimize the number of scanning steps for the second one. It is usually small for slicer-based IFS, which does not suffer from a similar trade-off and generally features sub-arcsecond narrow slits. A detailed discussion of the slit effect impact can be found in [48].

For fiber-based MOS or IFS, even in the extreme case of a point-like source at the fiber input, light is redistributed when propagating along the length of the fiber and output light usually almost fills uniformly (more precisely with an isotropic Gaussian function) the fiber end. There is thus very little slit effect; this has a cost, of course, in terms of substantial – say 15–25% – light loss by diffusion in the process. The LSF is then simply the convolution of the projected fiber core with the instrumental PSF. Usually, the projected diameter of the fiber on the detector is large enough, so that the spectral sampling satisfies the Nyquist criteria.

For lenslet-based IFS, the LSF is the convolution of the exit micro-pupil with the instrument PSF. The micro-pupils are, by definition, uniformly illuminated and thus this type of spectrograph does not suffer from a first order slit effect. A usually negligible second order effect is present; it is due to the geometric aberrations of the spectrograph camera in the imaging of the micro-pupils.

For scanning FP, the LSF is the convolution of the transmission of the etalon with the instrument PSF. The transmission of a plane-parallel etalon is given by the following Airy function:

$$T = \frac{1}{1 + F \sin^2(\delta/2)} \tag{8.5}$$

where δ is the phase difference and F the finesse of the etalon. As for lenslet-based IFS, there is no first order slit effect and the second order one, linked to the geometrical aberrations of the camera, is usually negligible.

For FTS, the theoretical LSF, as for a grating spectrograph, originates from the finite maximum path difference Δ, here between the two Michelson interferometer mirrors. The FTS LSF is a *sinc* function, which has undesirable oscillations; the first one has a negative amplitude of 22% of the peak and is situated at 3.2 half width half maximum. This is usually solved by multiplying the interferogram function by a suitable weighting function, a process called apodization since it removes the feet of the LSF. One widely used weighting function is a triangular function with the same Δ width. This results in a *sinc squared* LSF function, removing all oscillations at the cost of a spectral resolution loss by a factor 1.5. Note that this theoretical LSF mathematical shape is the same as that for a diffraction grating with the same maximum path difference, in this latter case between the opposite sides of the grating perpendicular to the grating grooves.

Some examples of typical LSF are given in Figure 8.3.

8.3 Noise Properties

We will differentiate between three components that add to give the photoelectron counts measured on detector pixels, namely *signal*, *background*, and *noise*. *Signal* is the count number coming from the object under study and *background*

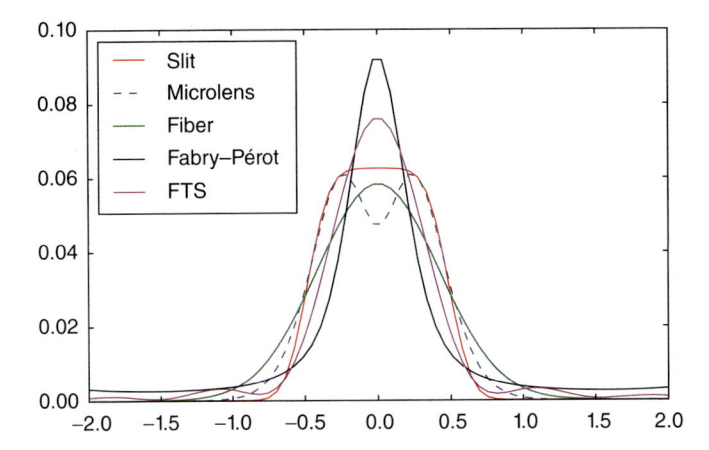

Figure 8.3 Examples of LSF for various types of spectrographs: slit-based, fiber-based, lenslet-based, Fabry–Pérot, and FTS.

from any other light flux component, while *noise* is the statistical fluctuation of the count numbers. For example, for the study of the central nucleus of a galaxy, the *signal* comes from the nucleus itself and the *background* both from night-sky airglow and the other components of the galaxy (disk and bulge) integrated along the line of sight. Note that the *signal* versus *background* distinction is somewhat arbitrary, as it generally depends on the exact science goal of the observation.

One inescapable noise component is photon noise (also called shot noise), which is the inherent photon flux fluctuation due to the quantic nature of light (Figure 8.4 panel b). Since the probability of a photon arrival on a detector pixel at any time is independent of the time since the last event, the probability $P_\lambda(x)$ to get x photons in a given integration time with a mean integrated flux of λ photons obeys the Poisson discrete probability distribution:

$$P(x, \lambda) = \frac{e^{-\lambda} \lambda^x}{x!} \tag{8.6}$$

Note that this is true only if the signal is expressed in photons (or photoelectrons). This distribution function has a variance equal to its mean value. For $\lambda > 10$, a very common occurrence, the rather unwieldy discrete Poisson distribution can be replaced by the continuous normal distribution function with the same mean value (λ) and same variance (λ). This gives

$$P(x, \lambda) = \frac{1}{\sqrt{2\pi \lambda}} \exp\left(-\frac{(x + 0.5 - \lambda)^2}{2\lambda}\right) \tag{8.7}$$

Another main source of noise is the additive readout noise. It is due to the amplifier processing chain and the conversion of electrons in analog-to-digital units (Figure 8.4 panel c). It is well represented by a normal distribution of zero mean value and variance σ_{rn}^2. The standard deviation σ_{rn} is expressed in electrons rms. Modern CCD controllers reach low readout noises, with typically $\sigma_{rn} \sim 2$–3 electrons rms. IR detectors have higher readout noise, in the 10–20 electrons rms range, but this is usually reduced to 6–9 electrons rms by using multiple

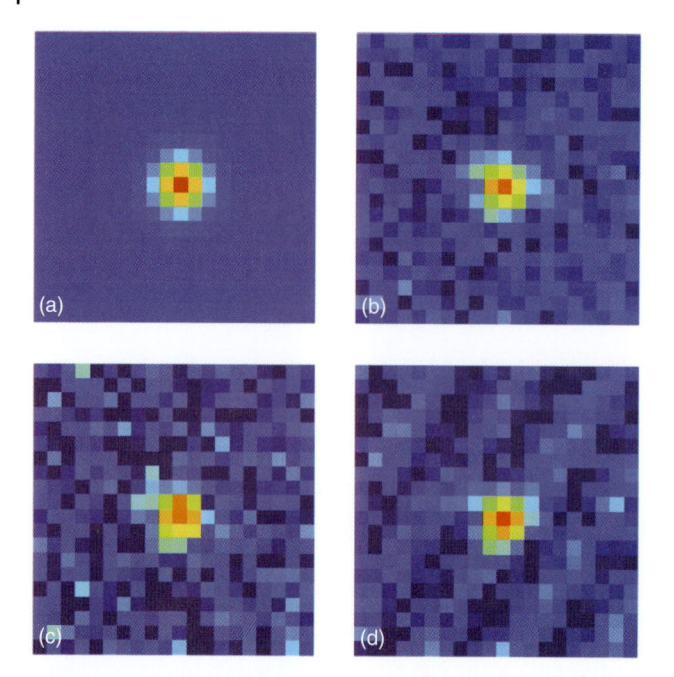

Figure 8.4 Examples of various detector noise patterns: (a) original source, (b) photon noise, (c) readout noise, (d) pickup noise.

nondestructive readout. Readout noise is generally the dominant source of detector noise, but there are other contributors. Here is a nonexhaustive list:

- Dark current is the electric current due to random spontaneous generation of electrons in the detector. It is expressed in electron per hour or per second. Dark current noise is its statistical fluctuation. The CCDs are always cooled to maintain their dark current at low values (typically a few electrons per hour) and the dark current noise contribution to the total noise budget is almost negligible. Although the new generation of (cooled) HgCdTe infrared devices have much lower dark current (~ 0.1 e$^-$ s^{-1}) than before, this is still 2 orders of magnitude higher than with CCDs.
- Pickup noise occurs when there is electromagnetic interference between the detector and other electromagnetic sources. It is visible as a regular pattern that comes in addition to the signal (Figure 8.4 panel e). Such noise can be suppressed by improving the ground isolation of the external sources of electromagnetic signal.

Given that the various noise sources are independent and assuming normal distributions, the total noise σ_N is just given by the square root of the sum of the variances:

$$\sigma_N = \sqrt{\sigma_{rn}^2 + n_{obj}\Delta t + n_{back}\Delta t + d_c\Delta t} \qquad (8.8)$$

where σ_{rn} is the readout noise, n_{obj} is the number of photons by second received from the object, n_{back} is the number of photons by second received from the background sources, d_c is the dark current, and Δt is the integration time.

If $n_{obj}\Delta t + n_{back}\Delta t \gg \sigma_{ron}^2 + d_c\Delta t$ the observation is dominated by the photon noise. This photon noise regime is optimal because the signal can be summed up a posteriori without noise penalty. If this is not the case, the observations are then in the detector noise regime.

For astrophysical imaging, the dominant source of noise is always the photon noise of the night sky emission. For extremely faint object spectroscopy at medium-high resolution ($Re \geq 5000$) this is not the case anymore and the detector noise stands out as the major limitation.

Cosmic rays-induced muons are another important source of noise, impacting the detector material and creating electron–hole pairs (Figure 8.5). These muons are produced in the hadronic cascade following a primary cosmic ray impact in the outer atmosphere. The number of pixels impacted by the cosmic ray depends much on the energy of the particle and the thickness of the photosensitive layer. For thin CCD the impact is generally limited to 2–3 pixels, but for red-sensitive thick CCD up to 1000 pixels can be wasted by a single event. The density of cosmic rays is also a function of the altitude; at sea level there are 20% less events than at 2000 m altitude. Contrary to the previous noise sources, cosmic rays are not an additive noise. Many spurious charges are created on each impacted pixel and they can be considered as lost for the observation data set. There is no other solution than to limit the exposure time and to obtain a series of exposures of the same target to get rid of most of the events. See Section 10.4 for a discussion on the optimum observing strategy. Because of their nondestructive readout capabilities, it is much easier to deal with cosmic rays in the case of IR detectors. What is actually measured on each pixel is charge build-up with time, as the array

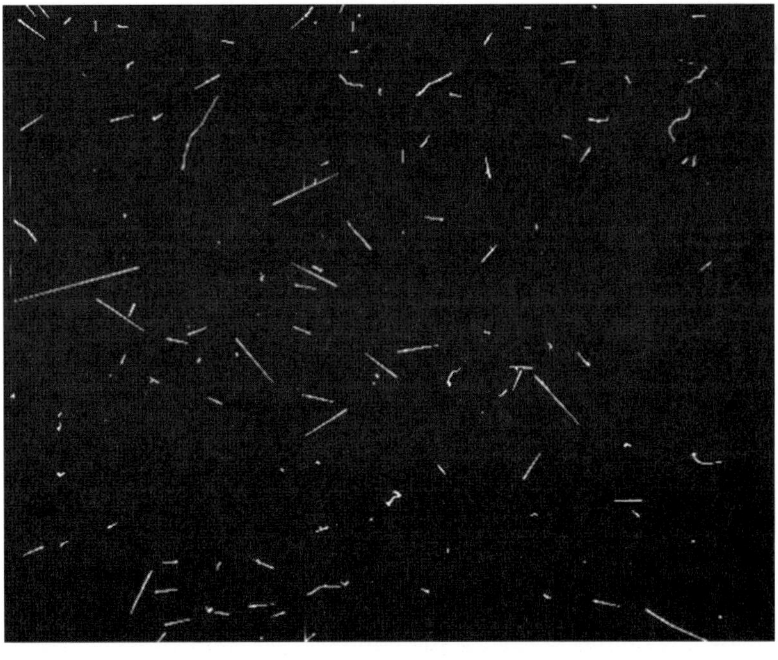

Figure 8.5 Example of impact of cosmic rays in a thick red-sensitive CCD.

is read every few seconds for the typically 5–10 min total integration. A one-off sharp jump on a pixel during this sequence is a clear cosmic ray impact signature and can be filtered out in real-time [87].

We quantify the relative importance of noise by the signal-to-noise ratio, usually written as S/N or SNR. If we just consider the photon noise, the SNR is given by

$$\text{SNR} = \frac{S}{\sigma_N} = \sqrt{n_\text{phot}} \tag{8.9}$$

where n_phot is the number of photons. The SNR is an important measure of the observation quality. The required SNR depends on the type of measurement we want to perform on the data. Signal detection typically requires an SNR of at least 3. For a normal distribution this corresponds to a probability of 0.3% of false detection. Because of all the complex effects that enter into the noise budget, the final noise probability function is rarely normal and it is safer to take some margin and to consider an SNR of 5 as a more reliable detection threshold. In most cases, however, we want to evaluate some physical parameter from the signal and thus require a higher SNR. Typical SNR values required are

- 5–10 for measuring the position of an emission line
- 10–20 for measuring a line width
- 50–100 for measuring higher moments of a line profile.

Integration time is the only handle to achieve the required SNR. Note that in photon noise regime, the SNR (Equation 8.9) is proportional to the square of the integration time and thus doubling the SNR requires four times more integration time. Prediction of SNR and integration time is the subject of the exposure time calculator and is discussed in section 10.3.

Accurate noise estimation is a key for getting reliable results. This is particularly true for faint source detection, which are always at the detection limit. It is usually relatively easy to get a good estimate of the noise property on the raw data, but it is much more difficult to get the same accuracy on the final data once it has been processed by the data reduction system. The various calibration (in wavelength, in sky coordinates, in flux) steps lead to many interpolations, which makes it difficult to track the variance information. The solution would be to compute and propagate the covariance matrix along the various steps but this is prohibitively expensive on the CPU and memory. Alternative solutions are being developed in the new generation of data reduction systems (see Section 11.3)

Finally, there are also a number of uncertainties that affect the observations and can be considered as noise, even if they cannot be easily quantified. These are all the systematics that are left by the imperfect instrument signature removal because of inaccurate calibration and data reduction. For example, imprecise sky subtraction will leave some residuals at the location of bright [OH] emission lines and wavelength calibration errors will produce errors in the measured kinematics. Another frequent source of problems is from inaccurate flat fielding. A strategy to minimize the impact of these uncertainties is discussed in Section 10.4. Because systematics are in most case due to a mix of various small, often nonlinear effects, they are very difficult to quantify and then to

8.3 Noise Properties

correct. They remain a source of uncertainty and the ultimate limiting factor when pushing instruments to their limits.

** Exercise 11 Noise Regime and Integration Time

This exercise illustrates the impact of the detector noise properties on the required integration time to achieve a given signal to noise. An astronomical source observed with an instrument at high spectral resolution produces $50 e^-\ h^{-1}$ at 6500 Å. The sky emission is $100 e^-\ h^{-1}$ at this wavelength.

1. Compute the SNR achieved in 2 h integration time in the case of an ideal detector.
2. For a $10 e^-\ h^{-1}$ dark current and $10 e^-$ readout noise, how long do we need to integrate to obtain the same SNR?
3. To limit the impact of cosmic rays and atmospheric change we perform a maximum integration time of 30 min. How many exposures are needed to achieve the same SNR in the case of the ideal detector?
4. Same question but now with the noisy detector.

Answer of exercise 11

1. Assuming no readout noise and no dark current, a simple application of Equation 8.8 gives $17.3\ e^-$ of noise and an SNR of 5.77 in 2 h.
2. From Equation 8.8 we can derive the integration time as a function of SNR:

$$\Delta t = \frac{1}{2}(Na^2 + a\sqrt{a^2 N^2 + 4\sigma_{rn}^2}) \qquad (8.10)$$

with $a = \text{SNR}/n_{obj}$ and $N = n_{obj} + n_{back} + dc$; this gives 2.6 h of integration to achieve the required SNR.
3. In the case of the ideal detector, there is no noise impact due to the splitting of exposure. The SNR is then achieved in four exposures.
4. The detector readout noise contribution must be taken into account for each exposure. The variance from the detector contribution in Equation 8.8 becomes $k\sigma_{rn}^2$ with k the number of exposures. Iterating on k, we found that 10 exposures or 5 h of integration are required to achieve an SNR of 5.89.

*** Exercise 12 Radial Velocity Accuracy versus Signal-to-Noise Ratio

To illustrate the relationship between signal to noise S/N and radial velocity accuracy, take a highly simplified triangular model of an absorption line spread over 5 detector pixels, with successive normalized fluxes $F = 1$, $(1 + k)/2, k, (1 + k)/2, 1$, where $k \leq 1$ is the normalized depth of the absorption line. The corresponding mean radial velocity is of course given at the middle of the half depth, $F = (1 + k)/2$, segment. The spectrograph dispersion is S in km s^{-1} pixel^{-1} and Σ is the continuum SNR.

1. Assuming pure photon noise, express the radial velocity uncertainty δv in km s^{-1} as a function of k, S, and Σ.
2. Comment on this result.

Answer of exercise 12

1. Photometric uncertainty on each point is δF. The line length has thus an uncertainty of $\sqrt{2}\,\delta F$, and the middle point uncertainty is half of that or $\delta F/\sqrt{2}$. The radial velocity uncertainty is $\delta v = (1/\sqrt{2}\,\delta F(\Delta v/\Delta F)$. Here, $\Delta F/\Delta v$ is the midslope of the absorption line, equal to $(1-k)/2S$. With the stellar continuum S/N Σ, the signal to noise at midpoint is $\Sigma\sqrt{(1+k)/2}$ and $\delta F = 1/\sqrt{(1+k)/2}$. Finally, $\delta v = 2S/\Sigma(1-k)\sqrt{1+k}$. δv is the radial velocity uncertainty in km/s, S is the dispersion scale in km s^{-1} pixel^{-1}, Σ is the continuum SNR, and k is the normalized line depth (k between 0 for a totally absorbing line and 1 for a virtually zero depth line).
2. This is highly schematic, but illustrates two fundamental points: the radial velocity accuracy scales with the signal to noise Σ of the continuum and with the line slope $(1-k)/2$. The former point is obvious. The latter is quite normal, as clearly there is no velocity information at the top and bottom of the line (a slight shift of the absorption line does not change the flux integrated on the corresponding pixel), and the best one is at the two maximum slope inflexion points (the pixel flux change is then the largest).

GOODSN-46-G141_16623

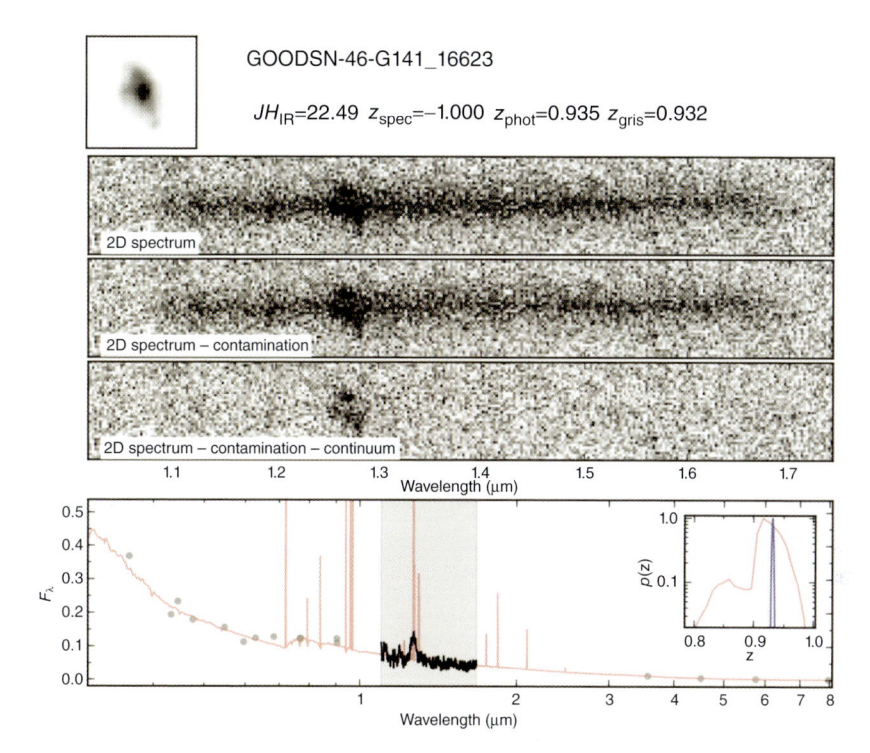

$JH_{IR}=22.49$ $z_{spec}=-1.000$ $z_{phot}=0.935$ $z_{gris}=0.932$

Top let: Infrared (F140W) Hubble Space Telescope direct image within the GOOD-South Field observed as part of 3D-HST survey [127, 128]. Middle: G141 grism spectra within the same GOODS-South pointing. Bottom: an extracted spectrum of a star-forming galaxy at $z = 0.9$ (in black) and the SED fit (in red). These observations were performed with WFC3 and ACS grism slitless HST MOS instruments. ([128]. Reproduced with permission of Pieter Van Dokkum.)

9

Impact of Atmosphere

9.1 Introduction

Interstellar and intergalactic space being essentially empty, light waves even coming from the far end of the Universe arrive relatively unimpeded in the solar system, except for wavelength-dependent absorption by occasional dust clouds inside galaxies including our own, which happen to be fortuitously located along the line of sight. After their up to 13-billion years journey, and in the last 50 μs or so, they are unfortunately heavily impacted by the Earth's atmosphere in a number of ways detailed in this chapter.

The most obvious impact of Earth's atmosphere is strong light absorption in many parts of the electromagnetic spectrum. In the domain covered by this book the situation is nevertheless quite favorable (see Figure 9.1), with very good transmission in the optical window (305–1100 nm) and fair ones in three near-IR windows, namely, the J-band (1150–1350 nm), H-band (1460–1810 nm), and the K-band (1930–2430 nm). Significant improvements can be obtained by going to a high-altitude dry site, with even wider and cleaner transmission bands, but the only full remedy is to observe from a high-altitude balloon, or even better from outer space.

Another important effect is the blurring of stellar images due to turbulent eddies in the atmosphere, the so-called astronomical seeing (Figure 9.2). This effect has a major impact on most astronomical observations from even moderate-sized telescopes, and in any case for virtually all observations with 3D instruments.

There are other atmosphere-related deleterious effects that play havoc with some specialized observation types. For instance, atmospheric turbulence creates also a scintillation effect: the light input collected from, say, a point-like star fluctuates in time within a large frequency range, a fraction of an hertz to about 100 Hz; this might be a strong limitation for the accurate measure of fast stellar brightness variations. In addition, atmospheric diffusion complicates the detection of very faint companions – including bona fide planets – around their much brighter parent star.

The following sections cover respectively (i) the basic seeing properties, (ii) their impact on 3D observations, (iii) real-time seeing correction, also known

Optical 3D-Spectroscopy for Astronomy, First Edition. Roland Bacon and Guy Monnet.
© 2017 Wiley-VCH Verlag GmbH & Co. KGaA. Published 2017 by Wiley-VCH Verlag GmbH & Co. KGaA.

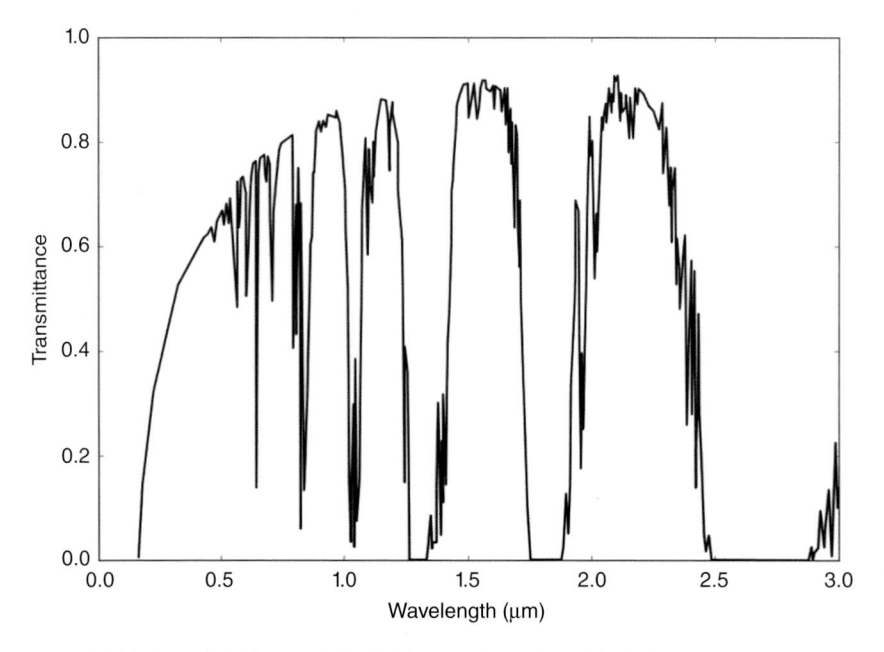

Figure 9.1 Atmospheric transmission in the optical and near-IR windows.

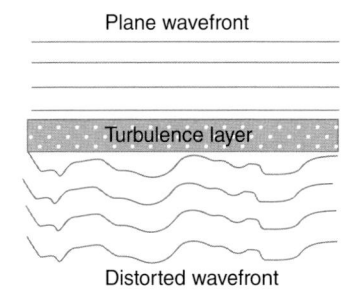

Figure 9.2 Astronomical seeing. Light wavefronts coming from a point-like astronomical object are almost perfectly plane. Images given by a "perfect" telescope would then be very small, limited by light diffraction to sub-0.1″ values for meters-size telescopes. Unfortunately, wavefronts observed at ground level are severely distorted by turbulent layers in our Earth's atmosphere. This degrades image quality to much larger values, almost never below 0.3″, even at the best world sites.

as adaptive optics techniques, and their impact on 3D observations, (iv) the other atmospheric impacts, and (v) the case for observing from the space above the Earth's atmosphere.

9.2 Basic Seeing Principles

9.2.1 What is Astronomical Seeing?

Any astronomical observation from the ground – as well as any Earth observation from space – suffers from the so-called seeing effect: a perfectly plane light wavefront coming from a point-like astronomical source is distorted by fast (a few millisecond timescale) temperature fluctuations naturally present in the

(a)   (b)

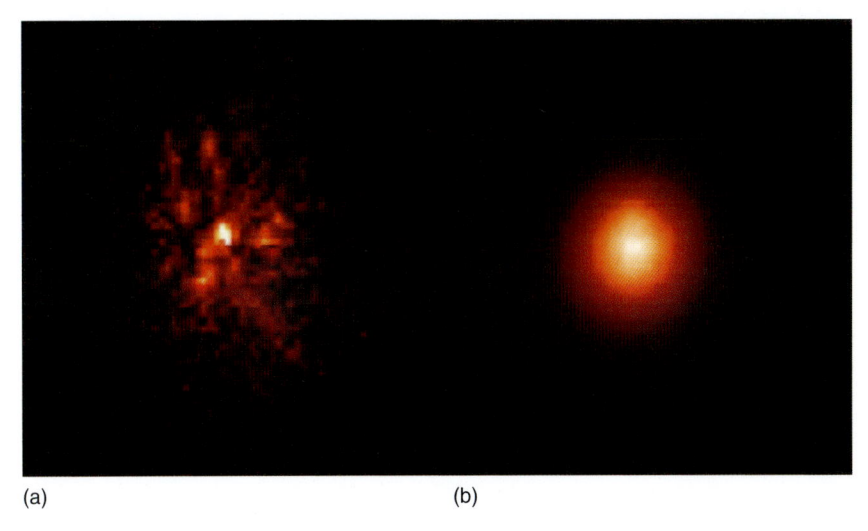

Figure 9.3 Turbulent Stellar Images. (a) Image of a point-like star taken through the Earth's atmosphere, with an exposure time short enough (a few ms at optical wavelengths) to freeze atmospheric turbulence. The irregular speckle pattern observed is due to light wavefront corrugations. Individual speckle size corresponds to the telescope diffraction limit, a few hundreds of an arcsecond for a large telescope at optical wavelengths. (b) Corresponding long time exposure image, that is, long enough to smooth out the dancing speckle pattern in (a), viz., a few seconds at optical wavelengths. Image size is at best a few tenths of an arcsecond across, even at the very best sites.

Earth's atmosphere. This atmospheric turbulence phenomenon creates a rapidly variable corrugated wavefront, represented by fluctuating light blobs (speckles) when imaged at the focal plane of the telescope. When integrated in time over a second or more, this gives the ubiquitous seeing effect: any point-like source image is heavily blurred in a more or less Gaussian disk, which represents this speckles pattern integrated over the observing time. This sequence is illustrated in Figure 9.3.

This is a major effect that strongly degrades the spatial resolution of virtually any astronomical observation: without it and for a perfect telescope, image sharpness would be limited only by the "finite size" of light, the so-called diffraction limit. For circular telescope of diameter D and light of wavelength λ, the image of a point source is an Airy function. The resulting blur diameter size is $1.22\lambda/D$: for an optically perfect 8-m telescope, which means a spatial resolution of around 0.016″ at a wavelength of 0.5 μm and 0.064″ at 2 μm. For ground-based optical-near IR observations, this is to be compared with median seeings around 0.6″–0.8″ at the best sites (median meaning that seeing is better than the median value half of the time, but worse the other half!). Also, the best seeings rarely reach the 0.25″ barrier, and then only for very brief (<1 min) integrations. This is in effect a huge increase in the etendue covered by an initial "point" source. To better appreciate its nefarious effect, note that image degradation impact goes actually as the square of seeing for any imager, and of course also for any bona fide 3D spectrograph.

9.2.2 Seeing Properties

On an empirical level, the local seeing S_0 is quantitatively defined as the Full Width at Half Maximum (FWHM) in arcseconds of the image of a point-like object at 500 nm obtained with a perfect large size telescope looking at the zenith. Except near the horizon, atmospheric turbulence follows the Kolmogorov model reasonably well, for which seeing is slightly wavelength dependent, with $S_\lambda \propto \lambda^{-1/5}$. Alternatively, seeing conditions can be defined, again at 500 m, by the Fried coherence length R_0: this is the diameter of a perfect telescope with an Airy disk of the same FWHM than the seeing disk. Physically speaking, this is the length scale over which turbulence becomes significant (quantitatively, for which the root mean square wavefront phase changes by 1 rad). These two parameters are directly related by R_0 (cm) $\sim 1/S_0$ ("). Typical median R_0 is ~ 15 cm at very good sites. One can visualize the telescope's primary mirror area A covered by N coherent zones of diameter R_0; the peak contrast of a stellar image at 500 nm, compared to the diffraction-limited one, is then divided by N. This is always a huge factor for a large telescope.

R_λ wavelength variation is much steeper than for S_λ, with $R_\lambda \propto \lambda^{6/5}$. For observations at an angle z from zenith, R_λ is multiplied by $\cos^{3/5} z$. This is an already significative degradation at 45° from zenith (a factor 0.86) and a strong one at 60° (a factor 0.66). When image quality is important – as it is for the vast majority of observations – it is thus strongly advisable to observe when objects are transiting near their highest sky location and much preferably choose only objects located much more than 30° from the horizon (getting, by the same token, better atmospheric transmission too).

As already stressed, only a few high-altitude sites offer a median night seeing in the optical domain substantially below 1". Seeing as good as 0.25" might happen, but during a few minutes only and on an exceptional basis. One (usually) major seeing component, the so-called ground seeing, is located in the first hundred meters or so above ground; the remainder is distributed over the upper atmosphere; a significant fraction is often located in a few high-altitude razor-thin (a few meters) inversion layers: they are responsible for the short-lived shaking occasionally felt in an ascending or descending passenger plane.

For daytime observations, with full Sun atmosphere heating, seeing is invariably much worse, 2–3" at best, the best period being the mornings, after the Sun is high enough above the horizon but before too much solar heating of the atmosphere has taken place. On the other hand, in the case of Earth's observations from satellites, image blurring (seeing) is smaller by an order of magnitude or so: this might look paradoxical since the light's journey down from the satellite to the Earth's surface is exactly the same as the one up from a ground-based telescope. The explanation is purely geometrical in nature: the nearly 10-km thick turbulent atmosphere lies far below (hundreds of km) the satellite, but just on top of the Earth surface being imaged, while the situation is opposite from a ground-based telescope looking up above the atmosphere. It is like taking a piece of ground glass to look at this page: if you put the glass in front of your eye, you cannot read this sentence; put it instead right on the page and you will have no problem.

9.2 Basic Seeing Principles

We see that in the seeing-limited case, unlike the diffraction-limited case, the spatial point-spread function (PSF) is independent of the telescope aperture (and only weekly wavelength dependent). However, for large aperture telescopes (say > 8 m diameter), the Kolmogorov turbulence model is not valid anymore and an additional parameter must be introduced: the finite outer scale L_0, which physically refers to the largest size of the turbulent eddies. At astronomical sites, typically $L_0 \approx 20$ m. The corresponding Van Karman turbulence model is used to estimate the PSF dependence. Compared to the Kolmogorov case (infinite L_0), it gives a smaller effective seeing. For example, with a small telescope FWHM seeing of 0.8" at 500 nm and $L_0 = 20$ m, the predicted FWHM seeing for an 8-m telescope is a significantly better 0.64".

Atmospheric turbulence also rapidly varies with time with the turbulence eddies both moving across the telescope and changing shapes. This turbulence time component is quantitatively expressed by its coherence time length τ_0, again at 500 nm. Wavelength (and zenithal distance) dependence is the same as for R_λ, with $\tau_\lambda \propto \lambda^{6/5}$. This very important parameter governs the response time required for getting improved images through the lucky imaging or adaptive optics techniques (see below). For a highly simplified model with the coherent time governed by frozen turbulence in a high altitude layer moving at the local wind velocity V, one gets roughly $\tau_0 \sim R_0/V$. $V = 180$ km h^{-1} and $R_0 = 15$ cm give $\tau_0 = 3$ ms. That is a very small value, typical of turbulence at that kind of short wavelength.

There are a number of techniques that aim at beating atmospheric turbulence to get much better image quality. The so-called lucky imaging [88], mostly done on intermediate size telescopes (2–5 m), breaks down a normal integration time of an hour or so in say a hundred thousand very short integrations (with a photon counting 2D detector), followed by a posteriori selection of typically the 10% best, and then co-added to finally get a vastly improved image. This is a rather powerful imaging approach especially in the optical range, but not well adapted to photon-starved 3D observations. The other main technique, adaptive optics correction, omnipresent on very large telescopes and especially upward of ~0.7 μm, "flattens" in real-time the turbulent wavefront and delivers it to any instrument, including of course 3D ones (see Section 9.4). For such techniques, both R_0 and τ_0 are the two key parameters that govern whether such vast improvements can be achieved and to what degree. Note that they can also be combined, again mainly for the optical domain.

There is actually a third key parameter, expressing how instantaneous seeing varies in the field of view. This effect is quantified by the coherence radius angle θ_0 again at 500 nm, with the same wavelength and zenith angle impact. It is a very important parameter as it governs the field that can be corrected either by lucky imaging or (first generation) adaptive optics. For a single dominant turbulence layer of height H above the telescope, one gets roughly $\theta_0 \sim (R_0/H)$. For $R_0 = 15$ cm and $H = 9$ km, this gives $\theta_0 \sim 3.3"$. That is a very small value, typical of turbulence at that kind of short wavelength.

In summary, at any time atmospheric turbulence is characterized by the three (evolving) parameters, R_0, τ_0, and θ_0 as per the insert: an R_0 diameter telescope looking at zenith would give a diffraction-limited image at 500 nm in an angular radius θ_0 field for an integration time τ_0.

ATMOSPHERIC TURBULENCE KEY PARAMETERS

D Telescope diameter; z zenithal distance; index 0 at 500 nm

Seeing FWHM angular size S_0 in "; $S_\lambda \propto \lambda^{-1/5}$
Diffraction-limited angular size $1.22\lambda/D$
Coherence length radius $R_0(cm) = 0.98/S_0$ ("); $R_\lambda \propto \lambda^{6/5}$

Coherence time τ_0; $\tau_\lambda \propto \lambda^{6/5}$
Coherence angle radius θ_0; $\theta_\lambda \propto \lambda^{6/5}$

$R, \tau, \theta \propto \cos^{3/5} z$

9.3 Seeing-Limited Observations

As seen above, even in the best seeing conditions, any moderate or large size telescope image resolution in the optical to near-infrared region is totally dominated by atmospheric turbulence. This impacts directly the spatial resolution of all 3D spectrographic flavors developed in this book. This is a major limitation, and getting minimal turbulence is arguably the number two requirement for site selection, number one being mostly cloud-free night conditions.

9.3.1 Seeing Impact on 3D Instruments

Modern telescopes use active optics techniques to achieve near perfect mirror optical shapes and alignment. A guide star image distortion is measured and integrated over ~30 s to average seeing effects. Commands are then computed and sent to hundreds of push/pull actuators located on the primary mirror and motorized motions of the secondary mirror in order to sharpen the images. In addition, the guiding star gives an almost instantaneous error signal used to get almost perfect guiding at time frequencies up to a few hertz. With these correcting systems, telescopes' intrinsic image quality is better than $0.1''$ and actual image quality is fully dominated by seeing.

As for any astronomical observation, a good knowledge of the spatial PSF is essential for the processing of the data. For natural turbulence only and under the Kolmogorov assumption, the PSF is the 2D Fourier transform of an $\exp^{-5/3}$ function, which is close to a Gaussian function. Residual telescope polishing errors add a faint $\sim 1/r^2$ halo. The resulting theoretical shape is also somewhat distorted by residual optomechanical errors such as imperfect focusing and instrument optical aberrations. It is thus necessary to derive the correct PSF from the data cube itself.

For direct imaging, the usual, and by far simpler, strategy is to use all stars present in the field as PSF proxies and go through a proper interpolation scheme to derive at least a rough evaluation of the PSFs all over the field. The same approach can be used with Fabry–Pérot type techniques; given the very small wavelength bin, the signal given by the stars is however much fainter than in imagery, and this approach is often ineffective. One saving strategy is then to use a better quality space-based or adaptive optics-corrected image of the same field with a known PSF taken in a similar (narrow) bandpass. A synthetic field image integrated over the same bandpass is then built from the scanning FP data cube, and the Fabry–Pérot spatial PSF is derived from the comparison between these two field images with different sharpness.

With their usually much smaller fields, integral-field spectrographs (IFS) PSFs can rarely be obtained from field stars, except when studying the close environment of a bright star. The most common strategy, as in the Fabry–Pérot case above, is then to use a better quality space-based or AO-corrected image of the same field with a known PSF taken with a filter, the bandpass of which is a subset of the IFS wavelength coverage. A synthetic field image integrated over the exact same bandpass is then built from the IFS data cube, and the spatial PSF is derived from the comparison between these two field images.

9.4 Adaptive Optics Corrected Observations

9.4.1 The Need for Overcoming Atmospheric Turbulence

As emphasized previously, image sharpness delivered by ground-based telescope is severely degraded by atmospheric seeing. The so-called adaptive optics (AO) techniques have been developed in the last 30 years or so, at first for defense purposes, then for astronomical and more recently medical imaging ones (in the latter case to allow better eye examination by correcting its optical imperfections). The approach is to overcome this vexing image quality limitation by *real-time* correction ("flattening") of the science target light wavefront delivered by the telescope to the instrument. AO correction's interest extends to any kind of observation, starting with direct imaging, but looks particularly promising when associated with 3D instruments, as they are able to make full use of the incredibly sharp 2D spatial quality delivered by this technique.

9.4.2 Adaptive Optics Correction Principle

In essence, an adaptive optics (AO) corrector is a real-time control system based on three basic components, as shown in Figure 9.4, namely, a wavefront sensor (WFS) that measures the wavefront deformation, a real-time computer (RTC) to compute the corresponding corrections, which are then sent to a deformable mirror (DM) in order to "flatten" the wavefront. A perfect control system would transform the corrugated wavefront delivered by a diameter D telescope in a perfectly flat wavefront that a perfect imager would convert in a diffraction-limited image, with an image core diameter $1.22\lambda/D$. To put in perspective, on an 8-m diameter telescope, this means an AO-assisted image core width of 63 mas (milli-arcsecond) at 2 µm and 25 mas at 0.8 µm, while the corresponding (median) seeing values at excellent astronomical sites lie around 500 mas at 2 µm and 600 mas at 0.8 µm. AO systems also correct the fixed

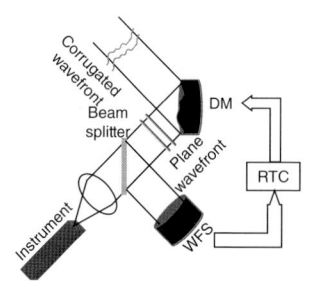

Figure 9.4 Principle of an adaptive optics control loop. The turbulent wavefront (top) falls on a deformable mirror (DM) and goes to a beam splitter. Light from the science object is sent to the instrument; light from a reference star in the field of view goes to the wavefront sensor (WFS). Signals from the WFS detector are processed by the real time computer (RTC), which sends appropriate voltages to the DM actuators to flatten the wavefront.

telescope wavefront errors (e.g., mirror's optical figuring errors) as well as slowly evolving ones (e.g., coming from differential mechanical flexures between the primary and secondary mirrors): this is not totally free however as this takes some of the highly limited deformation range of the DM actuators, leaving less for the much more demanding task of cancelling fast-evolving turbulence effects.

In practice, however, there are unavoidable residual correction errors that degrade the final image quality, as developed below.

Application of that technique with the first generation of astronomical AO systems, as schematically shown in Figure 9.5, is fairly straightforward. A small (~10–20″ diameter) field with both the science target and a fairly bright reference star is selected at the telescope focus, with the plane DM generally put on an image of the telescope pupil. The field is then sent to the science instrument, while the reference star is picked off – for example, with a dichroic beam splitter – and sent to the WFS. Wavefront distortion measures are sent to the RTC, the computed voltages of which are sent to a DM with typically hundreds of actuators in order to flatten the incoming wavefront. The AO-corrected field beam can input any kind of instrument, straight imagers of course, but also scanning interferometers, long-slit, multislit or integral-field spectrographs. Note that accurately flattening the wavefronts requires inter alia to get (i) a WFS with enough spatial resolution to derive finely the light wavefront 2D shape, (ii) a DM with enough actuators to closely antimimic the wavefront, and (iii) an overall loop (WFS measurements, followed by RTC computation, followed by DM deformations) fast enough to freeze atmospheric turbulence. These are heavy requirements that push even the simplest AO system on a large telescope in the multimillion USD/Euro range.

This basic AO system works in closed loop with the measuring device (the WFS) located *after* the correcting device (the DM) and thus looking at the previously corrected wavefront. This means that after the first few iterations, the wavefront sensor sees only small wavefront deviations that only need small differential corrections of the deformable mirror shape. This neatly minimizes the effect of any calibration errors of the three basic AO components (WFS, RTC, DM). Another plus of the closed-loop scheme is that any other source of wavefront aberration (e.g., telescope misalignment or imperfect mirror figuring) *before* the light is split between the science beam and the wavefront reference beam is equally corrected. Conversely, however, any differential aberration between the science beam and the wavefront sensor beam is *not* corrected: when, as is often

Figure 9.5 Single conjugate adaptive optics (SCAO). This is a highly schematic view of the simplest adaptive optics system. A single reference star in the observed field illuminates a single wavefront sensor. The real time computer sends commands to the single deformable mirror usually conjugated with the telescope pupil (hence the SC in SCAO), giving near diffraction-limited image correction in the immediate vicinity of the reference star. Typical correction parameters for an NIR AO system on a 4- to 8-m class telescope are response time a few milliseconds, number of corrected elements a few hundreds to a thousand. (Used Under Creative Commons License: https://creativecommons.org/licenses/by/4.0/.)

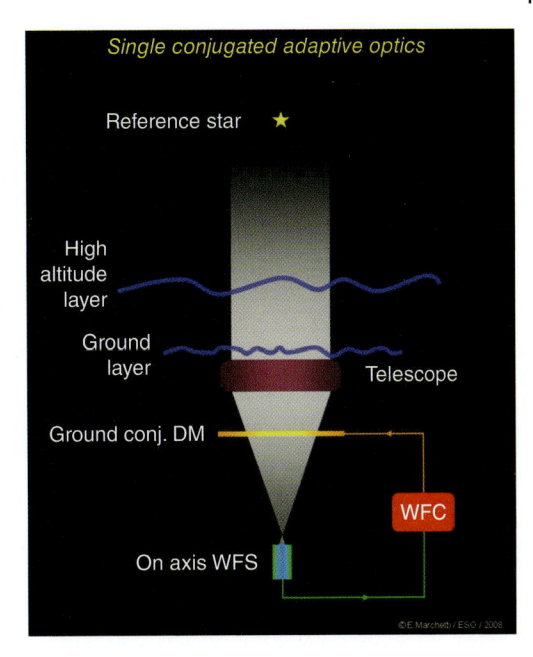

the case, the science instrument is sampling the images at the diffraction limit, that makes it hard to get good enough optical quality and small enough differential mechanical flexures between the science and the reference beams to fully exploit the image quality delivered by the AO system. It means that it is always better to put the WFS as close as possible to the science instrument, perhaps even in the instrument itself.

One good news with this basic AO scheme is that wavefront distortions in length units (say nm), due to atmospheric turbulence plus telescope imperfections, are achromatic, that is, independent of wavelength, longward of ~500 nm: that conveniently means that a given DM actuator deformation can correct the corresponding wavefront section at all wavelengths from the greenish optical to the mid-infrared domain. This is not true anymore below 500 nm, as atmospheric turbulence is then not only driven by temperature fluctuations (giving achromatic wavefront distortions) but also by water vapor fluctuations (giving wavelength-dependent wavefront distortions).

The bad news is that the inevitable remaining small wavefront distortions δW give dramatically different nefarious effects at different wavelengths, as image quality is driven not by residual wavefront distortions in length units but the corresponding phase errors in radians, $\delta\phi = 2\pi\delta W/\lambda$. The effect is that the image of a point source appears as a diffraction-limited core surrounded by a much fainter halo similar to the uncorrected seeing-limited image (see Figure 9.6). The correction quality is quantitatively expressed by its Strehl ratio S, which is the fraction of the total light that resides in the core (versus $1 - S$ in the halo). An approximate formula for phase errors smaller than 1 rad is $\ln S = -(\delta\phi)^2$, which varies dramatically with wavelength.

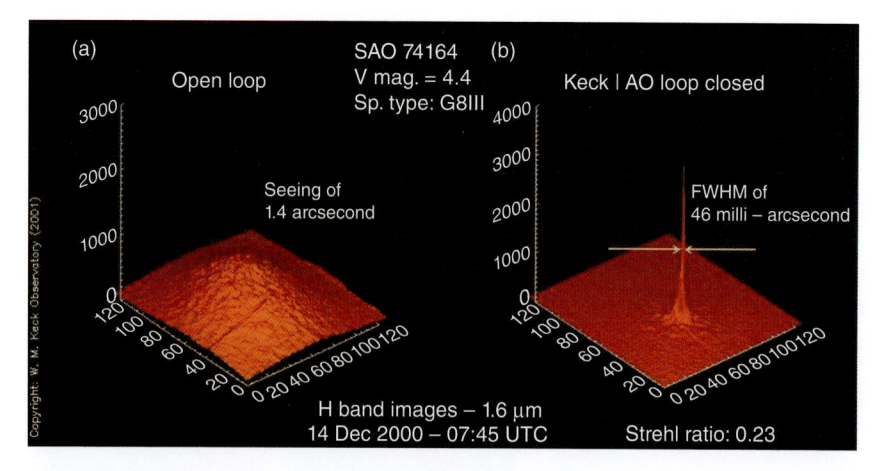

Figure 9.6 The figure shows 2D image profiles of a (point-like) star in the H-band given by the Keck1 10-m diameter telescope adaptive optics imager in 2001. (a) Natural seeing image, exhibiting a shallow light distribution with 1.4″ FWHM. (b) AO-corrected image with two basic components, a sharp peak (0.046″ FWHM) carrying 23% of the total star light, and a shallow halo, similar to the natural seeing PSF and carrying the remaining 77%. Note the spectacular contrast gain, ~200 between the two image peaks. (Courtesy W.M. Keck Observatory.)

9.4.3 Adaptive Optics Components

As indicated above, AO systems use three basic components, a deformable mirror (DM), a wavefront sensor (WFS), and a real-time computer (RTC). Considerable development has been made and is still going on on all fronts. This includes increasing DM actuator number and density, WFS detector pixel number, smaller read-out noise and faster read-out, as well as RTC computing speed. Here is a brief mid-2010 status:

- There are two kinds of deformable mirrors, "slope" mirrors that adapt their shape to cancel wavefront slopes and "curvature" mirrors that adapt to cancel wavefront curvatures. A "curvature" DM is made of two glued piezo-ceramic disks with opposite polarization. When a voltage is applied, one disk contracts and one expands, which causes a local curvature. Up to 500-actuator mirrors on a ~8 mm pitch have been fabricated. Three different technologies are used for "slope" mirrors: voice-coil actuator thin mirrors with up to 1177 actuators put on a 28 mm pitch, piezoelectric face sheet stacked arrays (Figure 9.7) with

Figure 9.7 Principle of the piezo deformable mirrors. A thin glass plate is bonded on a 2D array of piezoelectric actuators. High voltages (hundreds of volts) are applied to the actuators in order to deform the front glass plate up to a few microns, hundreds of times per second. (Reproduced with permission of Tokovinin.)

up to 1681 actuators on a 4 mm pitch, and electrostatic digital micromirror devices with up to 4092 actuators on a 0.4 mm pitch. Full stroke is ~50 µm for voice-coil mirrors, 10 µm with piezoelectric mirrors, 1 µm with electrostatic mirrors, which usually require a first AO stage to remove the large, high coherence radius wavefront distortions.

- Most wavefront sensors are of the Shack–Hartmann kind (Figure 9.8), well adapted to working in tandem with a "slope" mirror. A minilens array, illuminated by a reference star, is put on a pupil image. Each sub-pupil gives a star image on the detector. Their fast motions reflect the local wavefront slopes and are measured on the fly by a fast detector (usually a CCD/CID array) and sent to the RTC. A one millisecond delay can be achieved for even the highest present actuator number N, with high quantum efficiency and negligible noise (down to $1e^-$ RON at 1 kHz frame rate). For the curvature approach, its founder (F. Roddier) has developed a curvature WFS, based on comparing the illumination of two symmetrical in/out focus sections of the reference star beam. An array of avalanche photodiodes is generally used.
- RTC architectures use field programmable arrays and/or digital signal processors, and fulfill present needs. Computing power requirements scale however classically as the fourth power of the number of actuators (and of course as the inverse of the control loop time delay). They are also significantly higher for the advanced AO modes presented below. With the current slowing down of Moore's law, this would make it problematic to develop AO systems for the next generation of Extremely Large Telescopes, and even more when working in the optical region. Fortunately, smart real-time algorithms are being successfully developed with a less abrupt scaling law, speeding up the computations for all AO modes. Gains ranging from a factor of ten to a thousand, depending on the mode, have been reached (see, e.g., [89]).

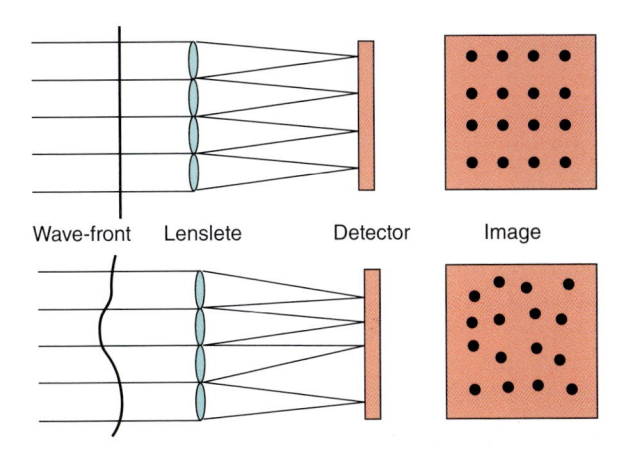

Figure 9.8 Principle of the 2-D Shack–Hartmann wavefront sensor. An image of the telescope pupil is paved by a 2D lenslet array, each giving an image of a reference star on the detector. *Top*: star images for a plane wavefront. *Bottom*: star images for a distorted wavefront. Locations of the multi-images of the reference star measured on the fly by the sensor reflect the instantaneous wavefront local slopes. (Reproduced with permission of Tokovinin.)

9.4.4 Adaptive Optics: The Optical Domain Curse

The crux of the problem, at say 500 nm wavelength, is that atmospheric turbulence (i) varies very quickly in time with an usually less than 1 ms timescale τ_0, and (ii) varies rather quickly in space, the coherent radius R_0 being rarely larger than 20 cm (see the turbulence parameters table in Section 9.2.2). For an 8-m class telescope, an AO system working down to 500 nm would thus require a deformable mirror with a very large number ($\sim$1600) of actuators and an ultra-fast AO control loop running well under 1 ms: working out the numbers shows that this would also require using one of the 6000 stars brighter than magnitude 6 (coincidentally visible to the naked eye) to feed the WFS with the required 10–20 photons per coherence area × coherence time for a good enough determination of the required AO corrections. Furthermore, for simple geometric reasons (see Figure 9.9), this AO correction would remain valid only in a very small coherent field θ_0 (also known as isoplanatic field) centered on the reference star and $\sim$2″ radius only, again at 500 nm. With these working parameters, total accessible sky area would be $6 \times 10^3 \times 4\pi$ square arcsecond only, or a fraction 1.4×10^{-7} of the total celestial sphere – in other words, a mammoth AO system with vanishing sky coverage. For a much smaller telescope, the AO system has less actuators in proportion to the telescope areas, but the poor sky coverage is exactly the same.

As of the mid 2010s, optical AO, even in its reddish part around 800 nm, is thus pretty limited to studies of the environment of very bright closeby stars, including searches for exoplanets (see Section 5.2), the parent star being then the obvious and perfectly located reference star. The situation improves dramatically in the near-IR, with τ, R, and $\theta \propto \lambda^{6/5}$. This gives a factor 6 improvement on each of these critical seeing parameters from 500 to 2200 nm. For example,

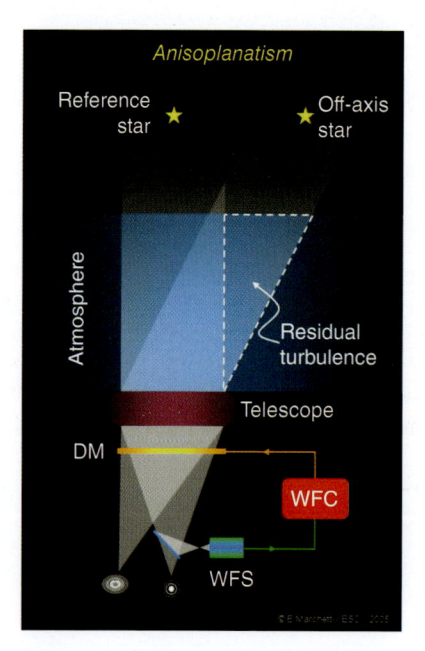

Figure 9.9 AO Isoplanatic field limitation for a single natural reference star. The figure schematically shows the negative impact of the natural reference star offset angle from the science target, as the wavefront sensed by the reference star beam is increasingly different from the science target ones as the offset angle increases, especially for the high altitude turbulence layers. (Used Under Creative Commons License: https://creativecommons.org/licenses/by/4.0/.)

on an 8-m class telescope, a 500 actuators AO system working with a few millisecond control loop will give a well-corrected image in a field 10–20″ diameter centered on a ≤ 13 magnitude star at 2200 nm on a good night, down to 1200 nm on an exceptional one. This is much better than at short wavelengths, but with still a small sky coverage, a few % of the celestial sphere only. On top, most of the accessible spots lie close to the plane of our galaxy, a good location to look at stellar formation, but definitely bad to observe far away galaxies. An exceptionally good AO night features a large coherent radius (R_0) and even more importantly a large coherence time (τ_0). In physical terms, this means nights with small overall turbulence integrated over all atmospheric inversion layers and located in the good season for which the jet stream, an extremely fast wind at the 12-km altitude level, is not present above the observing site.

In conclusion, the first generation of AO systems, even in the near-IR, suffered from extremely poor sky coverage (for lack of bright-enough reference stars) and small corrected fields (due to only on-axis wavefront correction). In practice, this severely limited the capability of adaptive optics systems to address cutting-edge astrophysical questions. This led in the last 20 years to the emergence of a small zoo of more sophisticated AO systems that aim to remove, or at least mitigate, most of these shortcomings.

9.4.5 Addressing the Lack of Reference Stars

The much too small sky density of bright-enough reference stars can be directly circumvented by creating an artificial star very close to wherever any science object lies (Figure 9.10) as first proposed in 1985 in the open literature[1] by Foy and Labeyrie [90]. In practice, this is done by shooting laser light collimated through a small – typically 50 cm diameter – telescope mounted and co-aligned with the big telescope. With the laser wavelength precisely tuned on the 589 nm sodium D_2 resonance line, naturally occurring sodium atoms trapped in a thin layer located 95 km above the Earth surface are excited and they send back light at the same wavelength in the form of a ~1.5″ diameter sodium reference "star." Unfortunately, 95 km (a little less actually since large telescopes are never located at sea level) is a far cry from infinity for an 8-m diameter telescope. This so-called cone effect, sharply limits the AO correction performance as shown in Figure 9.11.

The rather heavy solution to this conundrum is to deploy a small – ~20″ diameter – constellation of four to five lasers to sense the full atmosphere cylinder in the direction of the science targets with as many returning laser cones (Figure 9.12). With an optimum laser line shape, a 20 W continuous wave laser usually[2] gives bright enough reference sources for proper wavefront sensing under fair turbulence conditions even down to the reddish optical range, with say 10% Strehl at 650 nm. The laser guide star (LGS) returning beams are sent to as many wavefront sensors: this gives enough information for a real-time tomographic reconstruction of the wavefront at the center of the field (Figure 9.13).

---

1 but was already developed by the USAF for military purposes though.
2 Usually, because the 95-km layer is irregularly replenished by incoming meteoroid showers and its sodium atoms content could be occasionally depleted by a factor of 10 or so.

Figure 9.10 Upper atmosphere illumination by the ALFA Laser Guide Star system at the GEMINI-S 8-m telescope. The small (~1.5″ diameter) pseudo-star at the center of the figure comes from excited sodium atoms in the ~95 km high atmospheric layer above the Earth. The extended parasitic light at the bottom right comes from Rayleigh scattering of the upcoming laser beam in the atmosphere above the telescope, up to about 15 km. (Reproduced with permission of Gemini Observatory.)

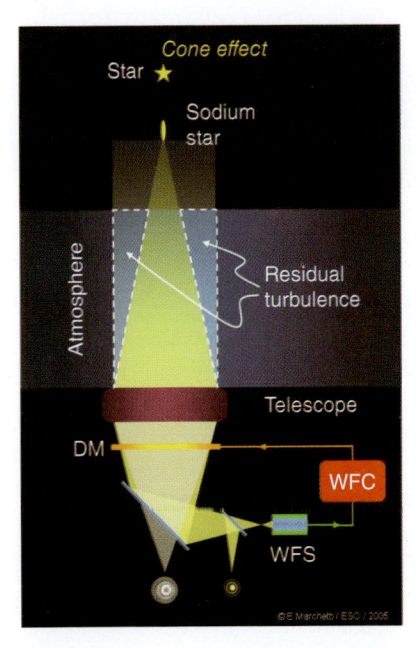

Figure 9.11 AO isoplanatic field limitation with a laser guide star. The Figure schematically shows the poor matching of the light cylinder from the science target (blue) by the light cone coming from a single laser guide "star" (yellow), except close to the ground. (Used Under Creative Commons License: https://creativecommons.org/licenses/by/4.0/.)

Figure 9.12 Multiple laser guide stars. The first ever laser guide star "constellation" launched in January 2011 at Gemini South. The image shows the 50-W laser beam as it shines upward toward the 95-km-high atmospheric sodium layer to create a pattern of five artificial guide stars (upper left) used to sample atmospheric turbulence for the Gemini Observatory GeMS adaptive optics system. The yellow-orange beam visible from the lower right to the upper left is caused by scattering of the laser's light by the Earth's lower atmosphere. (Used Under Creative Commons License: https://creativecommons.org/licenses/by/4.0/.)

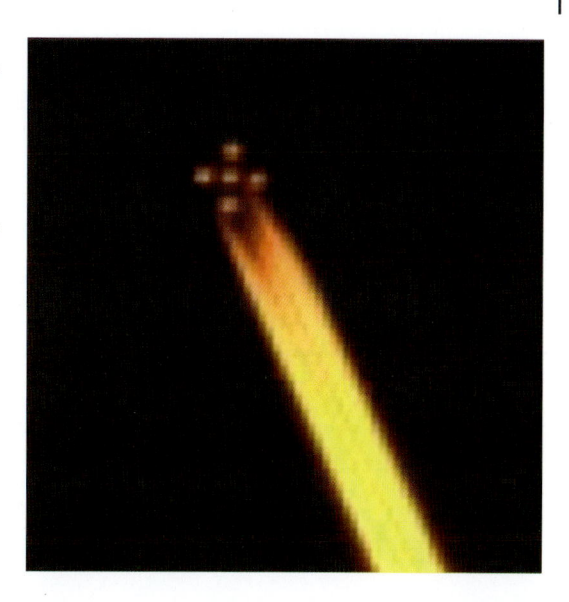

This approach has been accordingly christened laser tomography adaptive optics (LTAO). One complication is that star jitter – mainly due to atmospheric turbulence for small telescopes, to telescope vibrations for large ones – is not sensed at all by the LGS[3]: One thus needs a, fortunately faint (magnitude 17–18), natural guide star in the, again fortunately, large ($\sim 1'$ diameter), tip/tilt isoplanatic field to correct this effect, and preferably to provide accurate focusing at infinity. A $\sim 50\%$ sky coverage can be reached during good seeing conditions.

There are a number of practical difficulties to achieving routine operation of multilaser assisted AO systems, in particular (i) Mother Nature apparently dislikes laser operation at 589 nm wavelength and compact efficient reliable turnkey lasers [91] only arrived on the market[4] in the mid 2010s and are still expensive (hundreds of thousand Euros apiece); (ii) even thin upper atmosphere cirrus clouds might be enough to wash out the laser light too much for proper operation, especially at inauspicious times of low sodium upper atmosphere content; (iii) return light from each laser includes retro-diffused light from roughly the first 15 km above the telescope, which must be shielded especially when firing multilasers, and (iv) for safety reasons, it is mandatory to shut down the lasers when any aircraft (including eventual drug smugglers) is flying over the telescope. By the way, the latter problem is even worse for US-related facilities, which are in addition subject to prior approval of each planned laser pointing by the US Air Force Space Command.

Another more mundane difficulty is that laser "stars," unlike the real ones, are not located at infinity, only at a mere 92 km or so up from the telescope. At the telescope focal plane, this translates to a focus farther along the telescope axis

---

3 The atmospheric turbulence tilts responsible for star jitter bend indeed the laser beam on its way up, but work almost exactly in reverse when the sodium light comes back less than 1 ms later.
4 See eso1613 release on the ESO web pages for the first light of four such lasers at the VLT Observatory.

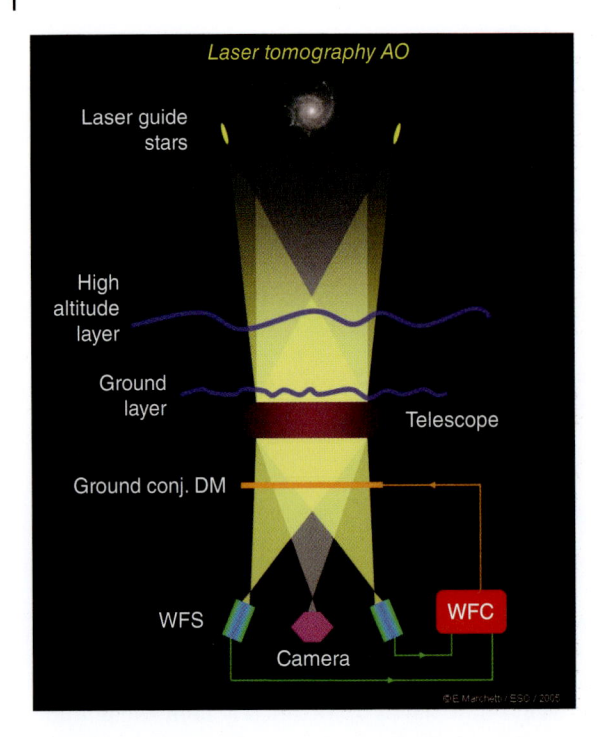

Figure 9.13 Laser tomography. The picture schematically shows the turbulence layers affecting the science object covered by two laser cones. In practice, fair corrections can be achieved with three to four 20 W, continuous lasers. (*Source:* http://www.eso.org/sci/meetings/2015/EriceSchool2015/Erice_Marchetti_1.pdf Used Under Creative Commons License: https://creativecommons.org/licenses/by/4.0/.)

than for natural stars. As the telescope tracks the science object, this defocusing varies with the elevation of the telescope, and optical trombones (variable path optical relays) are used to keep focusing the four to five LGS beams to their respective wavefront sensors. This is not trivial to achieve since much care is needed to avoid introducing differential aberrations between the science beam and the wavefront sensor beams.

9.4.6 Addressing the Small Field Limitation

The small corrected field limitation can be overcome by using more than one deformable mirror (DM), in practice two or three, each preferably located on the conjugate of a major turbulence layer (see Figure 9.14). Typical values for a telescope say located at 2.5 km altitude could be to put one DM conjugated to the ground (as in the classical single conjugate case), one to about 4 km above, and the last one to the 10 km level. Needless to say, this makes for definitely more complicated optical relays, wavefront systems, and real control algorithms able to disentangle the various layers contributions in order to send the proper commands to each corresponding DM.

This so-called multiconjugate adaptive optics (MCAO) technique originated from Rigaut and Ellerbroek circa 2000 [92] from Beckers [93] 1988 proposal. In the near IR region, it can cover a few (1–2) arcminutes diameter field with almost uniform ($\leq 10\%$) AO correction over the field. Note that the much simpler first generation AO technique described above is now retrospectively called single conjugate adaptive optics (SCAO). The technique was first put on the sky for day-time observations of the Sun. For night-time astronomy, the ESO MAD

Figure 9.14 Multiconjugate adaptive optics (MCAO). This is a highly schematic view of the MCAO principle. Multiple wavefront sensors, each illuminated by a natural or artificial star (hence the star-oriented label), are used here to sense separately the two main turbulence layers, at least one close to the ground and one 4–5 km above the telescope. Commands are sent to two deformable mirrors optically conjugated to these two layers, achieving diffraction-limited correction over a few arcminutes field. (Used Under Creative Commons License: https://creativecommons.org/licenses/by/4.0/.)

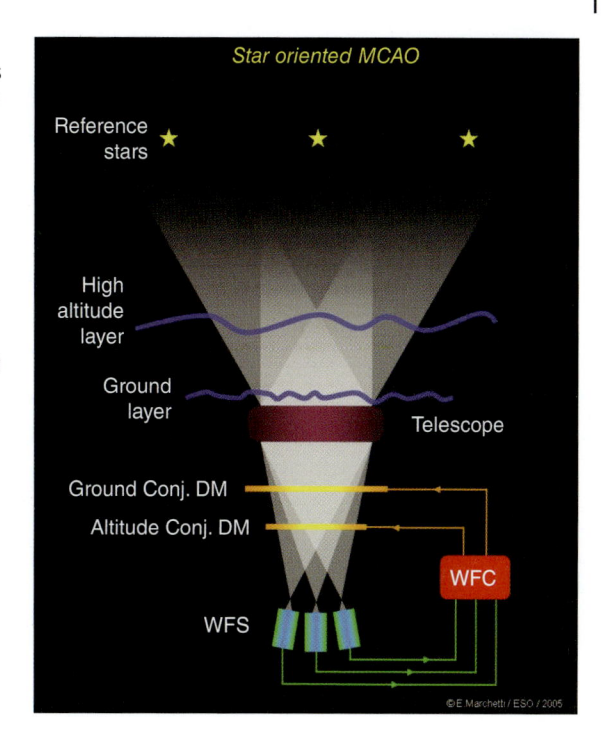

demonstrator was operated in 2007–2008 on the very few accessible science objects closely surrounded by two bright reference stars; the first fully operating GeMS MCAO system, complete with a laser guide star constellation started in 2011 at Gemini South [94].

9.4.7 Large Field Partial AO Correction

As noted above, during single conjugate adaptive optics (SCAO) observations, the quasi-Gaussian natural seeing stellar images normally shrink to a 5–10 times smaller coherent core surrounded by a still turbulence-size halo. Yet, occasionally, during such observations, the stellar images retain instead a quasi-Gaussian shape, but narrower by a factor of ~2 in area (or 1.4 in size). It took some time to realize that this happens when high-altitude turbulence is too fast for the SCAO system WFS to sense it: ground-layer turbulence being then typically 10–30 times slower (tenths of ms coherence time), it is the only atmospheric perturbation sensed by the WFS and then corrected by the DM. The resulting stellar images are still seeing-limited, hence roughly Gaussian, but with the ground-layer component erased. To achieve this feat consistently, multiple wavefront sensors are used to measure in real-time the ground-layer component only and send the appropriate commands to a single DM optically conjugated to the ground (Figure 9.15). This approach was first proposed by Rigaut [95].

The big plus of that approach is that this partial correction is valid over a substantial field, from around 1' across in the reddish optical to 6' across in the near-IR region. Since this is the ground-layer that generally features the largest Fried parameter R_0 and timescale τ_0, and always the largest aplanatic field

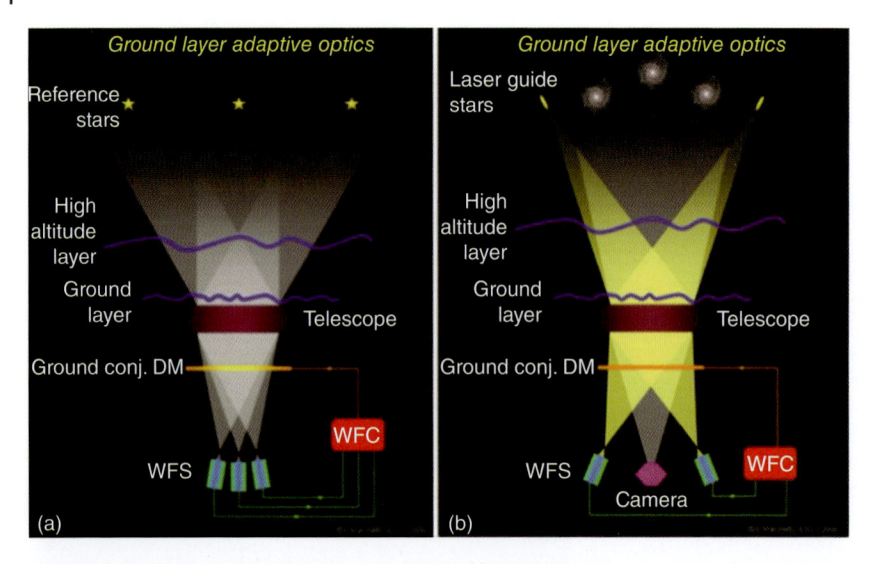

Figure 9.15 Ground-layer adaptive optics (GLAO). This is a highly schematic view of the GLAO principle. Multiple wavefront sensors, each illuminated by a natural (a) or artificial (b) star, are used to sense the turbulence layer close to the ground. Commands are sent to a single deformable mirror optically conjugated to the ground, thus erasing the ground layer contribution over a few arcminutes field. Courtesy E. Marchetti, ESO. (©The European Southern Observatory.)

(being by definition the closest to the telescope), it is anticipated that it could be possible to operate routinely with near complete sky coverage in the NIR and even, during the best AO periods, venture well into the optical domain. This so-called ground-layer adaptive optics (GLAO) technique can then be seen as essentially trading an actual telescope site for another one about twice better in terms of light flux concentration. To get near 100% sky coverage, one must still use multilaser guide stars. Sodium lasers can of course be used. Alternatively, since there is essentially no cone effect, cheaper pulsed Rayleigh lasers are also used instead, for example at the Large Binocular Telescope (LBT): with proper time gating, retro-scattered light from the atmospheric layers is used to derive the wavefront distortions.

9.4.8 AO-Based Scanning Interferometers

Some near-IR AO imagers feature an additional scanning Fabry–Pérot mode as a relatively easy way to study ionized gas regions in the light of a strong emission line at a spectral resolution of a few thousands (see, e.g., [96]). As AO-corrected images can vary dramatically over a time scale of a few minutes only, it is essential to achieve full scanning in such a short time and then to build the final 3D data cube over many iterations, in order to smooth out this deleterious effect. This could be easily reached using a photon-counting detector; it is much more difficult with the more classical CCD or IR array. In practice, this limits AO-based scanning interferometry to niche utilization on very bright regions.

9.4.9 AO-Based Slit Spectrographs

A number of near-IR AO imagers also feature an additional long-slit spectrographic mode as a relatively easy and quick way to probe interesting objects detected by the main imaging mode. Classical Nyquist sampling of the diffraction-limited images to retain the full image quality results in very narrow slits, well below 0.1″ on 8- to 10- m telescopes. The resulting small etendue makes it easy to reach moderate spectral resolutions (3000–5000) with relatively small efficient gratings; on the other hand, for an extended object including even highly distant galaxies, a large fraction of the targets light falls outside of the narrow slit and is wasted away. Besides, it is almost impossible to retrieve the spatial point spread function of the observation, including knowing precisely where the slit is situated on the object; this limits the physical value of the data to a few basic parameters.

The TMT 30-m diameter telescope project plans to go much further in this direction with a multislit version, the IRMS MCAO-fed cryogenic NIR instrument project [97]. With its J–H–K wavelength coverage and a 2′ diameter patrol field, its main science case is the identification of proto-galaxies in the early Universe from their redshifted ($z \geq 7$) Ly_α emission. As for the non-AO versions, this multiobject instrument with 46 slitlets will essentially be a survey facility. The main role of AO here is to increase by two orders of magnitude the contrast of the objects with respect to the bright night-sky emissions. Unlike 3D systems, and owing to its intrinsic 1-D spatial coverage and poor knowledge of the spatial PSF, it is not directed toward deriving accurate physical data on the targets, but to a rough characterization of ultra-faint objects. Given the very narrow width of the slitlets, this instrument requires an atmospheric dispersion compensator to accurately cancel the prism-like effect of the earth's atmosphere away from the zenith (see Section 9.5.2).

9.4.10 AO-Based Integral Field Spectrographs

With its 2D spatial coverage, integral field spectroscopy looks ideally suited to fully exploit the superb image quality delivered by AO systems. There is no need for multiple short observations to smooth out the observing conditions, object spatial information is fully registered, and limited if any atmospheric dispersion compensation is required as it just shifts the reconstituted images on the detector and can be corrected a posteriori in the data reduction phase. As illustrated in Section 4.4.1, there are a number of IFSs in operation on large telescope facilities that feature such a mode. They are mostly operating in the near-infrared domain, and because of their small field of view (see next paragraph) they only need "simple" single conjugate AO systems. Multilasers are however mandatory as most potential targets do not benefit from adequate reference stars. An example of the scientific potential of this approach is given in Figure 9.16 [130].

There are quite a number of provisos though with AO-assisted integral field spectrography (IFS). Firstly, as for any kind of IFS, there is a basic tradeoff between field pixels and wavelength pixels, which leads to a relatively small

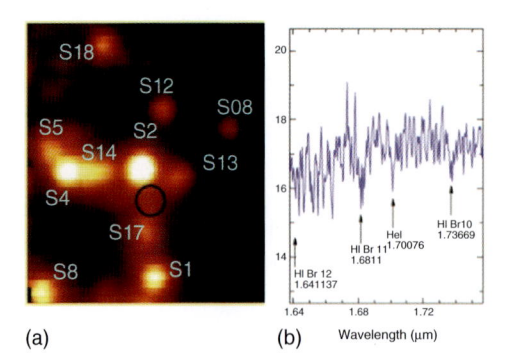

Figure 9.16 Sinfoni Galactic Center data. (a) Reconstructed K-band image of a small field around the center of our galaxy (open circle) with a spatial resolution of 0.075″. (b) Typical H-band spectrum of these central stars. Their hydrogen and helium absorption lines are used to get stellar radial velocities and physical parameters. (Reproduced with permission of Frank Eisenhauer.)

number of spatial pixels (spaxels). In the AO-assisted case, the use of Nyquist sampling (2×2 pixels) of the PSF diffraction-limited core inexorably leads to an extremely small field. For example, an 8-m mounted IFS with 10,000 spaxels, diffraction-limited, sampled at 2 µm, has a total field of $3″ \times 3″$ only. For any given instrument, the actual tradeoff much depends on its science drivers. For instance, an IFS dedicated to the study of compact star clusters or galaxy nuclei will always sample the field at the diffraction limit, in order to get the absolute best possible spatial resolution. Conversely, an IFS dedicated to the study of distant galaxies will generally use a coarser spatial sampling (say by a factor of 2–3), enlarging the field, for example, to $9″ \times 9″$, but most importantly concentrating the light enough on the detector pixels to reach ultra-faint targets.

An interesting case study is that of the MUSE wide-field *optical* IFS. In its main mode, it covers an $1′ \times 1′$ field with a rather small $0.20″ \times 0.21″$ pixel sampling. This is perfect for the absolute best atmospheric conditions at its Paranal location (∼0.35″ optical seeing), but an overkill for most of the time (0.66″ median optical seeing). To improve the spatial resolution of *inter alia* most IFS observations, the UT4 ESO 8-m Telescope is being converted by 2017 into an "Adaptive Optics Telescope" with a deformable 1.1 m diameter secondary mirror and a constellation of four sodium laser guide stars. With the addition of four wavefront sensors and an IR tip/tilt sensor in the MUSE instrument itself, the expectation is to achieve routinely by the end of the decade an improvement by a factor of ∼2 in seeing area through this GLAO approach.

Besides this GLAO-assisted main mode, MUSE also features an alternative "laser tomography AO" mode (LTAO) with near diffraction-limited sampling at 0.8 µm ($0.025″ \times 0.026″$ pixels), leading of course to a much smaller $7.4″ \times 7.4″$ overall field. Again, using the four telescope laser guide stars and the five MUSE wavefront sensors, LTAO correction is expected to provide fair diffraction-limited correction in the reddish part of the MUSE wavelength range, with $\geq 23\%$ Strehl above 750 nm in equal or better than median seeing conditions. This mode is expected to be in full operation in late 2017, and the jury is still out on the full scientific potential of this approach.

9.4.11 AO-Based Near-IR Multiobject Integral Field Spectrographs

Near-IR multiobject IFS are presently essentially galaxy crunchers devoted to the study of moderately distant galaxies in large (typically 6′ across) fields with a multiplex in the 20–30 range. They could easily – at least at the conceptual level– be fed with a typically 2′ diameter MCAO-corrected field, getting much better physical data on the science targets from the improved spatial resolution. This is a sound solution, albeit in search of a problem, since there are rarely enough bright-enough targets in such a small field, except in the rare distant clusters of galaxies.

A potential solution (Figure 9.17) has been proposed in 2004 by Hammer [98], albeit calling for a radical departure from any previous AO system: the concept is to work in a quasi-open loop rather than the fully closed loop shared by all present AO modes. Specifically, guiding and science probes are deployed in a large (6′–8′) patrol field. Each guide probe, with a ~2″ diameter field, gets its own DM and WFS, and is positioned on a natural or a laser guide star. Each science probe features a DM also, but no WFS. The RTC runs the n closed loops for the n guide probes and uses their real-time DM deformations to drive the science probes DM in open loop. The corrected science beam is then sent to the integral field spectrograph.

This so-called MOAO approach, from multiobject adaptive optics, imperatively calls for deformable mirrors for the science probes that can be accurately

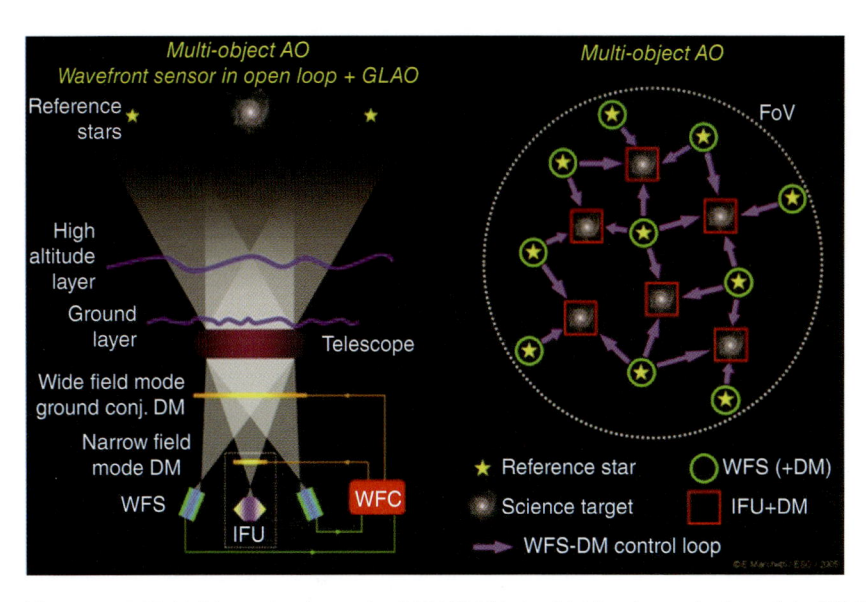

Figure 9.17 Multiobject adaptive optics (MOAO). This is a highly schematic view of the MOAO principle. The left image shows the correction principle for one science object (covered by an Integral Field Unit) flanked by two wavefront sensors. The right image shows the field of view coverage with open-loop deformable mirrors on the science objects and closed-loop deformable mirrors + wavefront sensors on reference stars. (Used Under Creative Commons License: https://creativecommons.org/licenses/by/4.0/.)

deformed in a known and repetitive way at a few per cent accuracy level. This is definitely not the case for the classical piezoelectric-driven DM that in particular exhibits hysteresis effects at the 15% level. This is no problem in ordinary closed-loop AO, but would be a disaster if used for the MOAO mode. Electrostatic-driven DM are used instead, and their stroke limitation is solved by using a "woofer" DM used in the GLAO mode. A feasibility pilot system has successfully proved the concept in 2013 in the Canaries, and Raven, the first MOAO demonstrator, had first light at Subaru in May 2014 (see Raven on Subaru Telescope web pages).

9.4.12 Deriving AO-Corrected Point-Spread Functions

As for seeing-limited 3D observations, a good knowledge of the spatial point spread function is essential for the scientific analysis of the observing data, that is, what the target observation is telling us in terms of dynamical/physical/chemical properties. This is unfortunately much more difficult to achieve for AO-assisted observations than for seeing-limited ones for three main reasons: (i) significant PSF variations in the field, if less so with the MCAO and GLAO flavors; (ii) strong PSF variations with wavelength irrespective of the AO flavor chosen; and (iii) the small fields of AO-assisted instruments that strongly limit the availability of "PSF" stars inside the observed field. In addition, a typical AO-enhanced PSF shape is intrinsically hard to measure: it consists of a narrow diffraction peak (Airy function) convolved with the instrument optical aberrations, plus an extended faint halo with close to a natural seeing Gaussian-like shape (Figure 9.6). Determining the shapes of these two components is usually not too difficult, but getting a precise value of their relative integrated flux contribution (the Strehl ratio S) is intrinsically hard. Given these difficulties, a relatively meager $\pm 10\%$ precision on the PSF shapes is considered as already superb.

9.4.13 Conclusion

Depending of the particular adaptive optics flavor used, the spatial resolution of AO-assisted 3D spectrography can be much improved, from a factor of 2 with ground-layer-corrected AO up to a factor of 100 with other AO flavors, leading to a much sharper view of small spots in the Universe. There are however a number of provisos, notably (i) the AO corrector part with, in particular, the deployment of multilasers is an expensive and complicated proposition, easily in the 10^+ million USD/Euro range; (ii) good AO corrections are at present mostly limited to the IR domain, and that only during good to superb seeing nights; and (iii) to effectively use the superb spatial resolution offered by AO correctors, the science targets need to be bright enough at the diffraction-limit level to give a detectable spectroscopic signal on the science detector, which is certainly not guaranteed when observing, for example, distant galaxies. In the same vein, suppose you have shot a soccer game with a high definition video camera, but with only a few photons per frame over the whole field: you will never be able to see the ball! Finally, determination of the spatially and even more spectrally variable spatial point spread function is usually achieved with limited precision. This directly impacts overall science output.

9.4.14 For Further Reading

To read more about the fascinating, if highly complex, adaptive optics field, and the many actors involved, there are a number of good reference books and reviews, the following in particular:

- Davies, R. and Kasper, M. (2012) Adaptive optics for astronomy., Annu. Rev. Astron. Astrophys., 50, 305–351.
- Roddier, F. (ed.) (2004) Adaptive Optics in Astronomy, Cambridge University Press.
- Tyson, R.K. (2000) Introduction to Adaptive Optics, SPIE Press, Bellingham, WA.

There are also at any time some good tutorials on the World Wide Web.

9.5 Other Atmosphere Impacts

Besides turbulence-induced blurred images, Earth's atmosphere offers a number of other nefarious effects discussed in this section. This includes light absorption over the electromagnetic spectrum, atmospheric refraction/dispersion, and night sky emission.

9.5.1 Atmospheric Extinction

As shown by Figure 9.1 in the introduction, Earth's atmosphere is transparent only in a limited number of windows of the whole electromagnetic spectrum. When observing conditions are optimal, the so-called photometric night (i.e., clear sky and limited water absorption), light extinction is a characteristic of the site and the elevation (h) at the time of the observation. The transmission $T(\lambda)$ at a zenithal distance (z) is related to the extinction values at zenith (E) by the following formulae:

$$\rho(\lambda, z) = 10^{-0.4E(\lambda)A_m(z)} \tag{9.1}$$

where $A_m = \sec(z)$ is the airmass, that is, the optical path length through Earth's atmosphere. An example of the extinction values at Paranal in the optical range is given in Figure 9.18.

9.5.2 Atmospheric Refraction

Light entering the atmosphere at elevation h (varying from 0° at the horizon to 90° at zenith) is refracted vertically as it passes through air layers of increasing density, and hence increasing index of refraction. At sea level and in the optical domain, this upward deviation goes from 0' at zenith to about 1' at 45°, 5' at 5°, up to 35' at the horizon. At the current altitudes of large telescopes – roughly in the 2400–4200 m range – these values are smaller by factors ranging from 1.3 to 1.8. Quantitatively, refraction ρ in arcminutes is given by the following empirical formula:

$$\rho = \frac{P}{101} \frac{283}{273 + T} \cot \frac{h + 7.31}{h + 4.4} \tag{9.2}$$

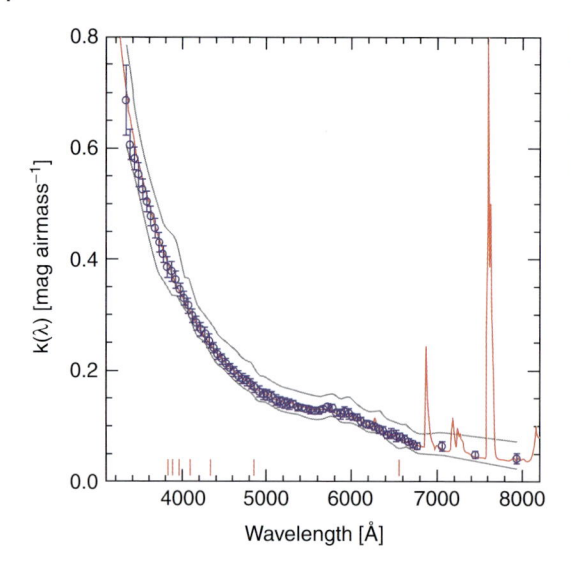

Figure 9.18 Measured Paranal extinction curve in the visible range [99]. (Used Under Creative Commons License: https://creativecommons.org/licenses/by/4.0/.)

Here, h is the apparent (i.e., after refraction) elevation in degrees, P is the ground pressure in kP, and T the temperature in $°C$.

This absolute deviation does not directly affect the performance of astronomical instruments. However, its differential values over the field can impact wide-field instruments. This impact is mild for imagers, scanning interferometers, and even less for integral field spectrographs because of their small fields: one needs only to correct a posteriori this slight distortion to get accurate sky coordinates over the field. The impact is stronger for multiobject instruments, as this distortion must be taken into account before observing, when positioning the slitlets or the fibers on the astronomical targets. In addition, for long observations, its differential value over time can be large enough to push some of the targets at significant offsets from the slitlets/fibers during integration. In the fiber case, this can sometimes be solved by slightly moving the fibers over time. Most often, it is simply mitigated by observing near the meridian, that is, when the field is close to its highest elevation in the course of the night.

As for any transmitting medium other than vacuum, air index of refraction n is wavelength dependent, as seen in Figure 9.19. This gives atmospheric dispersion, that is, round stellar images are transformed in elongated spectra in the vertical direction. In the optical domain, air optical constringence is about 89; hence, the angular length of 480–656 nm spectra is given by $\rho/89$, where ρ is the air refraction in arcminutes. For a 2400 m-high site and observations at 45° elevation, the resulting spectral length in arcsec is $60/(1.3 \times 89)$ or $0.48''$. This is not totally negligible for seeing-limited work, and is actually a large value when conducting adaptive optics-corrected observations. Air dispersion is much lower in the NIR and actually quite negligible for seeing-limited observations; this is not true however for AO-corrected observations, especially with large telescopes (and even more for extremely large ones).

Again, this is no problem for scanning interferometers or integral field spectrographs (except if working at the diffraction limit), as it can easily be corrected

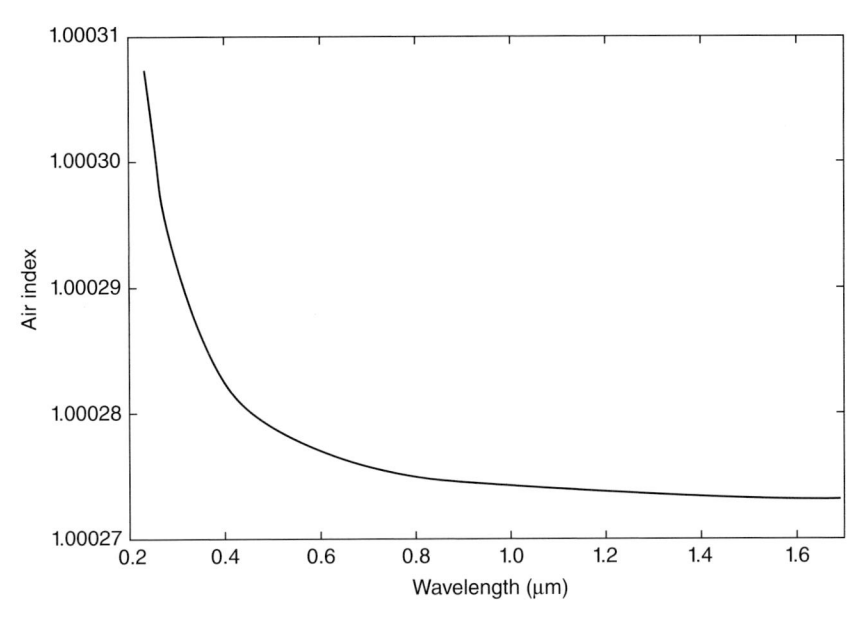

Figure 9.19 Air index of refraction as a function of wavelength for normalized conditions (sea level, temperature 20 °C).

a posteriori. It is, on the other hand, significant for multiobject instruments. A mitigating solution for the multislit flavor is to align all slitlets vertically. The full solution for both multiobject flavors – and for imagers as well – is to cancel atmospheric dispersion. This is usually done by putting two remotely controlled rotating prisms in the convergent beam before the instrument focal plane; their independent rotations give the two degrees of freedom necessary to apply the right correction amount along the right direction.

9.5.3 Night Sky Emission

The sky above the telescope is far from perfectly black even during the so-called astronomical nights (moonless nights with the Sun more than 18° below the horizon) and at the few sites with negligible light pollution from human activities. For the optical and nonthermal IR domains (below ~1.6 μm), atmosphere thermal emission is negligible and most of the light originates from chemical reactions in the upper atmosphere, roughly from the 80 to 300 km high layer. The major contributors are the oxygen and water vapor molecules, the sodium atom, and especially the OH radical. They are responsible for many narrow emission lines (<0.01 nm width). As can be seen in Figure 9.20 (optical) and Figure 9.21 (NIR), night sky emission is quite small at short wavelengths (300–600 nm) and progressively becomes huge for longer wavelengths. Note that these emissions can be readily canceled by observing from space.

The other significant contribution comes from scattering of solar light by thick dust disk located along the zodiacal plane (where solar system planets are located). Zodiacal light is maximal when looking in the zodiacal plane, but

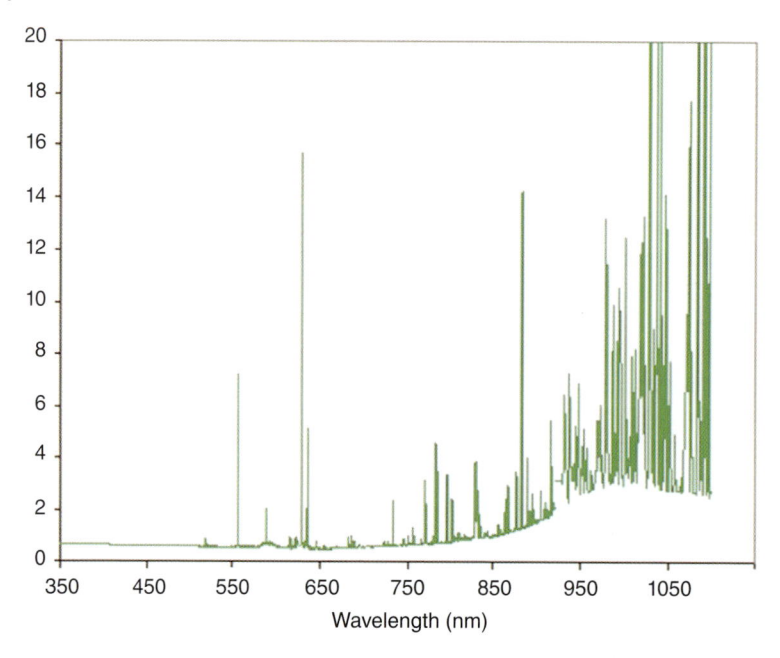

Figure 9.20 Optical night sky brightness at the Mauna Kea site in units of ph s^{-1} arcsec^{-2} m^{-2} nm^{-1}. (Reproduced with permission of Gemini Observatory.)

still present at its poles, its mean contribution at the ground being about 27% of the global sky brightness in the 300–600 nm region. It features the solar absorption lines spectrum and also contaminates space telescopes such as HST and JWST. It could be canceled only by going well out of the zodiacal plane, or at the edge of the solar system, both tremendously expensive propositions for a space telescope. Note that the darkest sky brightness from space occurs roughly between 1.5 and 3 μm.

9.6 Space-Based Observations

9.6.1 The Case for Space-Based Observations

In view of the many impediments to astronomical observations due to the Earth's atmosphere – and to a lesser degree to Earth's gravity – observing from space appears to be the perfect solution. One obvious and big plus is that the full electromagnetic spectrum is accessible, unlike the more or less dirty individual windows left for ground-based observations. Many space telescopes indeed work in regions fully inaccessible from the ground, from the UV to X-rays to γ rays, and on the other side of the spectrum in the medium and far infrared. Accordingly, and since some three decades, world astronomical resources are split about half–half between the building and operation of space and ground facilities. Interestingly enough, so are roughly the publication and discovery rates with either approach.

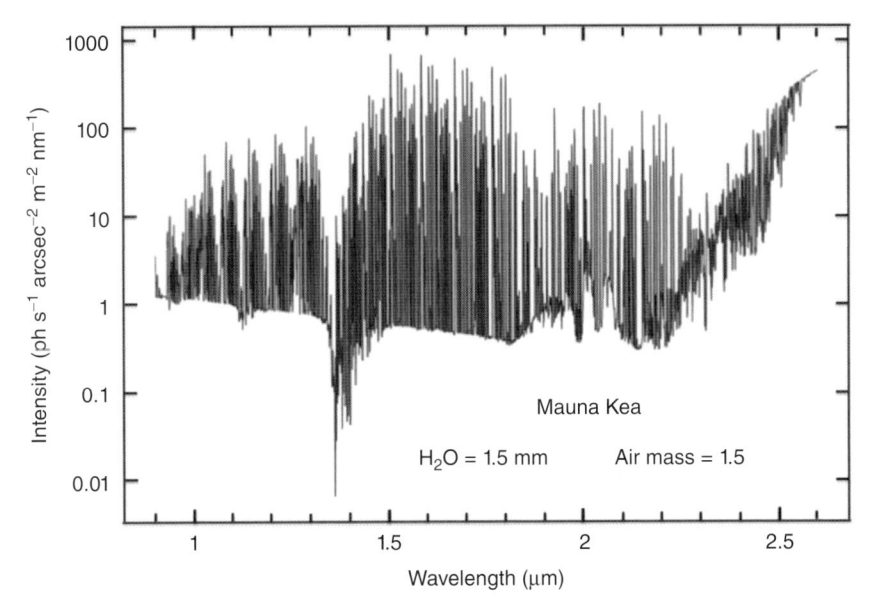

Figure 9.21 Near-IR night sky brightness at the Mauna Kea site in units of ph s^{-1} arcsec^{-2} m^{-2} nm^{-1}. Note the dense emission lines forest over a much fainter continuum. (Reproduced with permission of Gemini Observatory.)

There are other pluses with space operation, even when observing inside a reasonably good atmosphere window. The night sky above the Earth upper atmosphere is much darker than from the ground in the near-infrared region, roughly from 800 nm up. The telescope itself, when passively or actively cooled, is much darker in the thermal infrared domain, roughly from 1700 nm up. And, of course, there is no atmospheric turbulence anymore, and reaching diffraction-limited performance with close to 1 Strehl ratio now looks like a child's play, provided a couple of natural stars in the telescope huge patrol field are used for accurate on-field guiding.

This last point needs some qualification though. Reaching diffraction-limited performance requires extremely good optical figuring and near-perfect optics collimation and focusing. So, while the full paraphernalia of adaptive optics techniques, including laser guide stars, is blissfully avoided, there is still the need for a number of motorized functions, wavefront sensors, and even possibly deformable mirrors, albeit now working at low frequencies, in the <1 Hz range rather than the few kilohertz required for ground-based AO corrections. And, while wind-induced structure vibrations are not a factor anymore, for example, motor-induced fast vibrations can still be a significant nuisance.

9.6.2 Why all Telescopes are not Space Telescopes

In view of the above, one might wonder why all telescopes are not space telescopes, even when operating in the best atmospheric window (the optical ones). The rationale is actually quite straightforward: it costs almost 100 times more to build, install, and operate a space telescope with its instrumental package than for an equivalent size facility operating from the ground. This explains why, by the mid-2010s, there is only one 2.4-m diameter optical–near-IR space facility in operation, namely, the Hubble Space Telescope (HST). Along the same lines, its planned successor, the 6.5 m diameter near and medium IR James Webb Space Telescope (JWST), will likely be the only large space telescope competing directly with ground-based telescopes in the 2020–2040 period.

9.7 Conclusion

Atmospheric seeing severely limits the spatial resolution of all astronomical observations: even during the best periods at the best sites, its blurring effect is greater than the diffraction limit for telescope diameters greater than about 25 cm in the optical (at 0.5 μm) and 100 cm in the near-IR (at 2μm) regions. For present large telescopes in the 4–10 m diameter range, this is not a problem for the study of faint extended objects, since for optimum detection detector pixels must be illuminated at high focal ratios (F/1.5 to F/2.5), which anyway limits the best spatial resolution at the same 0.4″–0.8″ level than natural seeing. Note that at least in the near-IR region, this natural seeing limitation can be significantly improved (typically by a factor of 2 in seeing area) with the so-called ground-layer adaptive optics technique.

For bright enough objects, this limit is now successfully circumvented by adaptive optics techniques, albeit with large efforts and a number of provisos. It is now possible in the near-IR region (at 2 μm for good seeing nights; 1.2 μm for exceptionally good nights) to correct a sizeable field (minutes of arc) with a wide sky coverage ($\geq 50\%$), getting images with say 60% of the light in a diffraction-limited core at 2 μm and 30% at 1.2 μm. This opens relatively routine astronomical observations at the sub-0.1″ level, but requires a highly complex corrector, featuring multiple deformable mirrors and a laser constellation, and still needs AO-friendly nights.

For the next generation of extremely large telescopes (ELT) with diameter D of at least 24 m, even with the highest practical focal ratios (say F/1.25) and excellent seeing, the detector pixels will oversample natural seeing by big factors of easily 20–50. This is a terrible waste of the precious – and costly – detector pixels and of the huge optics required to feed them. It is thus essential to develop and deploy turnkey AO systems that are an integral part of the ELTs, instead of being optional correcting systems as is mostly the case in the present 8–10 m very large telescope generation. We are talking about heroic efforts ahead though, since the complexity of AO systems roughly scales with the number of coherent zones over the diameter D telescope primary mirror, that is, as D^2.

*Exercise 13   Seeing probability

1. Compute the probability of getting a DIMM seeing better than 1 arcsec at Paranal for the year 1999.
   Tips: ESO monthly seeing statistics are available at http://www.eso.org/gen-fac/pubs/astclim/paranal/seeing/statseeing.lis

2. If we assume the same seeing distribution, what median seeing can one expect for the observation of the Large Magellanic Cloud at the VLT in the I band?
3. In practice, we should expect a better seeing than the value previously computed. Why?

Answer of exercise 13

1. Using the table given at http://www.eso.org/gen-fac/pubs/astclim/paranal/seeing/statseeing.lis, the cumulative normalized distribution for 1999 gives a probability of 91% to get a DIMM seeing better than 1 arcsec at 5000 Å.
2. From the previous table we compute a median value of 0.92 arcsec for the DIMM seeing. This value is given at 5000 Å; in the I band, that is, at 8000 Å, using the Kolmogorov turbulence model, one gets $0.92 \times \left(\frac{8000}{5000}\right)^{-1/5} = 0''.88$. However, this is at zenith. The declination of LMC $-69°45'$ corresponds to $z = 45.1°$ at meridian at Paranal. The seeing value consequently increases to $0.88 \times \cos^{-3/5} 45.1 = 1''$.
3. As discussed in Section 9.2.2, the outer scale of turbulence at Paranal (approximately 20 m) is comparable to the primary mirror size (8 m) and cannot be considered as infinite. This will result in a better seeing than the estimated $1''$ value ($\sim 0.8''$).

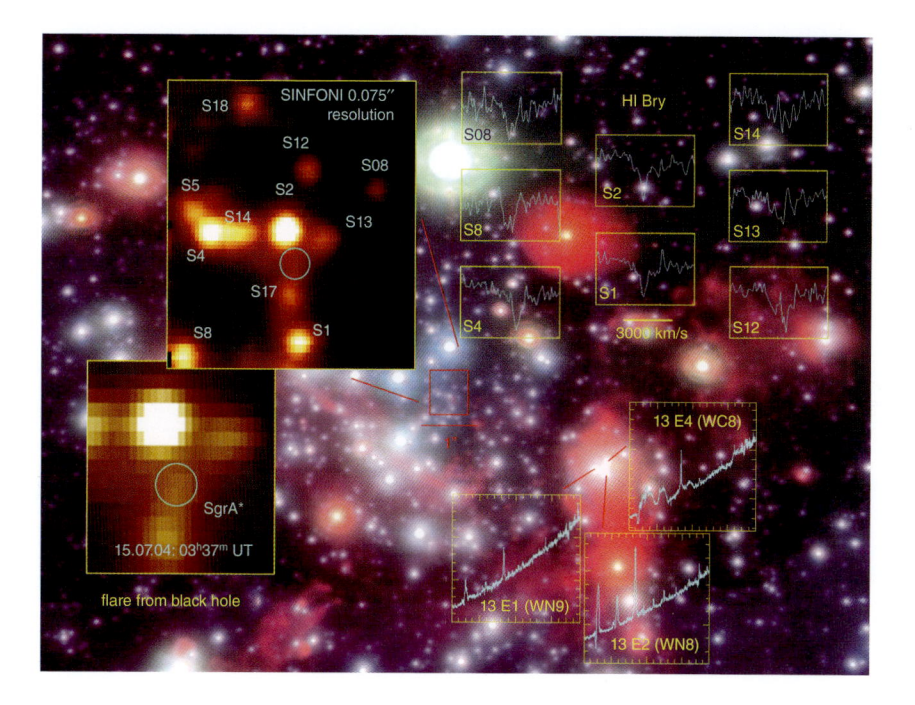

First adaptive optics assisted integral field spectroscopy of the galactic center supermassive black hole and its surrounding stars. The observations were done in 2004 with SINFONI, the world's first AO-assisted IFS on an 8–10 m class telescope. Zooming into the central light month of our Milky Way (top left), SINFONI can measure the spectra (top right) and radial velocities of the stars orbiting the galactic center, thereby providing together with high angular resolution imaging (background) the best evidence for the existence of supermassive black holes. It is also the first measurements of the spectrum of a so-called flare (left), the episodic outburst of infrared- and X-ray radiation from very close to the black hole event horizon. The bottom right shows exemplary spectra of the many very hot and massive stars of this region, which have formed about 6 Myr ago following the infall of a massive molecular cloud. (Reproduced with permission of Frank Eisenhauer.)

10

Data Gathering

10.1 Introduction

Telescopes and instruments are expensive and must be shared with many users. Thus, getting the scientifically most useful data from the minimum telescope observing time is key to a successful observing run. In this chapter, we review the various processes involved in planning the observations, estimating the integration time required to achieve a given signal-to-noise ratio (SNR), defining the optimum observing strategy, and finally running the observations at the telescope. Note that most aspects related to data gathering with 3D spectrographs are similar to the one involved for other observational techniques such as imaging. We nevertheless go to the full process in order to give a comprehensive view of data gathering with modern telescopes.

10.2 Planning Observations

The first – and easiest – task in building an observing program is to select the list of targets and find the appropriate time of the year for scheduling the observations. Given the target spherical equatorial coordinates (α, δ) and the observatory latitude (Φ), the source visibility is given by the classical formulae:

$$\sin h = \sin \delta \sin \Phi + \cos \delta \cos \Phi \cos(\text{LST} - \alpha) \tag{10.1}$$

where h is the altitude and LST the local sidereal time. In general, the visibility is given in airmass (A_m) rather than in altitude with the formulae $A_m = \sec(90° - h)$.

Objects will not generally be observable all night and it is always better to observe them at low airmass. The reason is that (i) sky brightness and absorption increase with airmass, (ii) seeing also deteriorates (e.g., a 0.7 arcsec seeing at airmass 1.0 will translate to a 1 arcsec seeing at airmass 1.5), and (iii) the impact of atmospheric dispersion is also more important in an instrument that does not have an atmospheric dispersion corrector (e.g., at 5000 Å the atmospheric dispersion is ~1.5 arcsec with respect to 7000 Å at an airmass of 1.5). On the other hand, if the instrument sits at a Nasmyth focus, objects cannot be observed if they transit too closely to the zenith. From (10.1) it is easy to find the optimal date to observe the targets and the duration where the objects will have the appropriate airmass to be observed in good conditions.

Optical 3D-Spectroscopy for Astronomy, First Edition. Roland Bacon and Guy Monnet.
© 2017 Wiley-VCH Verlag GmbH & Co. KGaA. Published 2017 by Wiley-VCH Verlag GmbH & Co. KGaA.

For AO observations the need of a bright enough guiding star is often a strong additional constraint, except when observing the close environment of a bright star. Even with laser guide star AO (e.g., LTAO, GLAO or MCAO; see Chapter 9), a so-called tip/tilt star needs to be used in the vicinity of the target in order to correct residual drifts of the field of view, which are unfortunately not sensed by laser "stars."

Another constraint is the moon's phase. Depending on the object brightness and wavelength of the observations, the observations must be scheduled in dark, grey, or bright time. Precise definition of dark, grey, and bright times is a function of the observatory. Here is the ESO definition:

- *Dark time*: When the fraction of lunar illumination is below 0.4, which corresponds approximately to ± 3 nights around the new moon.
- *Grey time*: When the fraction of lunar illumination is between 0.4 and 0.7 and the minimum angular distance of the Moon with the target is less than 90°.
- *Bright time*: When the fraction of lunar illumination is higher than 0.7 and the minimum angular distance of the Moon with the target is less than 60°.

In practice, this limitation does not apply to infrared observations, roughly above 800 nm, for which Moon's diffuse light is much fainter than in the visible.

Once the final list of objects have been selected pointing charts must be prepared (see Figure 10.1).

10.3 Estimating Observing Time

A key task of the observer is to get a reliable estimate of the integration time needed to achieve the science goals on the selected target. This is fundamental to the feasibility of the project and needs some attention. To perform this task, one uses an exposure time calculator (ETC), which is a simple tool to estimate SNR for a given target brightness taking into account the telescope and instrument characteristics and the expected observational conditions. Large telescopes all have built web-based ETC, which can be easily accessed by their community; see, for example, the ETC page on ESO web site at http://www.eso.org/observing/etc.

ETCs are instrument specific, but all are based on the same basic mathematical formulae that are discussed hereafter.

The source of interest or target is described by a light distribution $I_T(x, y)$ and a spectral flux distribution $F_T(\lambda)$. In general, we use simple light distributions such as point sources or infinitely extended sources. The spectral flux is also simply modeled by a continuum and/or an unresolved emission line. More sophisticated spatial and spectral distributions can also be used, for example, typical profiles for galaxies or typical spectral distributions for different types of stars.

The sources of noise are the detector read-out noise σ_{rn} and the photon noise due to the sky backgrounds $F_S(\lambda)$ and the detector dark current N_{dc}.

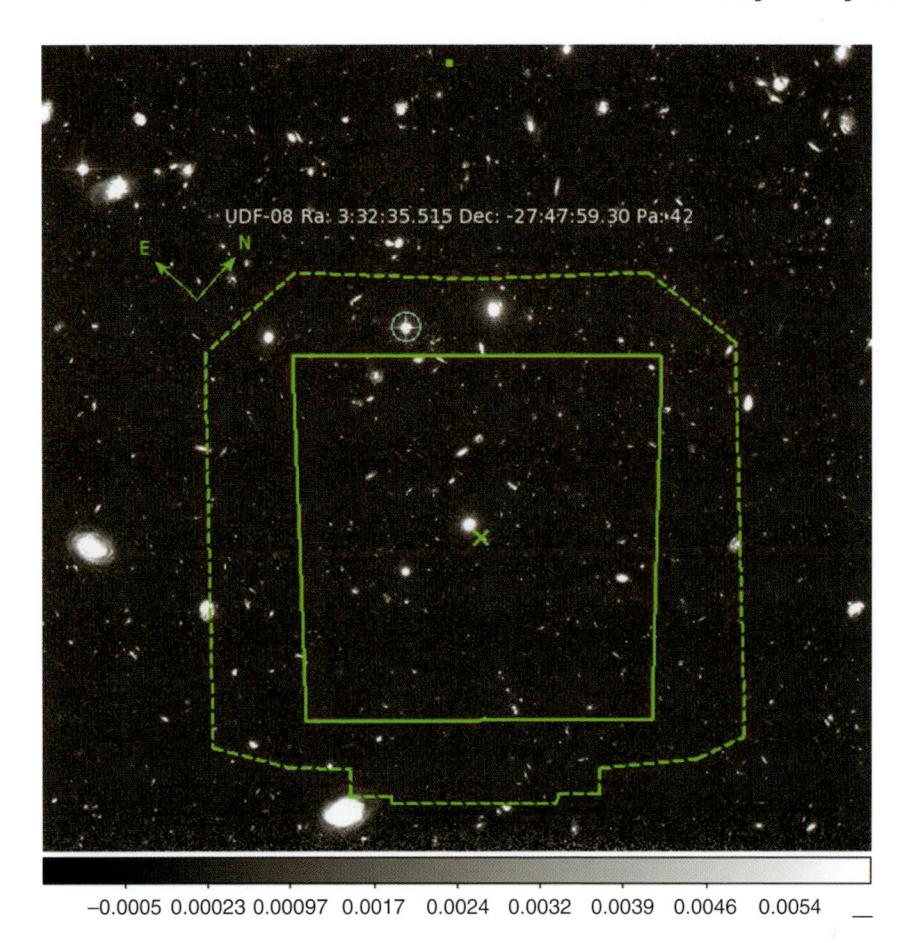

Figure 10.1 Example of a pointing chart used for MUSE observations of the Hubble Ultra Deep Field at the VLT. The MUSE field is visible (green rectangle), surrounded by the slow guiding area (dotted green curve). Note that the center of the field, the North-East orientation arrows, and the candidate guiding star are also shown.

The sole free parameter is the total integration time on the target. Note that the integration time is often split in a number of exposures n_e of shorter integration time t_e to limit the impact of cosmic rays (Section 8.3) or changes in the atmospheric properties (Section 9.1).

The SNR can be expressed as:

$$\text{SNR} = \frac{n_e t_e K_T F_T(\lambda)}{\sqrt{n_e t_e K_T F_T(\lambda) + n_e t_e K_S F_S(\lambda) + n_e N_{\text{pix}} \sigma_{\text{rn}}^2 + n_e t_e N_{\text{pix}} B_{\text{dc}}}} \quad (10.2)$$

F_T is target flux in erg s^{-1} cm^{-2} (CGS units) and $n_e t_e K_T F_T$ is the number of photon-electrons received from the target. K_T is given by

$$K_T = \frac{\lambda}{hc} \rho(\lambda, A_m) A_{\text{Tel}} T(\lambda) S_T(\lambda) \quad (10.3)$$

with λ being the wavelength (in cm), h the Planck's constant (6.626×10^{-27} erg s^{-1}), c the speed of light in vacuum (2.998×10^{10} cm.s^{-1}), $\rho(\lambda, A_m)$ the atmospheric extinction, which is a function of wavelength (λ) and airmass ($A_m = \sec z$), A_{Tel} the effective area of the telescope in cm^2, and $T(\lambda)$ the total throughput of the system (telescope + instrument, including detector quantum efficiency); $S_T(\lambda)$ is a factor that measures the fraction of the flux target captured in the selected area (usually given as a number of co-added spaxels and spectels).

F_S is the sky background flux in erg s^{-1} cm^{-2} arcsec^{-2} Å^{-1} and $n_e t_e K_S F_S$ is the number of photon-electrons received from the sky. K_S is given by

$$K_S = \frac{\lambda}{hc} A_{Tel} T(\lambda) \Delta_{spa} \Delta_{spe} \tag{10.4}$$

with Δ_{spe} the selected wavelength interval in Å and Δ_{spa} the selected area on the sky in arcsec^2.

σ_{rn}^2 and B_{dc} are respectively the detector readout noise variance and the detector dark current in photon-electrons by second and by pixel. N_{pix} is the total number of pixels that have been co-added to cover the selected wavelength interval and area on the sky.

Note that the total noise, as expressed in the denominator of (10.2), is the sum of three components: the photon noise of the object, the photon noise of the background and the detector noise, itself split in readout noise and dark current photon noise (see Section 8.3). Normal or Poisson noise distributions are assumed.

Let us take a realistic example to see how this works in practice. In the following, we will estimate the signal to noise achieved in a 30 mn exposure for a point source and a single unresolved emission line at 5200 Å with a total flux of $F_T = 3 \times 10^{-17}$ erg s^{-1} cm^{-2}. Observation is scheduled in dark time at an airmass of 1.1 and a seeing of 0.8 arcsec with the multiobject spectrograph VIMOS [100] at the VLT. The instrument configuration uses the low resolution grism R183 and a slit width of 1″. This translates to a spectel of 5.4 Å and a spaxel of 0.2×1 arcsec^2. The VLT effective area is $A_{Tel} = 4.85 \times 10^5$ cm^2. The transmission of the system, telescope included, is 0.18. The detector readout noise is $\sigma_{rn} = 5$e$^-$ and the dark current $B_{dc} = 7$ e$^-$ h^{-1} Using (9.1) of Section 9.5.1 and the extinction curve for Paranal, one can get the corresponding extinction at 1.1 airmass $\rho(5200$ Å$, 1.1) = 0.854$.

We start the computation with the faction of flux (S_T) of the target that is captured in the wavelength interval and in the selected spatial area. Along the slit length the flux of the target is spread over a few pixels because of the seeing. To retrieve most of the flux we will sum all the pixels that are within two times the seeing size. Assuming a Gaussian shape for the PSF this corresponds to 98% of the total flux. For the 0.8″ seeing we then need to sum over 1.6″, that is, 8 pixel of 0.2″. The flux outside the slit width is lost. For the assumed Gaussian PSF of 0.8″ FWHM the corresponding loss is 14.1%. Because of the low spectral resolution, a single spectel is enough to capture the totality of line emission flux. Then in total we have $S_T = 0.98 \times 0.86 \times 1 = 0.84$.

From (10.3) one gets the number of photon-electrons received from the target: $K_T F_T = 0.49$ e$^-$ s^{-1}, that is 880e$^-$ in total. Now let us compute the sky background. Taking the typical new moon Paranal sky brightness of AB = 21.8 magnitude arcsec^{-2} at this wavelength, one gets a background sky flux of $F_S = 6.25 \times 10^{-17}$ erg s^{-1} cm^{-2} Å^{-1} arcsec^{-2}. The spatial area is $\Delta_{spa} = 1.6$ arcsec^2 and the wavelength range $\Delta_{spe} = 5.4$ Å. Equation (10.4) then gives the number of photon-electrons by second received from the sky background: $K_S F_S = 12.3$ e$^-$ s^{-1}, that is, 22 083 e$^-$ in total.

Using the total number of pixels to co-add ($N_{pix} = 8$) Equation (10.2) gives an SNR of 5.8.

We now evaluate the relative impact of the different sources of noise. The previous example gives respectively 4%, 95%, and 1% for the object, sky, and detector variance contributions. In this case, the detector noise contribution is negligible and the data SNR is photon noise limited. In general, one should always try to be in this regime. This is the best regime as there is then no penalty when co-adding pixels in the spatial or spectral axis or splitting exposures.

Note that Equation (10.2) can easily be inversed to derive the exposure time t_e as a function of SNR S:

$$t_e = \frac{FS^2 + \sqrt{F^2 S^4 + 4 n_e N_{pix} \sigma_{rn}^2 K_T^2 F_T^2 S^2}}{2 n_e K_T^2 F_T^2} \tag{10.5}$$

with $F = K_T F_T + K_S F_S + N_{pix} B_{dc}$

Using the previous parameters, Equation (10.5) shows that an exposure time of 5355 sec is needed to achieve an SNR of 10. It is not recommended to perform such a long integration time in one go because of the impact of cosmic rays and atmospheric variability. Splitting the total exposure in three gives an integration time of 1795 or 5386 s in total. As expected, the difference with a long single exposure is very small because we are, in this case, in the photon noise regime. The situation would be very different when detector noise is not negligible (see an example in Exercise 11 of Chapter 8).

ETCs are useful tools to estimate the feasibly of a program and to prepare observing runs. However, it must be recalled that they are based on simplistic assumptions and they give only an indication of the exposure time to achieve the required target SNR. In the following we list the main reasons why the actual SNR could easily be quite different from the ETC predicted values.

- *Systematics*: By design, present ETCs do not take into account any systematics that can affect the data. There are instrumental imperfections (e.g., stability problems with the optics or the detectors) and data reduction ones (e.g., sky subtraction) that give rise to systematics in the data. Systematics can also result from calibrations that might not represent fully the night observing conditions (e.g., because of improper illumination of the internal calibrations). The impact of systematics is generally more important for faint sources detection. Self-calibrations can help reduce the impact of systematics but there are no instruments free of any systematics. When the level of residuals left by the systematics is significant, compared to the source signal, it is hopeless to increase the exposure time beyond a certain limit.
- *Noise properties*: ETCs assume that the noise follows a normal distribution and that each sample is independent. Depending on the detector properties,

a normal distribution is indeed often a good approximation of the true noise property at the detector level. However, noise properties of the final product given by the data reduction process might be very different. As discussed in Chapter 11 Section 11.3.4, efforts have been made to improve the noise handling in the new generation of data reduction pipelines, but they still have their limitations and the final noise statistics is in general not normal and not free from correlations.
- *Optimal extraction*: In most cases, ETCs assume a simple addition of spaxels and spectels to recover the flux of a source. In this case, the ETC gives a pessimistic value with respect to what can be achieved by more advanced schemes (e.g., variance weighted extraction) that increase the SNR of the extracted spectra.
- *PSF*: The simple Gaussian model of the spatial instrument PSF that is generally used in ETCs is a first approximation only. Actual PSFs tend to have broader wings than the Gaussian approximation and they are generally not constant over the field of view and with wavelength. In AO-corrected observations, PSFs are definitively not Gaussians and more sophisticated models are used.
- *Seeing*: ETC input seeing values are of course only best guesses of what will be achieved during the observations.
- *ETC input parameter accuracy*: The parameters used in the ETC might not be accurate enough. For example, having a sky model that is a good representative of the site characteristics is very important for NIR observations where strong and time variable OH emissions will be the main noise sources. Instrument throughput is also a critical item: sometimes, it can change with time due to, for example, long-term evolution of coatings or optics misalignments.

As can be seen from the above list, except for the optimal extraction, all other effects mentioned tend to lower the real SNR. For programming an observing run, it is then better to consider the ETC-computed SNR as an upper limit.

10.4 Observing Strategy

In the previous sections we have seen how to select the science targets and compute the required integration time with the ETC. But our observing program is not yet ready: as observers, we still have to face a number of strategic decisions: What objects to observe? In what order? What shall I do when the seeing is very good, of just bad? Shall I split the observations with shorter integrations? What calibrations should I take? etc.

One of the first strategic decision is whether to split the observation of a given field in a number of individual exposures. It is often advisable to have shorter exposures; multiple exposures of the same field help reduce the impact of faulty pixels and give a better chance to benefit for stable atmospheric conditions inside each sub-exposure. In infrared region where the sky subtraction is primordial, short exposures switching between the sky and objects are almost mandatory. In the visible, especially in the blue wavelength range, sky short term emission variations are less of a problem, but the longer the integration time, the larger the number of pixels impacted by cosmic rays. In space the environment is much more stable, but the impact of cosmic rays is also much larger, and splitting exposures is then mandatory. Deciding the optimum integration time for each individual exposure is largely a matter of tradeoffs. On one hand, we would like to get as many individual exposures as possible because (i) this gives a larger statistical sample that can be efficiently used to remove outliers, (ii) this limits the impact of cosmic rays on detectors, which cannot use nondestructive readout such as CCD, and (iii) this also allows to benefit from a more stable atmospheric environment. On the other hand, having too many exposures with short integration time impacts the efficiency of the observations by adding overheads. For example, if the total readout and exposure handling time is 2 min, making the individual exposure shorter will impact seriously the total overhead. Another reason to avoid making too short exposures is the impact of detector readout noise on the final SNR. As shown in the previous section, this happens when the detector noise starts not to be negligible with respect to the photon noise of the source plus background.

Another reason to split the observation in small pieces is to use a dithering strategy, that is, moving slightly the field between successive exposures. Same locations of the sky are then sampled by different parts of the instrument/detector system. This scheme can be a powerful tool to reduce the systematics by averaging the residuals coming from different parts of the system. How much and how to dither depends on the instrument characteristics and the science goals, but is also a matter of trade-off. If the instrument main systematics are on the detector pixels, an offset that would move by the diameter of a typical source would do the job. In some cases, the main source of systematics is not the detector pixel but some optical elements (e.g., fibers, lenslets, or slices) and thus different offsets should be considered. Note that using large offset will result in smaller field of view with optimal observing time. In other cases, such as multislits spectrography, one can only offset along the slit direction. With a squarish field of view, a good strategy is to rotate by multiples of 90° the field of view. This allows moving the objects over the instrument without any loss of the field of view. Such a recipe is used in MUSE [101] with an additional small offset to prevent the center of rotation of the field to stay at the same location. If many exposures are planned for a single field, it is recommended to implement a nonregular dithering pattern to avoid introducing spurious frequencies. A simple scheme is to perform a random draw of the location within a box of a given size, imposing a radius of avoidance around each point.

Dithering can also be used for better sampling of the spatial PSF. In this case, precise sub-spaxel offsets must be performed and an appropriate scheme to recover a finer final sampling executed during the data reduction process. This so-called drizzling scheme [102] has been successfully used for HST imaging to improve the spatial resolution of the WFPC2 and ACS images. It is much more challenging to use it for ground-based spectral-imaging system because of changing seeing conditions from one exposure to the other.

Another sort of dithering, called nod and shuffle, has been implemented in some instruments (e.g., LDSS AAT multiobject spectrograph [103]), PMAS integral field spectrograph [104]). This technique makes use of the capability of CCDs to move charges on the chip without reading the detector. Only a fraction of the detector area is used for imaging. After a first short integration, charges are transferred on the other part of the detector, which is used as a storage area. Then the telescope is offset to an empty location used as sky reference, and a sky exposure is taken. By moving back and forth the telescope and the charges on the detector, the sky and the objects are recorded through exactly the same optical path. This self-calibration approach allows a much better sky subtraction. The price to pay is that only a third of the detector area can be used for scientific data collection.

Mosaicking is used when the field of view to be observed is larger than the instrument field of view. The main question when preparing a mosaic is how to handle the edges. Depending on the instrument stability and how accurately it can be calibrated, this might require allowing some overlapping between fields. How much to overlap is also a trade-off; if one just wants to match the background level, a small fraction of the field of view will be enough. If relative astrometry is not accurate enough, one wants to be sure to identify spatial structures (e.g., stars) in common. Note that getting a homogeneous spatial resolution between the various mosaic fields is often difficult because of seeing time variation. When critical, one can build the mosaic from multiple observations of the individual subfields with short integration times and build a sequence that moves from field to field. This obviously impacts overheads.

10.5 At the Telescope

At the telescope (Figure 10.2), it is essential that the putative science or calibration targets are organized according to the various parameters that will be taken into account for real-time decision during the night. Here are the main parameters to be considered:

- *Visibility*: Detailed visibility curves (e.g., airmass as a function of time) should be computed for all observable objects.
- *Moon*: The important parameter is the moon phase and height above the horizon, as well as its angular distance from the object.
- *Seeing*: Because good and stable seeing is rare, using the best seeing only for observations that critically depend on it is obviously very important. Sorting objects by seeing categories (e.g., good, median, bad) is thus quite important to guide the real-time target selection.
- *Wind direction*: Pointing restrictions can occur when the wind speed is above a limit set by the observatory. In some sites the stronger winds blow preferably from a given orientation, depending on the season. If this is the case, it is important to identify objects that may be affected by these pointing restrictions.
- *Completeness*: Finally, it is also good to try to complete observations that have been already started but require additional exposure time to be completed. Keeping a record of what has been accomplished and what remains to be done is thus important.

Figure 10.2 The ESO VLT control room during day calibrations (Credits ESO).

- *Scientific priority*: Last, but not least, the relative scientific priorities of the targets are recorded. Conversely, it is a good idea to prepare a list of "fillers" that can be used when seeing is bad or when there are no observable high-priority objects at the time.

It is also important to carry a well-defined calibration plan. Calibration requirements depend on the instrument characteristics and stability, but also on the science objectives. For some science cases spectrophotometric accuracy may be very important, while for other programs a good astrometry (i.e., precise positions of the science targets in the field of view) will be mandatory. Some calibrations can be performed during the day using internal lights to illuminate a diffuser located inside the instrument (instrument flats) or inside the dome (dome flats). When feasible, this should be preferred to night calibrations to save the precious night time for science. This is, however, not always possible, the instrument characteristics may not be stable enough (e.g., flexures due to gravity change with rotation or systems sensitive to ambient temperature) and dedicated calibrations have to be performed during the night. Some "science" calibrations must also be performed on reference sky targets and thus should be inserted in the night program. This is the case, for example, for spectrophotometric standard stars, radial velocity standard stars and astrometric fields. The following list gives the usual mandatory standard calibrations:

- *Detector calibrations*: High signal to noise measurement of the bias, which measures the zero-level ADU, and of the dark current are required. A set of exposures with an internal lamp at increasing integration times might also be required to measure the detector linearity.

- *Geometrical calibrations*: Matching the image recorded on the detector with the sky coordinates is an important task, especially for integral field spectrographs that feature a complex reorganization of the field of view. This is also important for instruments where the field of view is split in physical subunits. This is generally done by imaging a mask with a known geometry in the instrument focal plane.
- *Flat-fields*: Getting an accurate flat-field, that is, the relative response of the instrument to a uniform illumination, is very important. Flat-field errors limit the accuracy of the sky subtraction and result in systematics. The use of internal calibration tungsten lamps is the best way to measure the high spatial frequency pixel-to-pixel flat-field correction. The black body response of tungsten lamps is a smooth function of wavelength and the lamps are bright enough to produce high SNR spectra. However, the illumination produced by an internal calibration unit never matches the sky illumination perfectly. To correct the remaining low frequency illumination errors, one can use the non-scientifically useful twilight period to obtain high SNR spectra of the sky. The resulting spectrum is a typical Sun spectrum, with the presence of many stellar absorption lines. Using it for pixel-to-pixel flat-field correction is difficult, but it is fine for correction of the remaining low frequency illumination errors.
- *Wavelength calibrations*: The mapping of detector position versus wavelength is done using gas discharge (or arc) lamps of various elements (e.g., HeCd, Ne, Ar). The lamp must be selected to have enough isolated bright emission lines in the wavelength range. A good recommendation is to have usable lines up to the edges of wavelength range to avoid inaccurate polynomial extrapolation of the wavelength solution. If the wavelength range is large, more than one lamp must be used. At high spectral resolution it may be difficult to find an appropriate set of lamps to cover a large spectral range. Recent alternatives have been developed using laser sources in the form of supercontinuum white light sources [105], which are able to produce regularly spaced bright narrow lines over a large wavelength range.
- *Spectrophotometry*: Flux calibration is performed by observing a spectrophotometric standard star with tabulated absolute flux as a function of wavelength. The best spectrophotometric standard stars are white dwarfs, which have almost featureless smooth spectrum spectra.
- *Astrometry*: Observation of a reference astrometric field (e.g., open cluster or multiple stars) is required to obtain the absolute coordinates of the observed fields.

During the night, the program will be implemented according to the external atmospheric conditions, which must be carefully monitored. The most important – and the most difficult – goal is to use the seeing conditions efficiently. Some sites are equipped with a differential image motion monitor (DIMM), which uses a dedicated small telescope to infer the seeing from the monitoring of bright star motions [106]. Note that DIMM inferred seeing is generally different from the true seeing measured at the instrument focal plane. For large apertures (e.g., an 8 m class telescope) the outer scale length of the atmospheric turbulence (typically 10–30 m) cannot be considered infinite with respect to the primary mirror size (see Section 9.2.1). In this case, the seeing value seen at the telescope focal plane is better than the one given by the DIMM seeing monitor. The other main reasons for such discrepancy are the optical aberrations within the instrument, the effect of guiding errors, and so on. Note that DIMM values are given at the zenith for a given wavelength (often taken at 5000 Å) while observations are generally performed at different airmasses and for various wavelength ranges. An alternative to DIMM values is to use the telescope auto guider information. It has the advantage to take into account the finite outer scale effect and has the same airmass as the observations.

At some observatories such as ESO or GEMINI, most of the observations are not performed in visitor mode but in service mode. In this mode, the observations are centrally scheduled and executed by a dedicated team of astronomers for the community. Observations are organized in a series of observing blocks with attached constraints (e.g., seeing). This so-called observing queue is then used with a dedicated software to optimize the use of telescope.

Once on-sky data have been obtained, the observer's next task is to check if the data quality is appropriate for the science objective. It may be difficult, and possibly not desirable, to perform full data reduction of the freshly obtained data sets (e.g., when the associated calibrations are not yet completed), but a quick reduction might be good enough to assess the data quality. Final assessment generally depends on the science goal, but getting an idea of the data integrity, the seeing, and the SNR for the target is already a good step at this stage. Depending on such assessment, the target may have to be re-observed if data quality is too poor. Results of these investigations, as well as any events that have occurred during the exposure (e.g., abrupt change of seeing, transmission change due to clouds, telescope guiding errors), should be recorded for future data reduction and analysis tasks.

10.6 Conclusion

From the scientific idea that leads to a successful proposal and ultimately a published paper, there are many inevitable steps that need careful attention. We have seen in this chapter how to prepare and optimize the observations and how to perform them at the telescope.

Space observations have the advantage that they are performed in very stable and predictive environmental conditions. Because of weight (and cost) constraints, instruments in space telescopes have generally a limited number of modes and benefit from well-defined calibration plans. They are also by design operated remotely by professional astronomers. In that respect, performing space-based observations is much easier than performing ground-based observations that suffer from changing atmospheric conditions. On the other hand, ground-based instruments usually incorporate many different operational and instrumental modes, which introduce a lot of flexibility and may provide a better match to the required performances to achieve the science goals. The negative side of this flexibility is that they are usually more difficult to operate and calibrate, leaving unknown systematics in their data, which are often difficult to remove at later data reduction stages.

Today ground-based large telescopes are also largely operated remotely in queue scheduling by professional observers. The observations are standardized with precise calibrations. This leads to a more efficient use of these costly machines. It is then possible today to get cutting-edge scientific data without having been in contact with the telescope and the instrument used. This now often includes collecting and reducing archived data coming from observations required by other people for similar or even totally different science goals. However, it is recommended that one confronts real observing conditions at least once. This gives a better understanding of the processes involved and can help improve the observational strategy for future observations.

*Exercise 14   Observation scheduling

1. Find the optimum date to perform faint object spectroscopy at visible wavelength in the Hubble Deep Field South (22 h 32 mn, −60° 33′) at the VLT (24.6° S, 70.4° W) in 2014. Tips: use the ESO ephemerides at http://www.eso.org/sci/observing/tools/calendar/skycalc.html.
2. Using 1.5 maximum airmass, how long can the field be observed during the selected night?

Answer of exercise 14

1. According to the ESO ephemerides, the LMST is 22:31 on 8–9 September 2014 at the middle of the night. The Hubble Deep Field South (HDFS) will then pass at the meridian. However, because we want to perform faint object spectroscopy at visible wavelength we need to observe in dark time. The dates 8–9 September are full moon days and thus not useable. The nearest new moon dates are 24/25 August with LMST 21:31 at midnight and 23/24 September with LMST 23:30 at midnight. These two periods will thus be optimal for HDFS observations.
2. Using the airmass definition, we derive the corresponding altitude

$$h = 90 - \arccos \frac{1}{1.5} = 41.8° \tag{10.6}$$

the time the object is above this airmass can be directly derived from Equation (10.1); LMST $- \alpha = \pm 3.14$ h. The object is then observable between 19.39 and 1.67 h LMST. We check that this time slot is indeed included in the night time given by the ephemerides [17.25–03.25] hours. The available observing time is then 6.28 h.

**Exercise 15   Optimum slit width

The plan is to observe a spatially unresolved faint emission line galaxy with a slit-based MOS. The instrument has a fixed instrumental setup (grating, optics, and detector) but the slit widths are configurable.

1. For a given seeing, and the blue part of the spectrum (no bright sky lines), describe the impact in the emission line SNR when opening the slit width.

2. Repeat the discussion in the red part of the spectrum with numerous bright [OH] lines.
3. Why it is not recommended to open the slit much larger than the FWHM of the seeing, even in the blue part of the spectrum?

Answer of exercise 15

1. A larger slit width will reduce the light loss outside the slit, although the gain is negligible when the slit becomes large enough. For example, assuming a Gaussian PSF, the slit loss is given by

$$F(w,\sigma) = 1 - erf\left(\frac{w}{\sqrt{2}\sigma}\right) \tag{10.7}$$

with w being the slit width and σ the Gaussian seeing (FWHM/2.3548). The corresponding slit loss when the slit width is two times the seeing FWHM is less than 2% and already negligible.

At the same time, the number of spaxels to sum up in the direction perpendicular to the slit increase and thus the readout noise will be larger. If the source and/or the background are bright enough, the detector readout noise will be small with respect to the photon noise and the additional noise due to the spread of the signal over many spaxels will have little impact. The spectral resolution will decrease linearly with the slit width. When the LSF becomes larger than the intrinsic emission line width, the contrast of the object emission line decreases with respect to the sky background with a negative impact on the SNR.

2. The difference when operating at wavelength where bright [OH] sky lines are present is the strong increase in the background when the spectral resolution decreases due to the slit opening. The sky background can easily be multiplied by a factor of 10 or more when the resolution decreases with a strong negative impact on the SNR.

3. As already discussed in the first question, there is no much gain in opening the slit width larger than two times the seeing FWHM . The other reason is the slit effect (see Section 4.2 in Chapter 4), which introduces systematic errors in the wavelength calibration and in the measured line shape.

11

Data Reduction

11.1 Introduction

The goal of the data reduction system is to remove the instrument signature and provide fully calibrated spectra, images, or data cubes. Depending on the instrument type and layout it can be more or less complex. Providing a user-friendly, good data reduction system is an integral – and often crucial – part of any successful instrumental development. Even instruments with clever instrument design, careful manufacturing, and fine alignment will not produce good science if they do not come together with a properly designed, realized, and validated data reduction pipeline.

The data reduction software package validation usually comes at the end of the instrument building and integration process. Unavoidable delays in these phases usually shorten the time devoted to data reduction software system testing. As a consequence, quite a number of instruments start operating with an insufficiently tested pipeline. The general science user, if he or she is not a member of the instrument building team, will then most likely encounter many difficulties in dealing with the software and getting its science done. This is a very bad start that could have a devastating impact on potential customers. In some extreme cases, these difficulties have remained unsolved, leading to early demise of the instrument.

A way to anticipate and to shorten the time needed between instrument final integration and start of operation is to build an instrument numerical model and/or use a prototype that shares the main instrument characteristics. Such tools allow performing early testing of the integrity and validity of the data reduction scheme.

Even the best data reduction software carefully designed and realized by experienced astronomers and professional software engineers and carefully tested with advanced instrument numerical models will have to be tuned and adapted to the real characteristics of the instrument. Deep knowledge of the instrument behavior will take time – it is more a matter of years than months for a complex instrument – and software development is expected to continue during its whole operating life, at least at a minimum level.

Optical 3D-Spectroscopy for Astronomy, First Edition. Roland Bacon and Guy Monnet.
© 2017 Wiley-VCH Verlag GmbH & Co. KGaA. Published 2017 by Wiley-VCH Verlag GmbH & Co. KGaA.

In this chapter, we will present the basics of data reduction software that are common to most of the instruments presented in this book. We then discuss the specificities related to each instrument species and present some more advanced concept. Finally, a detailed example in the form of the MUSE integral field pipeline is described.

11.2 Basics

In the following, we describe a generic data reduction process largely valid for most of the instrument types described in this book (Figure 11.1). More specific and detailed data reduction processes will be described in the following sections.

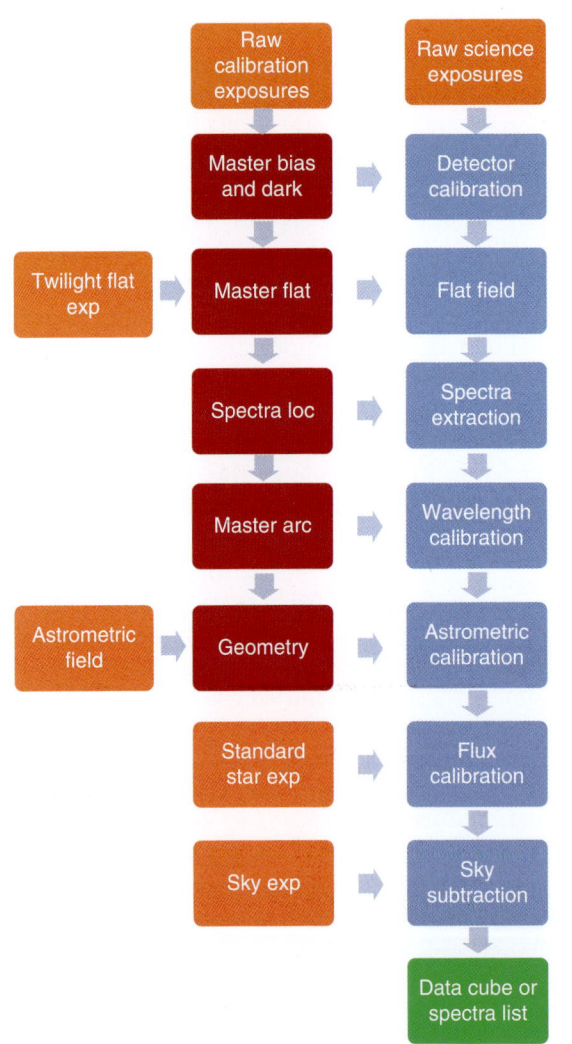

Figure 11.1 Schematic of MOS or IFU data reduction process.

The input for the data processing is a set of raw files: the science and the corresponding calibration raw files (see Section 10.5). Depending on the instrument and the type of science project, the data volume may be more or less big: for example, a typical night with long exposures with a multiobject spectrograph produces typically ~100 Mb of raw data, while a series of short integrations with MUSE can produce up to 250 Gb in a single night.

The process starts with detector calibration. A bad pixel mask, which flags all dead, hot, and nonlinear pixels, is created using dark and flat frames. A high SNR master bias frame is created by averaging a set of raw bias exposures after a sigma-clipping cycle to remove possible outliers. Similarly, a master dark frame is created using a series of dark exposures. The bias and the dark current are then subtracted from each science raw exposure. If the bias level is not stable enough, it might be necessary to use the prescan and/or overscan pixels directly from the science raw exposure to derive a more accurate value. A series of raw flat field exposures are then bias subtracted, combined, and normalized to produce a high SNR master flat field. The science exposure is then divided by this master flat field to remove the inhomogeneity in the detector response and possible light vignetting in the optical train. Twilight exposures can also be used to provide additional low spatial frequency corrections.

The master flat field is also used to define the location of the spectra on the detector. The process must take into account the distortion of the optics, either using a physical model of the instrument with some fitting parameters or measuring the spectra location directly on the detector. The first technique is generally more robust since it is using some a priori information to predict spectra location on the detector. The second technique has the advantage of being model independent but is more sensitive to noise and outliers.

The next step is wavelength calibration. For a grating or prism spectrograph, a high SNR master arc exposure is produced using a series of arc lamp individual exposures, which are combined after bias subtraction. Using the computed spectra location, the arc lamp spectra are then extracted and the location of the emission line peaks derived. These locations are then compared to a table of wavelengths of the main emission lines for the chemical elements present in the lamp. Matching pattern algorithms are used to identify the appropriate wavelengths. The pixel–wavelength mapping is then derived.

Geometrical calibration is performed using the appropriate raw exposures (including astrometric calibrations if available), and the spatial location of each spectra is saved in an appropriate reference table.

Science spectra are then extracted and rebinned in wavelength using the measured pixel–wavelength relationship. Empty parts of the field, or specific offset exposures, are used to extract the sky spectrum, which is then subtracted from the science spectra.

The same process is used for the spectrophotometric standard star observation. The wavelength calibrated standard star spectrum, summed over an area large enough to get the total flux, is compared to the reference absolute flux table. The ADU–flux conversion is then derived and used to calibrate in flux (e.g. $\text{erg s}^{-1} \text{ cm}^{-2} \text{ Å}^{-1}$) the science spectra.

For a MOS-type instrument, the result of the data reduction process is a set of n spectra $S_k(\lambda)$ and their corresponding spatial location α_k, δ_k, with $k = 1, \ldots, n$. For an IFU-type instrument, there is an additional step. The spectra are interpolated on a regular spatial grid resulting in a 3D data cube $C(\alpha, \delta, \lambda)$.

11.3 Specific Cases

In this section, we discuss the peculiarities of data reduction for the various instrument types described in this book. We also discuss enhanced data reduction schemes that go beyond the basics described in the previous section.

11.3.1 Slitless Multiobject Spectrograph

Data reduction of slitless MOS (Section 2.2.2) is quite complex given the source and background overlaps at the detector plane. This has been the subject of a major development in the 2000–2010 period for the slitless multiobject modes of the Hubble Space Telescope and in particular for its ACS (UV-Optical) and NIC-MOS (near-infrared) cameras. A typical data-reducing algorithm incorporates the following main steps:

- Each slitless data cube is accompanied with a direct image, preferably taken with the same instrument without disperser and in roughly the same wavelength range.
- A standard extraction SW package (SExtractor [107] in that case) then produces a photometric/morphological catalog of the objects automatically identified in the field.
- Together with photometric and spectroscopic calibrations, this provides the necessary information to relate each spectrum to its corresponding object and proceed to an optimum extraction (i.e., weighting each spectrum pixel contribution in order to maximize signal-to-noise ratio), calibrate the spectra in wavelength and intensity, compute their spectral resolutions, and quantitatively assess contamination due to overlapping spectra.

Needless to say, we are talking about hundreds of thousand lines long data reduction packages that took years to develop, test, and refine.

11.3.2 Scanning Fabry–Pérot Spectrograph

Data reduction of scanning Fabry–Pérot (see Section 3.3) data is very different from grating spectrography because of the way the wavelengths are recorded. A set of exposures are obtained with different width of the etalon optical cavity. Each exposure corresponds to a specific interferometric pattern in the form of circular rings corresponding to the various interference orders. The analysis of a calibration monochromatic source such as an arc lamp allows deriving the corresponding transformation between (x, y, n) and $(\alpha, \delta, \lambda)$, where x, y are the detector coordinates, n is the exposure number (corresponding to a given etalon width), (α, δ) are the sky coordinates, and λ is the wavelength. This is called the phase correction.

All science exposures are then processed to produce a data cube. Because of the complex mixture of wavelength, spatial coordinate, and exposure number, any variation of atmospheric conditions will impact the recovered spectral distribution in a nonhomogeneous way. The difficulty is then to take into account possible changes of image quality and atmospheric transmission between exposures. This can be in principle solved by repeated fast scanning of the wavelength range so that fluctuations of the atmospheric conditions are well averaged. This is, however, not always possible because of the impact of readout noise and the increasing overheads.

An alternate way is to produce a same or higher spectral resolution spectrum in the same wavelength range of a bright enough star in the field, for example, with an integral field spectrograph: this spectrum is then used as a spectrophotometric "ruler" to correct frame to frame transmission variations in the FP data. Frame to frame variations of the star profile could in principle be used to also correct image quality fluctuations, but requires an even brighter field star, not necessarily available.

11.3.3 Scanning Fourier Transform Spectrograph

Similarly to scanning Fabry–Pérot spectrographs, scanning Fourier transform spectrographs (see Section 3.4.2) have a different way to record the variation of flux with wavelength. A set of exposures are performed with various optical path differences, which results in an interferogram at each spatial location. Phase correction derived from a reference calibration is used to transform (x, y, n) to (α, δ, ν) where x, y are the detector coordinates, n is the exposure number (corresponding to a given optical path difference), (α, δ) are the sky coordinates, and ν is the light frequency. A Fourier transform of the interferogram then gives the flux distribution at each detector pixel as a function of the wavelength, resulting in a data cube $(\alpha, \delta, \lambda)$.

11.3.4 Getting Noise Variance Estimation

Because of strong OH atmospheric emission lines above roughly 630 nm, the sky subtracted signal of an object fainter than the sky will have an SNR that features strong variations with wavelength. Quantitatively speaking, at the location of a bright sky emission line, the noise can easily be 100 times larger than a few Å away. Getting a good estimation of the noise on each spectral pixel is then very important for faint object spectroscopy.

In principle, it is quite easy to get an estimate of the noise variance in the raw science data. As presented in Sections 8.3 and 10.3, it is, to the first order, the sum of the photon noise variance and the detector readout noise. If we approximate the photon noise variance by the number of detected photons, we have an approximation of the true variance at each detector pixel. It is relatively easy to propagate this variance information along the various transformations implied by the data reduction scheme described in the previous section. However, this works only if the points of measure are independent. This is true, for example, for bias and flat-field correction, which works at the pixel to pixel level. But for wavelength calibration, an interpolation is required to put all extracted spectra along the same grid and the independent point assumption is not valid anymore. This is also problematic for the transformation from pixel to sky coordinate.

The correlation introduced by the interpolation scheme can be taken into account by propagating the covariance matrix rather than the simple variance. In practice, however, the size of this matrix is often too large. Let us take, for example, the interpolation on sky coordinate with a seeing of 0.8" with a sampling of 0.2"; the covariance matrix, even restricted to the non-negligible values, will be 8^2 or 64 spaxels. Adding the wavelength interpolation will add a factor of 5 at least, resulting in a covariance data cube 2 orders of magnitude larger size than the data cube itself.

To limit noise estimation problems resulting from these multiple interpolations, new data reduction schemes have been developed (e.g., MUSE [108] and KMOS [109]), which keep the raw pixel information until the last data reduction steps. Within this approach the various coordinate transformations are applied at the detector pixel level, not modifying the independent flux values except for simple scaling factors. The data reduction process results in a large table of calibrated flux values sample on an irregular grid. One can then work directly on this table, or perform a single interpolation to compute a regular data cube $(\alpha, \delta, \lambda)$. In the latter case, the correlated noise problem is still present, but it avoids the multiple interpolation process contribution.

11.3.5 Minimizing Systematics

The goal of the data reduction process is to remove the instrumental signature in order to get fully calibrated source signals in physical units. Even with much care to get the correct calibration files and a bug-free data reduction software, there is always some instrumental signature left in the processed data. This residual signal is called the systematics.

As discussed in Section 10.4, a dithering strategy can help mitigate the impact of the systematics by averaging them over many exposures. When enough exposures are performed one can use a better combination scheme than just simple averaging. The median has the advantage to be less sensitive to outliers. A hybrid method is to remove outliers by using sigma-clipping around the median value and then performing an average of the clipped data.

In imaging a successful technique called super-flat-fielding has been developed to improve flat-field correction to much better than the 1% level. When a large number of different observations are performed in similar conditions (e.g., integration time, moon brightness) and if the fields are relatively sparse, it is possible to combine all exposures in such a way that objects are averaged out, leaving a combined sky exposure. The deviation from flatness of this combined sky exposure can then be used to correct the science exposure. The advantage of this method is that it uses science exposures with similar illumination and dynamics to create a self-calibrated flat field, which is generally much better than flat fields obtained from a calibration exposure. This approach can also be used with integral field spectrographs. It works as for imaging, except that it results in a flat-field data cube rather than a simple flat-field image.

11.4 Data Reduction Example: The MUSE Scheme

The previous sections may give the feeling that performing data reduction is straightforward. This is only true with respect to the general principles. The devil is in the details. In reality, performing state-of-the art data reduction of data taken with any of the instrumental flavors described in this book is not easy and requires, even for the experienced user, some attention and a lot of perseverance (Figure 11.2).

As a specific example, we detail in this section the data reduction flow using the MUSE IFU pipeline (Figure 11.3). This section is not intended to be a cookbook for MUSE observers, which can be found at www.eso.org/sci/facilities/paranal/instruments/muse/doc.html, but it should allow the reader to get an idea of the whole process. The MUSE pipeline has been optimized to deal with the large data volume provided by the instrument and to keep track of the data variance. The detailed process is described in [108].

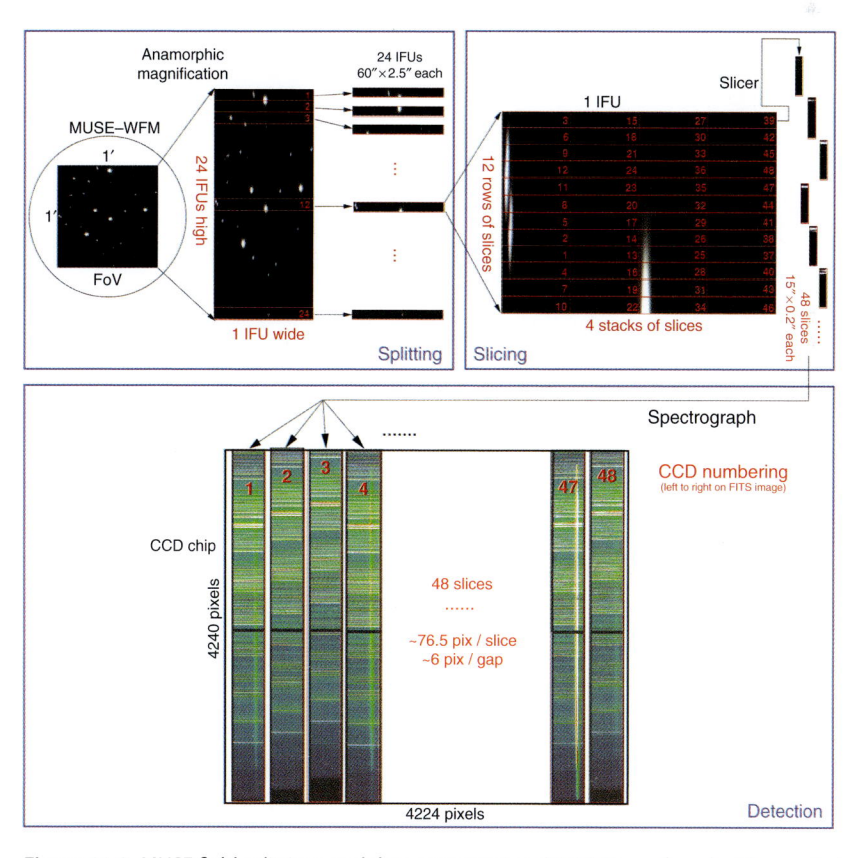

Figure 11.2 MUSE field splittings and data organization. ([108]. Reproduced with permission of Peter Weilbacher.)

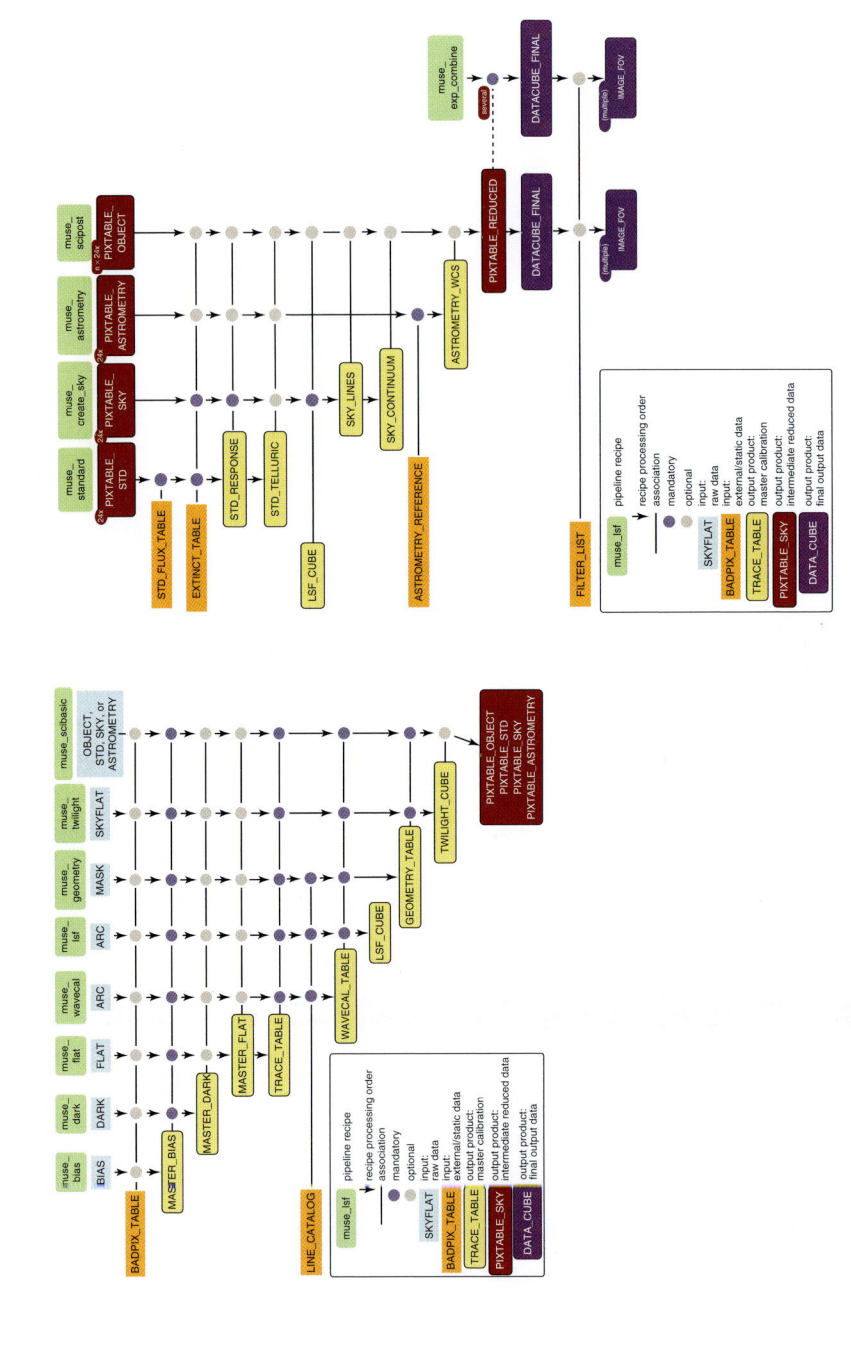

Figure 11.3 MUSE pipeline schematic. ([108]. Reproduced with permission of Peter Weilbacher.)

11.4.1 Detector Calibration

As described in Section 5.4.1, MUSE has 24 channels, each one featuring a CCD with 4096 × 4112 active pixels. To minimize readout time, each detector has four readout amplifiers that are used in parallel. The effective detector area is thus split into four quadrants. Overscan regions are added at the edge of each quadrant, giving a cross-pattern visible on each raw frame.

The daily standard set of calibrations include 11 bias exposures. A bias is a zero-integration time exposure, which gives the number of counts of the offset voltage applied to the detector. An example of a bias raw frame is displayed in Figure 11.4a. Note that each detector must be considered separately. The gain, bias level, and noise property are thus different from quadrant to quadrant.

The basic idea of the bias recipe is to average the 11 bias to provide a master calibration bias file with high SNR and subtract it to the science exposures. However, the bias structure is not perfectly stable with time and the overscan regions on the detector must be used to renormalize each bias before performing the combination of the 11 exposures. In the first year of operation we have discovered that in some quadrants, the bias level features a significant vertical ramp-up, which means that it takes a number of pixels from the edge of the detector before reaching its constant level. This ramp-up was not stable with time and to take it into account a polynomial fit is performed on the overscan region for each quadrant and subtracted to the bias exposure. This process removes the bias level leaving only its 2D structure, which is thought to be stable enough to be averaged over the 11 bias exposures. The overscan regions are also stripped out by the recipe. A sigma-clip is then performed to remove deviant pixels before doing the combination. Hot pixels characterized by systematic high values are also identified and marked as bad pixels. These computations are parallelized for the 24 channels in order to speed up the processing.

Note that this problem of bias stability was of course not foreseen in the design of the pipeline and has been discovered because of left-over systematics in the resulting science exposure. However, given the time variability of the problem, the fact that it was not systematically present on the 96 quadrants, and the size of

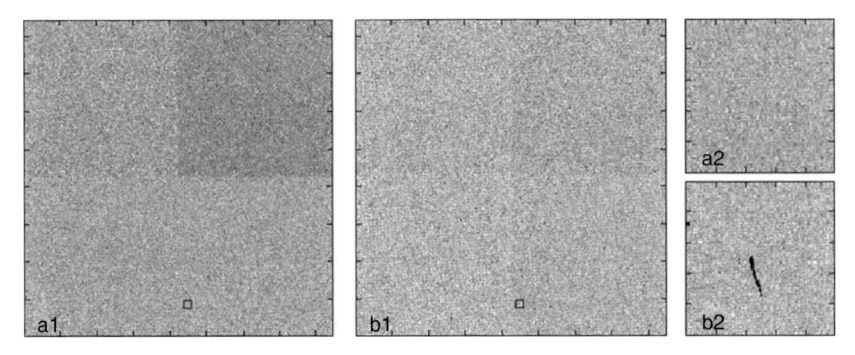

Figure 11.4 MUSE raw detector calibration exposures: (a1) channel 12 of bias exposure, (b1) 30 mn dark exposure of the same channel. A zoomed window, displayed as a red square in (a) and (b), is shown in a2 (bias) and b2 (dark). Note the impact of cosmic ray in the dark exposure.

the MUSE exposure with some 400 million pixels, it was not so easy to find the source of the problem and it took some time before a solution could be implemented. This illustrates the importance of quality control during an instrument development, including the need of quality control parameters to monitor what is happening at each stage of the data reduction process.

A number of dark exposures (i.e., integration with shutter closed) are also regularly obtained to measure the dark current properties of the detectors (see Figure 11.4b). The dark subtraction recipe starts with removal of the bias from the exposures. Similarly to the master bias process, the bias vertical structure is removed by polynomial fitting on the overscan region of each individual dark exposure. Then the master bias is subtracted to remove the remaining 2D bias structure. The exposures are then averaged after sigma-clipping of the deviant pixels and scaled according to the exposure time. Hot pixels are also identified and marked as bad pixels. The resulting master dark exposure contains the dark current level and spatial structure for the 96 quadrants.

The MUSE e2v CCD detectors have a small dark current of the order of 1 electron per hour. It is thus very difficult to get a master dark exposure with good enough SNR. In general, dark current contribution is neglected in the processing. However, dark exposures are useful to monitor the detector characteristics. For example, the problem of inaccurate bias subtraction reported in the previous paragraph was more visible in the processed master dark than in the science frame.

11.4.2 Flat-Field Calibrations and Trace Mask

The daily standard set of calibrations also includes 11 flat-field exposures. These exposures are obtained with the calibration unit and two tungsten lamps with different color temperature in order to have a good signal over the wide spectral range of MUSE. An example of a raw flat field exposure is displayed in Figure 11.5a.

The first step of the flat pipeline recipe is identical to the master dark recipe: a combination of the 11 exposures into a single high SNR master flat after removal

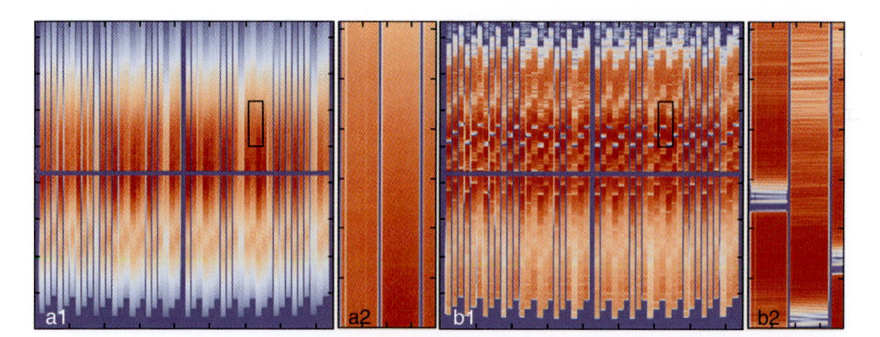

Figure 11.5 MUSE raw flat fields calibration exposures: (a1) channel 12 of internal tungsten lamp, (a2) zoom of the small window in a1, (b1) twilight sky exposure, and (b2) zoom of the small window in b1. In addition to the sun typical absorption spectral features, one can see the strong telluric absorption in the red due to the Earth's atmosphere.

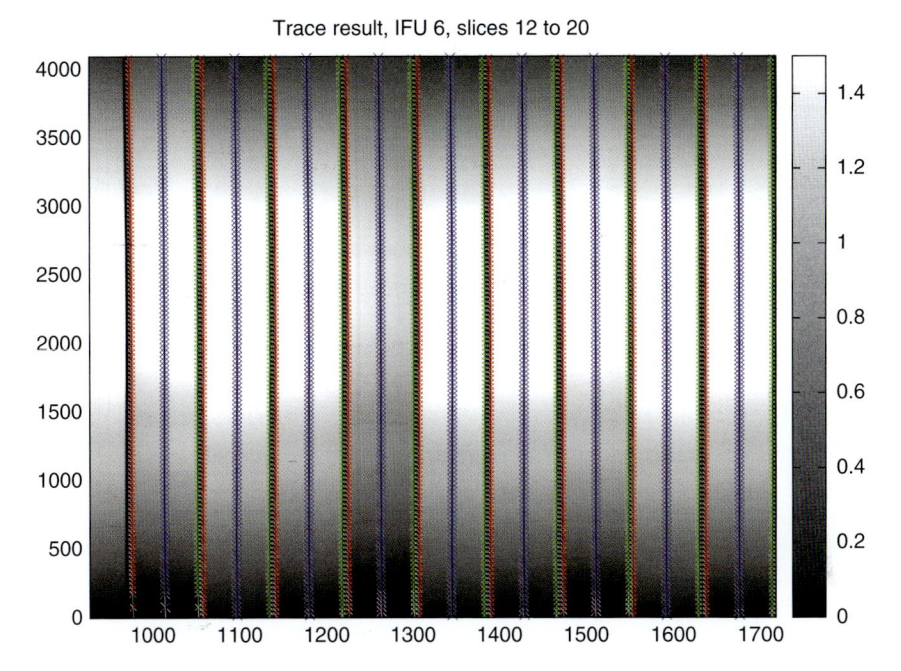

Figure 11.6 Example of trace mask. The green and red curves give the edge location of each slice on the detector. ([108]. Reproduced with permission of Peter Weilbacher.)

of the bias level. Dark pixels are also identified and marked. The second step is the location of the slices on the detector. To trace their positions, a flux threshold is applied to locate their edges. This is repeated along the wavelength direction. Deviant positions are identified and a polynomial fit of the edge positions of each slice is then performed and saved in a table called the trace mask. A typical result is shown in Figure 11.6. The central width and the average flux for the 1152 slices (48 slices times 24 channels) are also saved as quality control parameters.

These two set of parameters are essential to monitor the alignment of MUSE and the impact of temperature variations. Channel misalignment of the splitting and relay optics in the CCD row direction result in truncation of the slice width on one side of the channel and flux variation between the upper and lower slices due to misalignments in the column direction. A regular monitoring of these parameters is performed as a system health check.

The master flat field allows efficient removal of the pixel to pixel flat field response. It is, however, not fully representative of the on sky flat-field response; the reason is that the calibration unit is not a perfect match of the telescope pupil, and the integrating sphere flux is not perfectly isotropic. In addition, we found that it is slightly vignetted on one edge by some mechanical element. The solution is to use an exposure performed during twilight, which is much more representative of a night exposure. The disadvantage is that the sun spectrum with its numerous absorption lines is not smooth with wavelength (see Figure 11.5b for an illustration). In the MUSE pipeline we use a combination of the two: the internal flat field for the high frequency pixel-to-pixel variation and the twilight flat for the low frequency spatial and spectral variation.

Because of temperature changes during the night, slight variations of slice illuminations can occur. This effect can be important for large temperature variations (e.g., a few degrees). To correct for these, one could just perform a series of flat field exposures and compute a master flat field that matches the temperature of the science exposure. This would add too many overheads and is in fact not really needed. Instead, a single short flat exposure is obtained at regular intervals during the night or when the temperature changes. This flat exposure is then used to correct the slice illumination on the associated science exposure. This process is described in more detail later in Section 11.4.5.

11.4.3 Wavelength Calibrations

The last element of the daily calibration set is the wavelength calibrations. Three different arc lamps are used to get full coverage of the MUSE wavelength range (4650–9300 Å): a mercury–cadmium (HgCd) lamp, a neon (Ne) lamp, and a xenon (Xe) lamp. See Figure 11.7a for an illustration. Each lamp is exposed separately with different integration times to avoid saturation of bright lines and to limit lines blending. Exposures are repeated five times to improve the SNR on faint emission lines. The resulting 15 exposures are processed by the wavecal pipeline recipe.

For each lamp, the arc exposures are bias subtracted and averaged. The arc spectrum is extracted at the center of each slice using the previously computed trace mask. Arc emission line positions are detected using SNR thresholding and identified using pattern matching with the corresponding input reference lamp emission catalog. A Gaussian fit is performed to obtain the precise center of the line and its centering error and FWHM. This is repeated for each column of the slice and for each lamp. The final wavelength solution for each slice is derived from a two-dimensional polynomial fit of all positions of all lamps and their corresponding catalog wavelengths. This fit is iterated to reject outliers until full convergence. An example of the resulting solution is displayed in Figure 11.8.

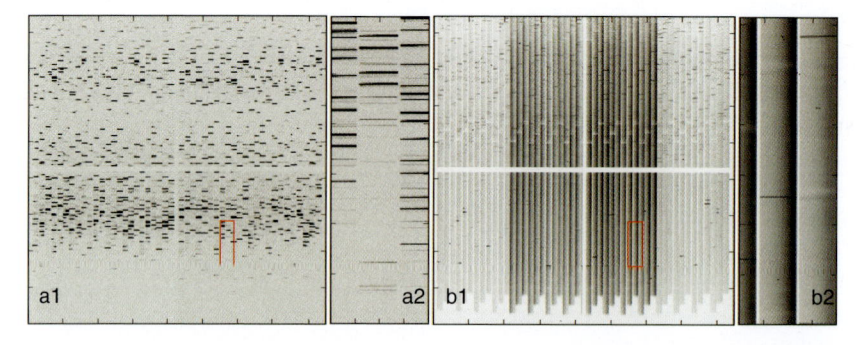

Figure 11.7 MUSE raw wavelength calibration exposures: (a1) channel 12 of the internal Ne arc lamp of 0.7 s exposure (zoomed view of the window marked in red is shown in a2). (b1) Example of a science 400 s exposure centered on NGC 3379 elliptical galaxy nuclei. Note the [OI] brightest sky line that can be seen on top of the galaxy bright continuum.

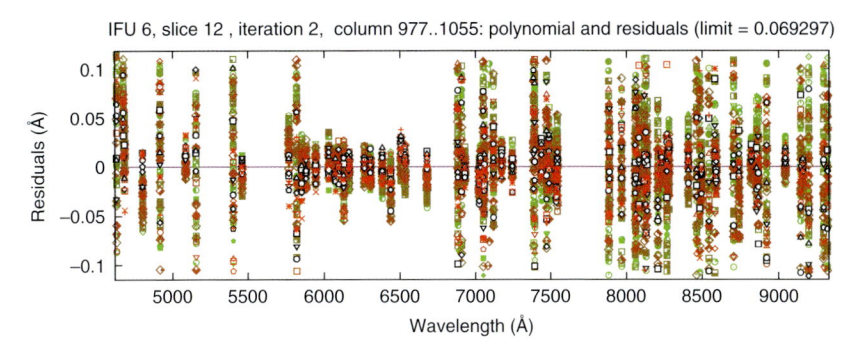

Figure 11.8 Example of MUSE wavelength calibration result. Here we see the residuals for one slice. ([108]. Reproduced with permission of Peter Weilbacher.)

This wavelength solution is used to calibrate the sky exposures in wavelength. Temperature variations during the night can introduce some shifts in the wavelength calibration. Fortunately, in most cases the brightest sky emission lines in science exposures can be used to compute on the fly the small resulting offset (a fraction of pixel) to the calibration solution. See Figure 11.7b as an illustration.

A Gaussian fit gives the line properties and in particular the FWHM of the spectral line profile. This so-called line spread function or LSF is an important parameter of the instrument. In some cases, a more precise estimation of the LSF than just a Gaussian fit is needed. This is the case for example, for accurate sky subtraction or when working with science objects with unresolved emission lines. A dedicated recipe performs this task. This is not easy for MUSE because of its poor sampling of the LSF ($\sim$2 pixels only).

11.4.4 Geometrical Calibration

The geometrical calibration is the process by which the position of the slices on the detector are related to their location within the field of view. This is not trivial given the two splitting stages of the instrument: the first one splits the field in 24 channels and for each channel there is an additional splitting in 48 slices performed by the image slicer (see Figure 11.2). Differences in the manufactured system as well as in alignment and tilts of the numerous optical elements involved result in a complex geometry. Failure to match this geometry would result in poor image reconstruction. Such difficulties are well known for integral field spectrographs of any kind and often, when not well enough calibrated, have resulted in the degradation of final image quality.

In MUSE the 1152 slice locations in the field of view are calibrated using a special sequence involving a multi-pinhole mask in the calibration unit focal plane. A large number of exposures are obtained with the mask being vertically shifted by small steps. The software measures precisely when the pinhole enters and leaves the slices by monitoring their illumination. Together with the known basic properties of the mask, the offsets of the shifts, and the instrument design, this data allows determining the relative location (x and y position) and angle of each slice of each channel of MUSE. Data taken with another mask after the exposure sequence are used to verify the output of the procedure.

11.4.5 Basic Science Extraction and Pixel Tables

After completion of all these calibrations, the next step is to remove the instrument signature from the science raw data still leaving the atmospheric signature (e.g. atmospheric dispersion, sky emission and absorption), which will be removed at a later stage. For now, we concentrate on removing the part due to the instrument itself.

A major goal of MUSE is extremely faint source spectroscopy. In this respect, getting the signal is not enough as accurate noise knowledge is equally critical. As discussed in Section 11.3.4, in order to propagate this noise information along the various data processing steps it is mandatory to avoid any resampling that would add correlated noise to the dataset. For MUSE we have then adopted a one-step resampling scheme, which used the concept of "pixel table." This table contains a list of all illuminated pixels (the data values, the bad pixel status, and the variance) of one exposure together with the coordinates (x, y, and λ).

The creation of the pixel table is performed by the *scibasic* pipeline recipe. The input of the recipe is a raw science exposure (see some examples in Figure 11.9). The raw data is trimmed to remove the overscan region, bias subtracted and flat fielded using the master flat and the twilight flats. Each pixel value, converted from ADU to photoelectrons using detector quadrant gains, pixel location on the detector, and the channel number are saved to the pixel table. The corresponding slice number is also saved. Using the geometrical and the wavelength calibration solutions, the corresponding sky coordinates and wavelengths are added to the table. Optionally, the wavelength calibration offset is derived from a bright sky line Gaussian fit and added to the wavelength solution.

Noise variance is estimated as the sum of the readout and the Poisson noise. It is properly propagated using the various mathematical operations involved in the process (e.g. conversion, bias, and flat-fielding). This approach assumes that the CCD pixels are uncorrelated, which is true to first order, and that Gaussian error propagation is applicable.

Correction of illumination effects due to temperature change is then performed using the appropriate single flat field obtained at a similar temperature during the night. The total flux (summed over the slice and wavelength) of each slice is computed and compared to the corresponding value of the master flat field. The pixel values are then renormalized by the ratio of these two values. The output of the *scibasic* pipeline recipe is a pixel table; see an example in Table 11.1.

11.4.6 Differential Atmospheric Correction

As discussed in Section 9.5.2, observations performed away from the zenith suffer from atmospheric dispersion and may result, if not corrected, in variations of the image centroid with respect to the wavelength. Given the absence of atmospheric dispersion corrector for the MUSE wide field mode, its wide spectral range, and its blue spectral limit (4650 Å), the impact starts to be not negligible at air mass >1.4 and must be corrected a posteriori by the pipeline. The pipeline uses an

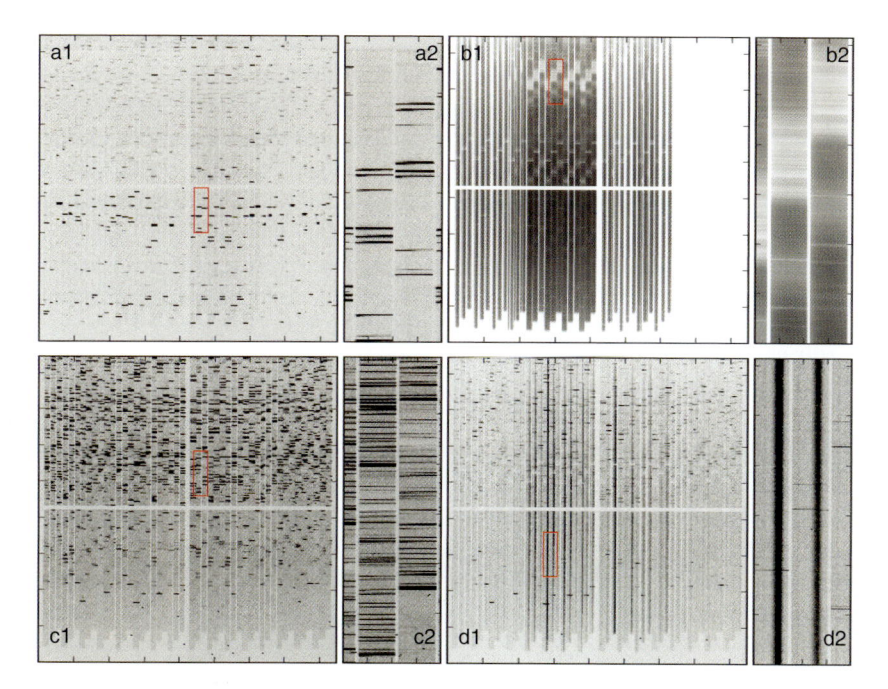

Figure 11.9 Examples of raw science exposures. For each object, one full channel is shown plus a zoom on a specific region (red rectangle). (a) The planetary nebulae NGC 3132 (60 s exposure). Note the bright emission lines all over the field. (b) Planet Saturn (1 s exposure). The deep CH_4 absorption bands can be seen in the red region. (c) A 25 mn exposure in the Hubble Deep Field South. The raw image is dominated by the bright sky OH emission lines. (d) The brightest stars in the globular cluster NGC 6752 (120 s exposure) are clearly visible with their strong continuum.

Table 11.1 Example of a pixel table content created by the *scipost* pipeline recipe. XPOS and YPOS are the spaxel coordinates (here as α, δ offset from the field center) and LAMBDA is the wavelength in Å.

XPOS	YPOS	LAMBDA	DATA	DQ	STAT	ORIGIN
4.740865707	−0.000191712	4749.747	96.415	0	751.245	1510314049
4.744184017	−0.000192353	4749.808	1.181	0	38.406	1527091265
4.747479916	−0.000192996	4749.869	42.447	0	344.738	1543868481
4.750754356	−0.000193641	4749.931	14.472	0	138.085	1560645697
4.754006863	−0.000194288	4749.992	63.546	0	492.823	1577422913
4.757237911	−0.000194937	4750.054	58.388	0	448.521	1594200129
4.760447025	−0.000195588	4750.116	41.018	0	331.388	1610977345
4.763635159	−0.000196241	4750.178	35.509	0	290.086	1627754561
4.766802311	−0.000196896	4750.240	22.667	0	199.526	1644531777
4.769948006	−0.000197553	4750.302	37.454	0	429.687	1661308993

Corresponding data and variance values are in the DATA and STAT columns. DQ is the data quality flag (0 means the voxel is ok). ORIGIN stores the channel number and the voxel location on the detector. The full table contains 323 920 150 similar lines.

algorithm from Filippenko [110], which is based on Owens's formula [111] to convert relative humidity to water vapor pressure. This model gives good results in most cases. If a very accurate correction is needed and if there is a star bright enough in the field, the alternative is to fit the star centroid with wavelength and apply this computed empirical correction to the data set.

11.4.7 Sky Subtraction

Sky subtraction is a critical step. A good sky calibration is a prerequisite for long exposures on faint objects. This is especially relevant for deep field observations, especially above 7000 Å where the OH emission lines are bright and numerous. If the objects are small and sparse enough, one can use the many empty spaxels to estimate the sky contribution. In the more tricky case of nearby objects larger than the field of view, the only way to estimate the sky contribution is to use an offset exposure on a nearby empty field. This additional exposure can be short, given that many spaxels can be used to produce a high S/N sky spectrum. While the continuum component of the sky spectrum is stable enough to be estimated before or after the science exposure, this is not generally the case for the OH emission lines that exhibit strong variations on shorter time scales. Fortunately, nearby objects are often bright and thus sky subtraction is not as critical as it is for deep field observations.

The field of view of MUSE is small enough that the sky contribution can be considered as constant over the field of view. Thus, one could think of just averaging the spectra at the empty spaxels and then subtracting this mean spectrum to all spaxels of the field of view. This simplistic approach is, however, not accurate enough and leaves strong residuals. The reason is that, even if this intrinsic sky contribution is indeed fairly constant over the field of view, line spread functions (LSF), because of optical aberrations, change slightly from place to place. This does not impact the sky continuum subtraction because its variations are smooth with respect to the LSF. But this is not anymore the case for the emission lines that are unresolved and thus are just equal to the local LSF, which in the case of MUSE is just 2 pixels wide and thus a bit undersampled. Imperfect peak subtraction leads then to strong oscillations if the LSF variations are not taken into account.

The other difficulty in the proposed scheme is to identify blank sky regions properly. One can look at the white light image but a spatially extended structure with emission lines but no or little continuum would not show up. There is thus a risk that the blank sky region defined from a white light image will then subtract spatially extended emission lines.

The sky subtraction scheme implemented in the MUSE pipeline [112] tries to circumvent these difficulties by using a detailed model of the LSF variation across the field of view and a physical atmospheric model of the sky contribution. The LSF model is derived by fitting the high S/N lines profiles obtained with the arc lamp exposures. Different parametric and non-parametric models have been used. The physical model of the sky uses the the atmospheric emission lines properties that originate from molecular transitions [113, 114]. The lines are grouped by the originating molecules (OH, [O I], Na I, O2) and the upper

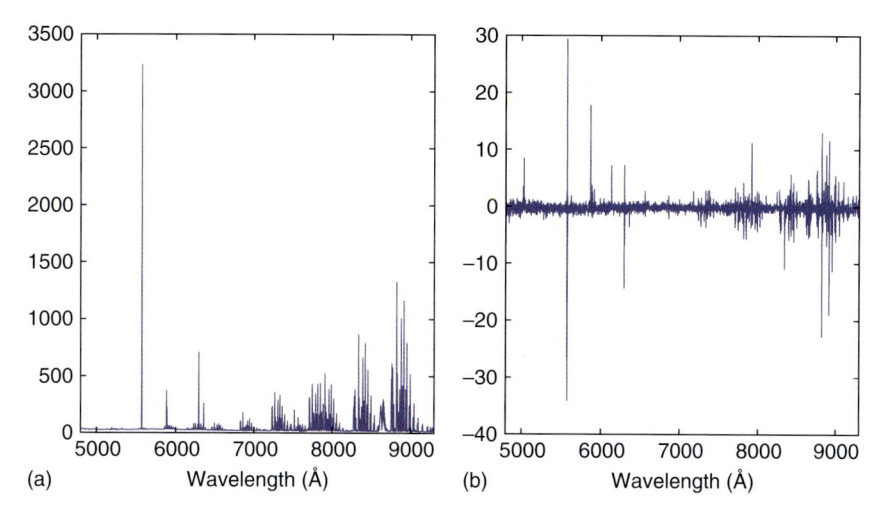

Figure 11.10 Example of sky subtraction in a deep field 25 mn exposure. Spectrum of an 'empty' location averaged over a 1 arcsec aperture before (a) and after (b) the sky subtraction. Note that while proper sky subtraction indeed removes its average contribution, it does not (and cannot) remove its Poisson noise contribution. Flux are 10^{-20} erg s^{-1} cm^{-2} Å^{-1} units.

transitional levels into twelve groups. Within each group, the emission flux ratio is fixed. This is used by the fitting process, together with the LSF model, to derive the sky spectrum, before LSF convolution. This fitting of the sky physical model does not consider other emission lines than the known OH molecular and atomic lines and thus is robust against accidental removal of astrophysical diffused emission lines, except of course if they fall at the same location as the atmospheric lines, but in this case they would not be detectable anyway. The pipeline recipe operates at the pixel table level to avoid any interpolation. An example is given in Figure 11.10.

11.4.8 Spectrophotometric and Astrometric Calibrations

The next step in the calibration process is to remove the atmospheric and telescope signature. Spectrophotometric calibration is used to convert the measured counts in proper flux units, for example, erg s^{-1} cm^{-2} Å^{-1} in the CGS system. A spectrophotometric standard star with proper absolute spectrophotometry tabulated flux must be observed. The pipeline computes the total flux of the star by integrating over its spatial extend and derives, by comparison with the tabulated reference values, the instrument response as a function of the wavelength. Correction for air mass is applied and this instrument response can then be used to convert detector counts to flux. The response curve must be smoothed to avoid the addition of spurious high frequencies in wavelength to the object spectra. It must also be interpolated across the absorption telluric bands of the atmosphere. There are two main broad absorption features in the wavelength range of MUSE. This results in a telluric correction spectrum that is applied as well to the object data cube. An example of a response curve and a telluric correction is given in Figure 11.11.

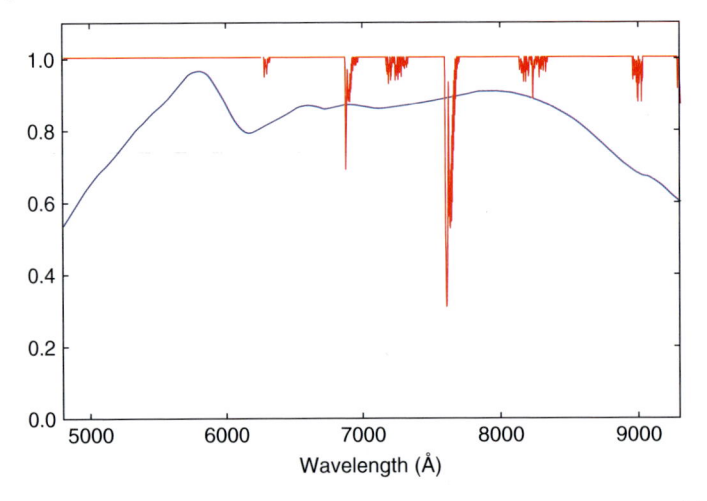

Figure 11.11 Spectrophotometric calibrations: instrument response (in blue, arbitrary units) and telluric corrections (in red).

This calibration is of course valid only if the observations are obtained in the same *photometric* conditions as the standard star observation. Given that atmospheric conditions are by essence variable, it is recommended to observe at least one spectrophotometric standard star for each night and preferably two. Ideal time is just after sunset or before sunrise, when the sky is too bright for doing scientific observations but dark enough for a short exposure on a bright standard star.

An ideal standard star should have its spectrophotometry known at least at the same spectral resolution as the instrumental setup, should be bright enough, but not too much to avoid too short integration time, and – most importantly – its spectrum should be as featureless as possible. The reason for the latter requirement is that most stars have, like the sun, many deep absorption lines. The sharp features associated with the absorptions lines make them not well suited to derive a smooth response curve and inappropriate to derive the atmospheric telluric absorption correction. White dwarfs have the appropriate smooth spectra (see an example in Figure 11.12), but they are generally faint and only a few have been observed with the required accuracy.

The astrometric calibration recipe is used to remove the remaining field distortions due to the instrument and the telescope itself. An astrometric field of stars of precise known location is observed. Given the relatively small field of view of MUSE, we need a reference field dense enough to get a sufficient number of bright point sources in the MUSE field, but not too dense to avoid crowding effects. The best fields are the outskirts of globular clusters that have been observed at high resolution by the HST (see an example in Figure 11.13). In a typical field, we get up to 30 suitable stars, which is sufficient to obtain a good astrometric solution. The process reconstructs white-light image from the processed data cube, and detects and fits positions of the stars found in the field. The positions are then

11.4 Data Reduction Example: The MUSE Scheme | 231

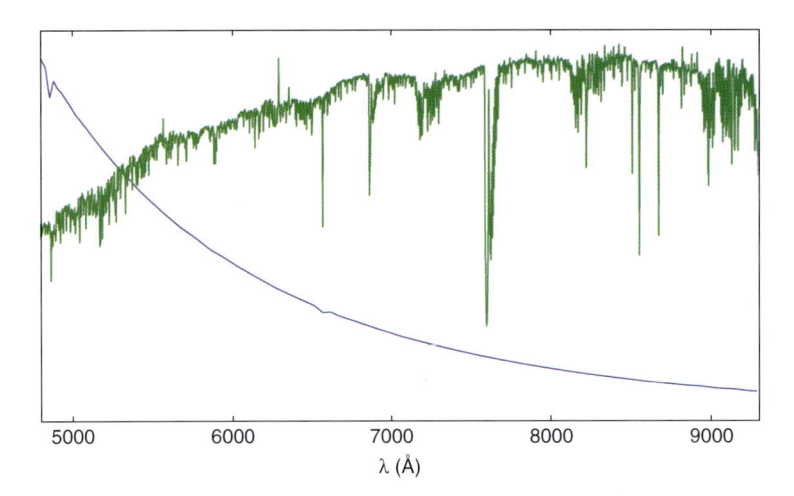

Figure 11.12 The spectrum of the sdO type HD49798 standard star (in blue) compared to a classical K star (in green).

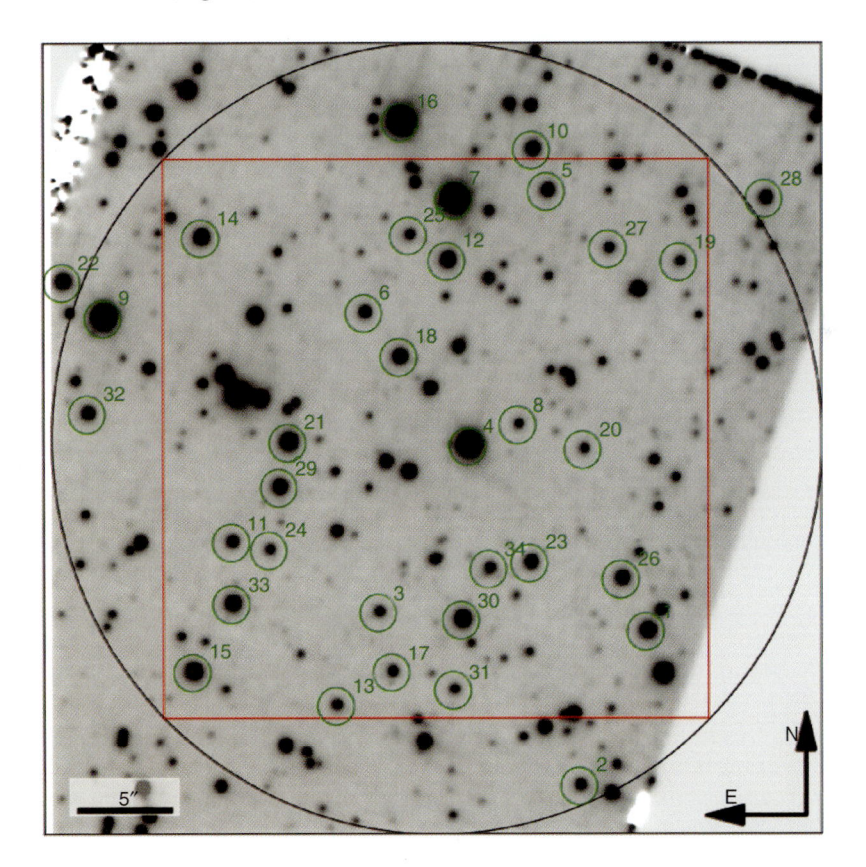

Figure 11.13 The outskirts of the NGC 3201 globular cluster used for astrometric calibration (courtesy Sebastian Kamann). The MUSE field of view is displayed in red. The isolated stars (with less than 1% contamination from close neighbors within 2 arcsec) are identified as green circles overlaid over an HST image of the cluster. (Reproduced with permission of Sebastian Kamann & NASA.)

matched against the reference catalog and a 2D polynomial fit is performed to reproduce the distortion. This solution is then applied to the science exposures.

The astrometric observations must be performed in good seeing conditions to obtain an accurate mapping. The solution is stable enough and these observations do not need to be done very frequently. It is better to run the astrometric and the geometric recipes together: the geometric one can be performed the day following the night observation of the astrometric field. Both are critical to obtain a good and precise image reconstruction. Note that the astrometric calibration is also able to take into account other low frequency field distortions that may have been left out by the geometric process. However, errors in individual slice positions will not be corrected by the astrometric calibration process.

11.4.9 Data-Cube Creation

The final step is to build a regularly sampled 3D array in the $(\alpha, \delta, \lambda)$ system coordinate. This so-called data cube is the final product of the whole process. It is saved in the usual multidimensional FITS format commonly used in astronomy. The first plane contains the data array in erg s^{-1} cm^{-2} Å^{-1} units, and the second ones the estimated variance array.

The scheme used to transform the irregularly sampled values in the pixel table in a regularly sampled data cube is an adaptation in three dimensions of the drizzle method developed for Hubble Space Telescope imaging [102]. This method identifies the pixel table data points that overlap with the final voxel and performs a weighted sum of these voxels. Each input pixel table data is assigned a weight proportional to its intersection with the output voxel in the 3D space. Figure 11.14 visualizes the drizzle method in two dimensions, with three input data points that overlap onto one output pixel. The parts of the input data that overlap the output pixel and therefore contribute to it are drawn as opaque regions. Before performing the drizzle scheme a sigma-clipping is performed to reject possible outliers.

The estimated variance associated to the pixel table data is also interpolated the same way, assuming independent Gaussian noise propagation. The main difficulty is that this process creates a correlated noise that can no longer be characterized by just a variance value. In principle, we might take this into account by propagation of the covariance matrix. But even reduced to non-negligible values, this matrix is still large, of the size of the spatial PSF and LSF ($5 \times 5 \times 3$), and it would be computationally prohibitive to propagate it. As can be seen in Figure 11.15, the effect of the correlated noise is clearly visible on a single exposure. When combining multiple exposures, the variance map is less structured because of smearing out due to the number of samples involved in the drizzle, but the effect is still there and the variance is not any more a good evaluation of the noise.

An alternative would be to keep the final pixel table and perform the analysis directly on the table. In practice, this is unrealistic because of its very large size. Owing to its structure, a pixel table for a single exposure takes 8.5 Gb, already 3

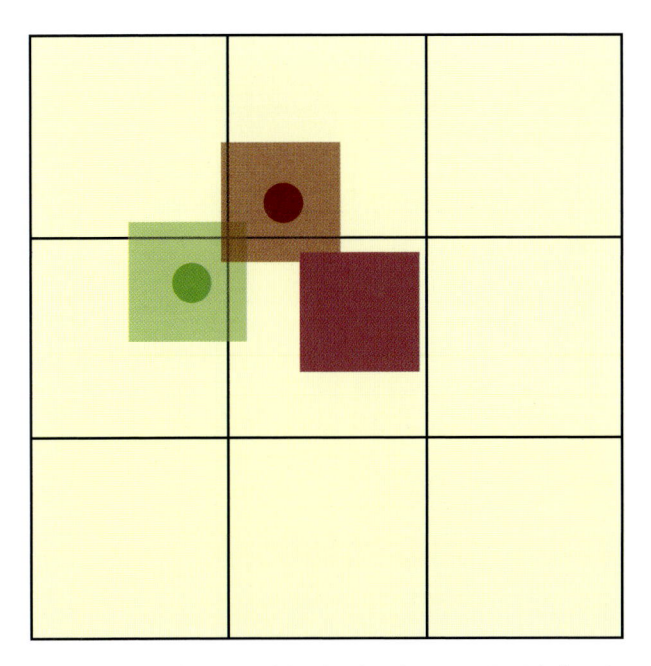

Figure 11.14 Schematic of the drizzle scheme in 2D with three input data points that overlap onto one output pixel. The parts of the input data that overlap the output pixel and therefore contribute to it are drawn as opaque regions. (Reproduced with permission of Peter Weilbacher.)

times bigger than a data cube, and it grows proportionally with the number of subexposures. But the size is not the unique difficulty, as most of the visualization, post-processing, and analysis techniques assume a uniform sampling along both the spatial and spectral axes. All matrix operations belong to this class of signal processing: for example, the classical fast Fourier transform (FFT) used to perform efficient convolution belongs to this category.

11.4.10 Data Quality

As can be seen from the previous sections, many steps are performed to move from the raw data to the final data cube. The MUSE pipeline has been optimized to run efficiently, taking benefit from parallel processing on multicore workstations. With enough computing power, the creation of a single data cube from the raw data is a matter of an hour or less after all the calibrations have been processed. The merging of many exposures is also a demanding task, but on the whole, a fully reduced data cube can be produced in a few hours. It is of course always possible that some mishap happens during one of the many processing steps, often resulting in artifacts in the final data cube that are hard to recognize. Nor, given the huge overall data volume, could users inspect all intermediate data processing sets.

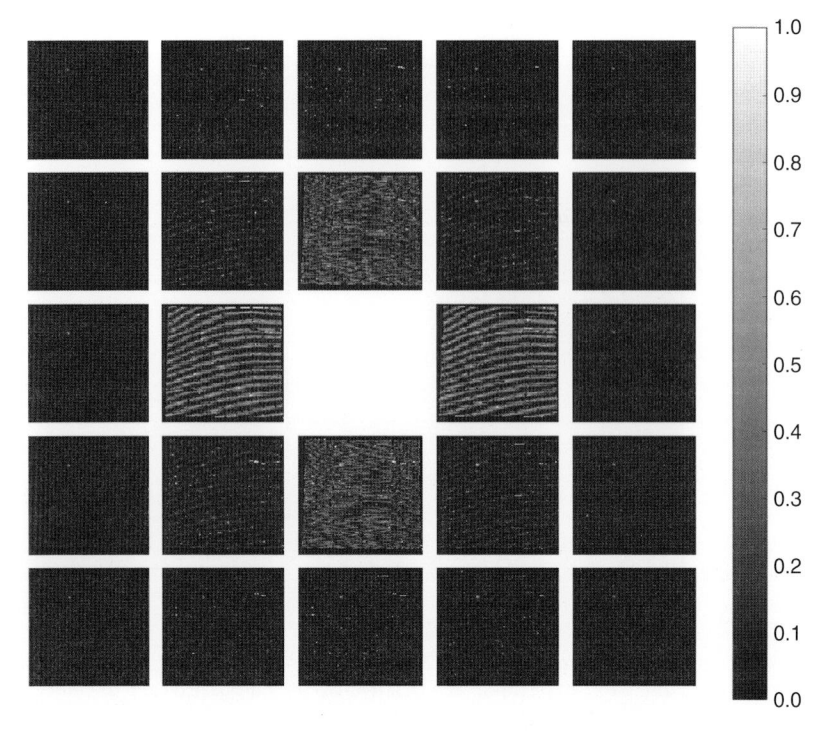

Figure 11.15 Spatial correlated properties in the MUSE data cube after drizzle interpolation. Each image shows the correlation map with its ±1, ±2 spaxel neighbors. (Reproduced with permission of Jean-Baptiste Courbot.)

To help in this respect, each recipe produces a number of quality control parameters that are attached to its output, in general, as header information or additional files. For example, some statistics (e.g., mean, median, standard deviation, min and max) for each slice of the wavelength calibration is saved on the header of the master wavelength calibration file. This is more manageable than the large number of polynomial coefficients, which are contained in the wavelength calibration table, but still holds 1152 values. The total number of quality control parameters is thus very large and tools to perform graphical representation of the relevant parameters have been developed by ESO as part of the health check control of the MUSE instrument (Figure 11.16).

One important aspect is the measurement of the data cube spatial PSF, and to a lesser extent, the spectral LSF. While the latter is supposed to be stable enough to be inferred from calibration data, the former is highly variable, in large part because of the large fluctuations of the seeing. As discussed in Section 8.2.1 there are many other effects that can also impact the final PSF. The PSF estimation is relatively straightforward when a point source bright enough is present in the observed field. As shown in Figure 11.17, a MOFFAT model is a good approximation of the PSF shape provided that the FWHM is allowed to change with wavelength as predicted by the turbulence atmospheric model (see Section 9.2).

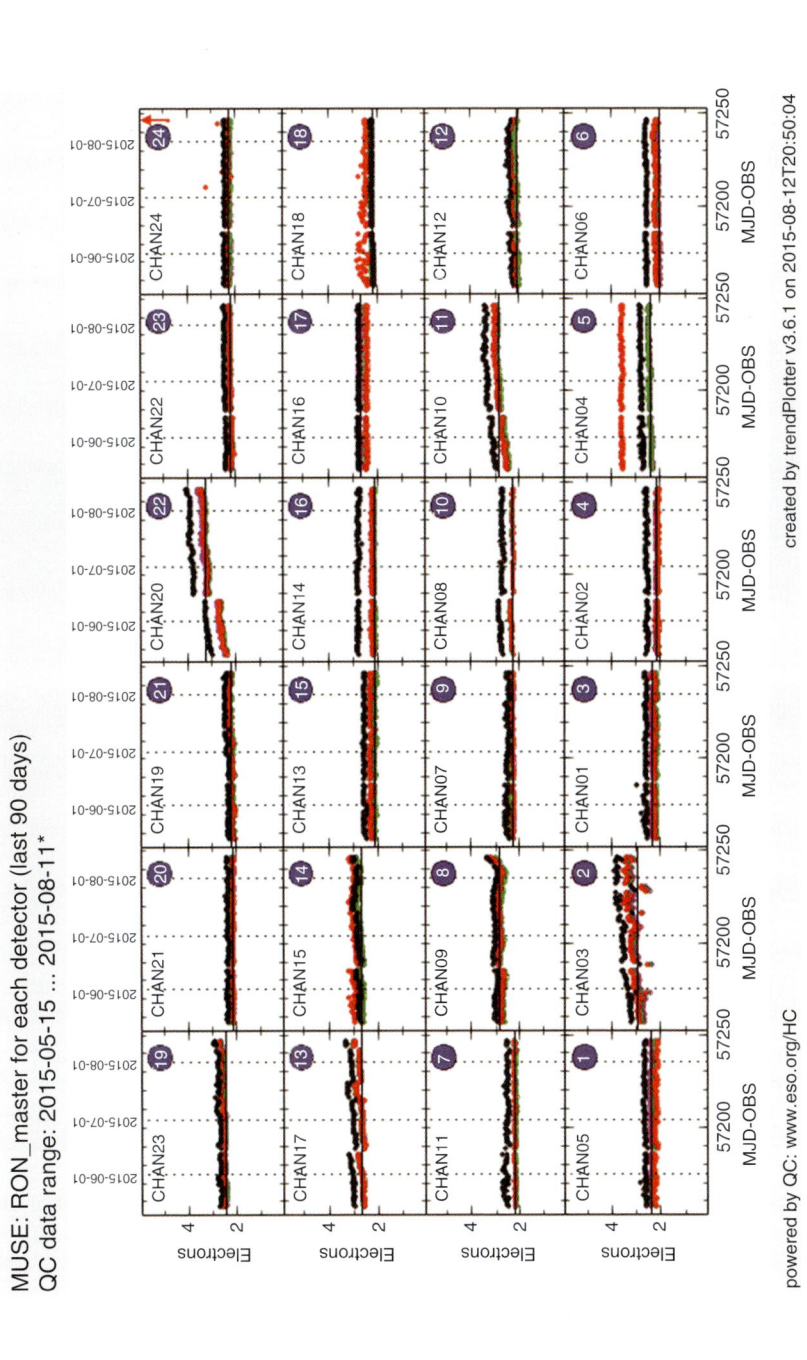

Figure 11.16 Monitoring of the readout noise levels of the 96 quadrants of MUSE detectors. (Used Under Creative Commons License: https://creativecommons.org/licenses/by/4.0/.)

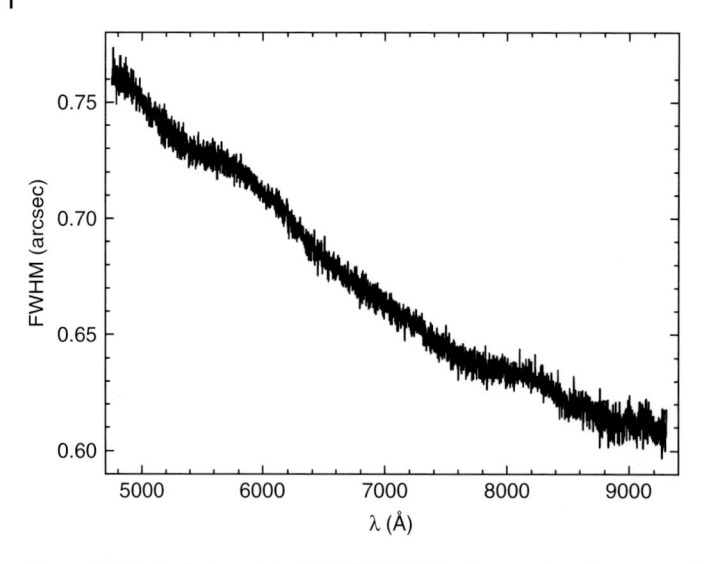

Figure 11.17 Evolution of the MOFFAT FWHM with wavelength measured on the deep observation of the Hubble deep field south with MUSE [101].

11.5 Conclusion

We have seen in this chapter the basic characteristics of the data reduction software. We also have investigated in more detail the pipeline used for the MUSE integral field spectrograph. However, all instruments have their own idiosyncrasies and the potential user must read carefully the up-to-date software documentation. Except for instruments that have been in operation for a long period and with an active software development, it is likely that the pipeline has some bugs or hidden "features" that may introduce some inaccuracies in the data products. Even for a bug-free system, the data reduction system is based on a number of assumptions and approximations that are not necessarily fulfilled. In addition, instruments evolve with time. For example, detector characteristics in space show rapid evolution due to the hard particle environment. Time evolution may also come from misalignment due to small changes in the optomechanical characteristics, especially for moving elements. Such changes may need some adjustment of the software.

In general, it is not recommended to use data reduction pipelines as black boxes. A minimal knowledge of the instrument and software characteristics is required. Critical assessment of the pipeline results should always be the rule. Playing with the pipeline parameters and trying other methods, including independent software tools, will help the user to check the reliability of the dataset provided by the pipelines.

12

Data Analysis

12.1 Introduction

The final product of the data reduction described in the previous chapter is in most cases a data cube (except for MOS spectrographs, which deliver a set of spectra) expressed in physical units (flux versus wavelength over a 2D field given in sky coordinates). The process that leads then from such a reduced data set to the science result is called data analysis. This is a very broad subject, highly dependent on the science case, which goes far beyond the scope of this book. In the following sections we will focus on a few generic processes related to the handling and visualization of data cubes, a subject that goes well beyond purely astronomical pastures. Note that data analysis of a set of individual data as in the case of multiobject spectrography is standard practice in astronomy and is not covered below.

12.2 Handling Data Cubes

Throughout this book we have used the term data cube to refer to the final product of 3D spectrographs. In general, a data cube can be described as a set of data points regularly spaced along three independent axes (x, y, z). There is, however, an important difference between our astronomical data cubes and common-sense data cubes. The latter have three homogeneous dimensions, similar to the three dimensions of space of a classical 3D object while the data cubes discussed at length in this book have inhomogeneous dimensions $(\alpha, \delta, \lambda)$: the first two dimensions (α, δ) are the spatial coordinates, in angular units (e.g., arcsec) or physical units (e.g., parsec), while the third dimension λ is in wavelength units (Å) or velocity (km/s). This makes an important difference in practice. For example, most of the 3D data analysis or data visualization softwares developed for homogeneous data cubes will not be very useful in our case, or for example, for Earth remote sensing hyperspectral data cubes. Let us take a simple example: an effective way to scrutinize a 3D object is to rotate it and look at it along various perspectives. This would not make much sense with an astronomical data cube as it would mix wavelength and spatial coordinates in a bizarre way.

Note also that the data cubes produced by the various 3D instruments described in this book have very different geometry (see Figure 1.5). Scanning Fabry–Pérot and Fourier transform spectrometers produce *flat* data cubes with much larger spatial dimensions than the spectral one. The reason is that the spectral dimension is obtained by scanning with time and thus is limited in size by the telescope time spent on the object to well under a hundred spectral pixels. The two spatial dimensions correspond to the detector size: with a single 4k × 4k detector and Nyquist 2 × 2 sampling, this gives 2048 × 2048 spatial pixels. Integral field spectrographs on the contrary produce long *tubes* with the spectral dimension in general much larger than the spatial dimensions. The mainly geometrical reason is that usually one detector axis is used to get the 1D spectral data (2048 Nyquist sampled spectral pixels for a 4k × 4k detector) and the other one to get the 2D spatial field (1024 Nyquist sampled spatial pixels for a 4k × 4k detector, or a meager 32 × 32 pixel field). Note that in the case of MUSE, this lack of field is well mitigated by the use of 24 detectors working in parallel.

Given the specificities of our data cubes, there are two major ways to handle them: as a collection of monochromatic images at various wavelengths or a set of spectra at each spatial location.

12.2.1 The Spectral View

This is the classical way to handle integral field data cubes. Each spectrum of the data cube is analyzed in sequence and an image of the given astrophysical quantity (line flux, line radial velocity, line velocity width, …) is produced. One can then use existing data analysis spectral software which has been developed to perform spectra data analysis: for example, the RVSAO IRAF packages [115] can be used to obtain radial velocities from spectra using cross-correlation and emission line fitting techniques.

Let us take the simple example of ionized gas velocity field of a galaxy. The precise location of an emission line peak is computed by centroiding or Gaussian fitting and the radial velocity derived using the Doppler–Fizeau shift. Each spectrum produces a single number: the radial velocity V_r in km s^{-1} units. Recording the velocities with their spatial coordinates produces then a 2D array $V_r(\alpha, \delta)$, which is the velocity field of the galaxy in the selected emission line. In practice, the ionized gas does not extend everywhere in the field of view, being usually concentrated in regions of intense star formation. In low flux regions the resulting velocity field will be noisy and must be post-processed by classical imaging techniques to improve its SNR. For example, running a 2D median filter on the velocity field image helps remove outliers if at the expense of the spatial resolution. An alternative is to replace each spectrum by the average of its spatial neighbors. This improves the resulting SNR of the derived radial velocity field, obviously again at the cost of less spatial resolution. Note that these two ways of improving the velocity field quality are not equivalent because they imply nonlinear processes such as Gaussian fitting. Improving the SNR of the spectra before invoking the fit is always better than trying to fix the poor quality of the velocity field by image post-processing.

12.2.2 The Spatial View

To produce a set of monochromatic images from the data cube, one can use the many powerful image processing softwares lying around. For example, we can use the well-known SExtractor software [107] to identify individual sources at various wavelengths within the data cube.

In some cases, we can process each individual image to compute a single quantity and to check how it changes with wavelength. The result is then a single plot. For example, let us compute the centroid of a spatially non-resolved double star. If the two stars have quite different spectra, the computed centroid will change with wavelength. This is the simplistic version of a more elaborate 3D de-blending software that separates overlapping stars by using their differential spectral information.

12.2.3 The 3D View

It is always possible to use a combination of the two approaches. For example, one can spatially convolves each image of the data cube to improve the SNR and then processes each spectrum of the resulting data cube to extract the relevant quantities.

However, the most powerful techniques are the ones that make use of the full 3D informational content of the data cubes. Let us see an example. If we want to search for sources in a data cube we may use a software to automatically detect sources at each image plane. This produces a series of catalogs, which we will try then to merge to identify the sources that appear at many wavelengths, such as the continuum sources. However, because each image plane is processed independently there is no guarantee that the continuity between two successive images is maintained.

Let us try another, but still non-3D, way. We can process each spectrum of the data cube and search for emission lines. Then if we find the same emission lines in neighboring spaxels we then decide that they belong to the same source object. But again, given that each spectrum is processed independently, the fit may converge to different solutions in case of low SNR and thus the spectra will not be identified as belonging to the same object.

In most cases, the best is to use the full 3D information. A simultaneous fit or detection of all the voxels in the $(\alpha, \delta, \lambda)$ planes does a better job in identifying the faint objects because it makes use of the full information content. In particular, it makes better use of the natural correlations that exist in a data cube. In any real object the physical content cannot change completely abruptly from one voxel to the other. Often this is intrinsic to the object itself. But even for an unresolved point source with a single narrow emission line, atmospheric seeing and finite instrument spatial and spectral resolution produce a number of correlated voxels, not just a single one. Using full 3D software techniques helps benefit from these natural correlations and produces the best and less biased results. These techniques are, however, more complex to develop and also more computationally expensive, especially with large data cubes.

12.3 Viewing Data Cubes

The first thing that observers do when the data cubes are freshly delivered by the data reduction software is to look at them. Since present computer screens do not yet have full 3D capabilities, a data cube viewer is needed.

Now that 3D spectrographs are becoming more and more common in astronomy, widely used image visualization softwares start to incorporate data cube visualization. This is, the case of the SAOImage DS9[1] and Aladin [116, 117] softwares. See Figure 12.1 for an example of data cube visualization in DS9. The easiest way to visualize the data cube is to look at successive monochromatic images. Usually, there is just a cursor to move to create a sort of movie of the data cube with wavelengths. This functionality is present in all data cube viewers. Although it is a very simple capability it is quite effective because it uses the human brain's ingrained capability to detect motion and identify differential changes from a series of images displayed successively. For example, this makes quite easy, and fun to boot, identifying the velocity field of ionized gas emission in a galaxy.

The second most used capability is to look at individual spectra at any spatial location. It generally takes only pointing the cursor at a location identified in an

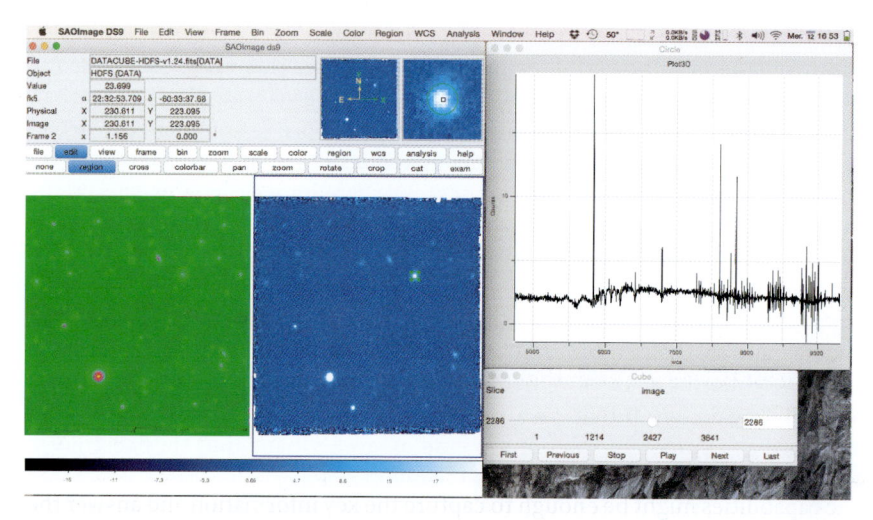

Figure 12.1 An example of a DS9 data cube view. The main DS9 window shows two frames: the right one (green color scheme) is the traditional white light image coming here from the MUSE Hubble deep field south observation. One monochromatic plane image (or slice) of the corresponding data cube is shown on the right frame with a blue color scheme. The wavelength (index 2286) has been selected from the cube with the slider window at the bottom right. Using the region type cursor, a small aperture was selected on the image (see the zoom window at the upper right) and the corresponding spectrum is shown on the right window (Plot3d).

---

1 SAOImage DS9 development has been made possible by funding from the Chandra X-ray Science Center (CXC) and the High Energy Astrophysics Science Archive Center (HEASARC). Additional funding was provided by the JWST Mission office at Space Telescope Science Institute to improve the capabilities for 3D data visualization. It is available at http://ds9.si.edu.

image plane to display the corresponding spectrum. In some cases the SNR of an individual spaxel is too low and it is better to have the possibility of getting a higher SNR spectra by averaging over an aperture. Most softwares incorporate also this capability. Monochromatic layers may also have poor SNR and it is also very useful to be able to sum over wavelength to improve the displayed image. There are also some softwares that have been developed specifically for data cube visualization. This is the case of QfitsView[2] [118]. These softwares generally incorporate more features and are more efficient to handle data cubes (Figure 12.2).

More general 3D visualization tools, not specially devoted to astronomy, can also be used. See, for example, the VisIt[3] software. It offers many different ways to view 3D information but not all of them are adapted to the specific 3D information of astronomical data cubes (see Section 12.2).

Another common way to use image views of the data cubes is to select different wavelength parts of the data cubes and build RGB color images of the combination. This is similar to what is regularly done with images obtained through three different filters, except that it can be expanded to more complex combinations than just three colors. A good example is shown in Figure 12.3 where the Orion nebulae MUSE data cube has been used to derive a number of color images showing the physical property of the gas in the nebulae.

What is currently missing in most softwares is the handling of the variance information. The variance data cube that is sometimes provided by the data processing pipelines can of course be visualized in the same way as the main data cubes themselves, but it is not yet possible to display, for example, a monochromatic image with its associate variance as a data point can be displayed with its error bar. There is however another way to get similar information, which is to display the corresponding SNR data cube obtained by dividing the data by the square root of its variance.

12.4 Conclusion

In this chapter, we have seen the premise of the data analysis, how to view and inspect the data cubes using data cube viewers. This first step allows playing with the data cubes, inspecting them by moving into the wavelengths space, identifying structures, and looking for the corresponding spectra. In some cases, these basic capabilities might be enough to capture the key information and answer the scientific questions addressed. But in general, more quantitative analysis will be needed and users have to develop some scripts to perform the work. The easiest approach is to use the spectral view of the data cube and use some existing software to perform the appropriate computation on each individual spectrum and then to build images of the derived quantities. But for low SNR or highly

---

2 QFitsView is a FITS file data cube viewer. It is developed at the Max Planck Institute for Extraterrestrial Physics by Thomas Ott. It is available at http://www.mpe.mpg.de/~ott/dpuser/qfitsview.html.
3 VisIt is VisIt is an open source, interactive, scalable, visualization, animation, and analysis tool developed by the Lawrence Livermore National Laboratory. It is available here: https://wci.llnl.gov/simulation/computer-codes/visit/.

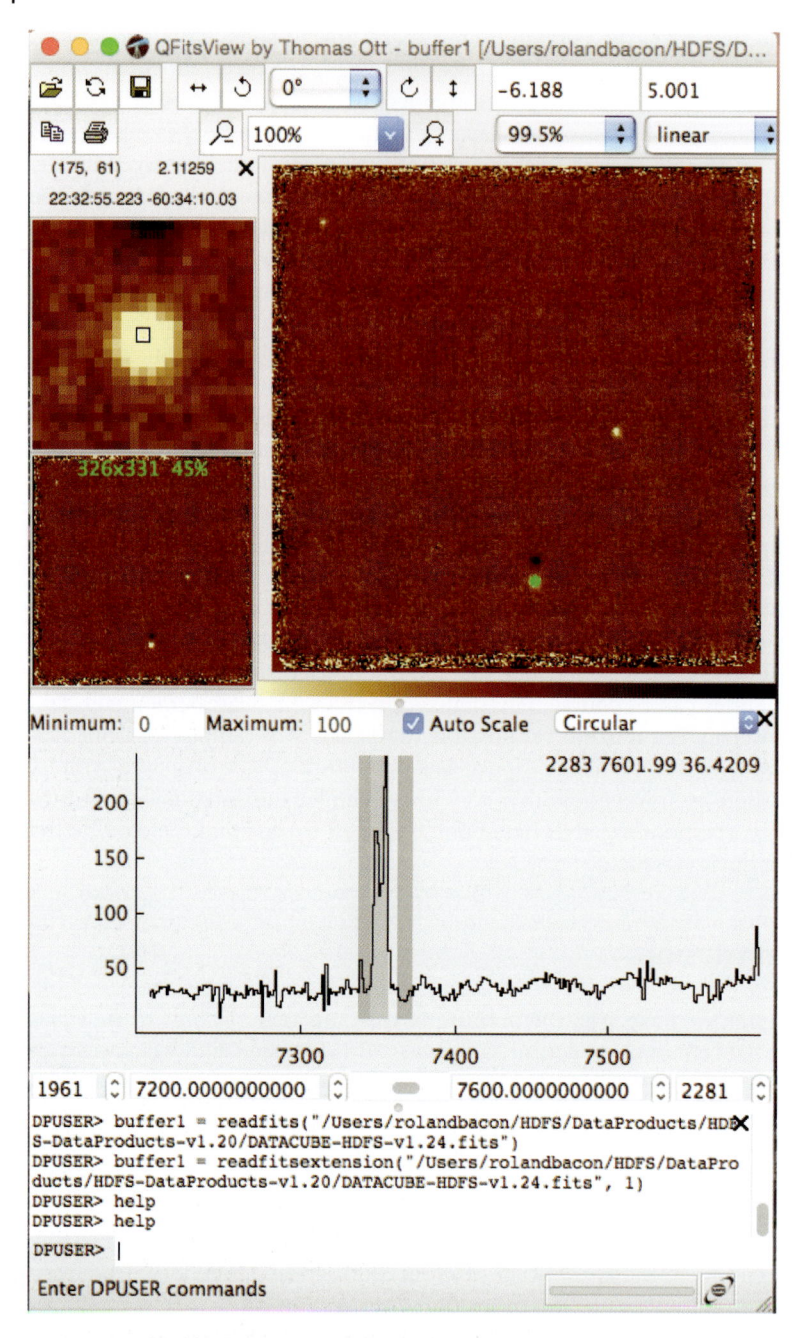

Figure 12.2 QfitsView example session. The upper window shows the image plane view and the lower ones the spectral view. In this example, we have used the QfitsView capability to perform continuum subtraction by selecting the line and continuum emission wavelength range in the spectra window. The resulting narrow band image is displayed on the image view where only emission line objects at the selected wavelength show up.

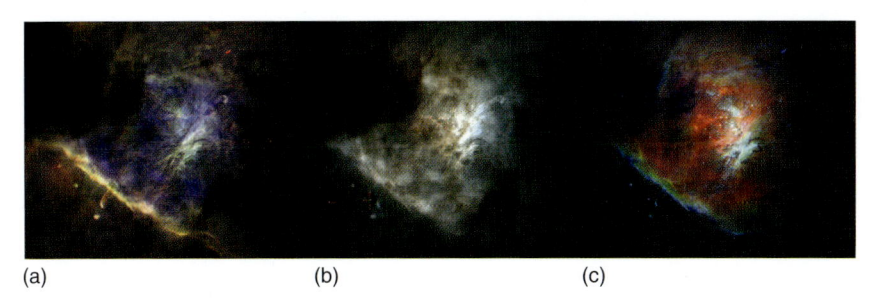

Figure 12.3 Different views of the Orion nebulae MUSE data cube. (a) Color composite using three emission lines fluxes, in blue: $H\beta$, green: [N II] 6584, and red: [S II] 6731. (b) Color composite, showing emission line fluxes in red: Paschen 9, green: $H\alpha$, blue: $H\beta$. (c) Enhanced color image, showing the three ionization levels of oxygen: red: [O III] 5007, green: [O II] 7320, blue: [O I] 6300. (Reproduced with permission of Peter Weilbacher.)

complex data this might not be enough and the user will have to deal with the three dimensions of the data cube in a global way. This requires much more work, but can exploit the full 3D content of that data cube and provide more robust and efficient analysis than the simplest approach. However, before diving too rapidly in the often treacherous waters, it might be useful to consider the software development cost versus the scientific added value.

12.5 Further Reading

- Andreon, S. and Weaver, B. (2015) *Bayesian Methods for the Physical Sciences*, Springer Series in Astrostatistics, ISBN: 978-3-319-15286-8.
- Starck, J.-L. and Murtagh, F. (2006) *Astronomical Image and Data Analysis*, Astronomy and Astrophysics Library, Springer-Verlag, Berlin, Heidelberg, ISBN: 978-3-540-33024-0.
- Wall, J.V. and Jenkins, C.R. (2012) *Practical Statistics for Astronomers*, Cambridge Observing Handbooks for Research Astronomers, ISBN-13: 9780521732499.

13

Conclusions

13.1 Conclusions

So far, we have mainly presented the 3D instrumental approaches that have enjoyed considerable development and are currently in operation on most observational facilities worldwide. Yet, these relatively few success stories hide the many other species that have been proposed, most often prototyped, and possibly used, but only to vanish more or less quickly from the scene. A casual look at the tables of content of any of the biannual SPIE conferences on ground-based and space astronomy instrumentation is enough to find a bewildering variety of newly advertised approaches, ranging from small variants to radical departures from mainstream instrumentation. Actually, there have been so many different instrumental offsprings put forward over the last 40 years or so, that, without this high attrition rate, they would by now all but clog every observing facility in sight.

This state of affairs testifies to the creativity of the community, but also generates substantial waste in monies and human investment, plus occasional psychological drama. This justifies looking carefully at the big question "what makes an instrument successful?" as the theme of this concluding part. There are many facets to that question that are briefly explored below, covering general-use versus team-use instruments, science drivers, serendipity potential, observing parameters space, reliability and observing performance, data reduction and analysis pipelines, competitiveness and uniqueness, and synergy with other facilities.

The general-use versus team-use operation dichotomy is a far-reaching one that reflects on many aspects of an instrument's success.

13.2 General-Use Instruments

The life of a general-use instrument is anything but a ride along a long, calm river. In the halcyon days of up to the 1970s, it was enough to offer some basic facility, say long-slit spectra over the photographic wavelength range, out of less than a handful of choices. Provided the instrument worked at all – quite a number of good-looking spectrographs ended stillborn, for example, because

Optical 3D-Spectroscopy for Astronomy, First Edition. Roland Bacon and Guy Monnet.
© 2017 Wiley-VCH Verlag GmbH & Co. KGaA. Published 2017 by Wiley-VCH Verlag GmbH & Co. KGaA.

of unmanageable mechanical flexures – its support by the almost captive clientele of its home telescope was ensured. With increased competition between telescopes, equipped with large and eclectic instrument suites, all catering for largely the same customer pool, this is not true anymore, and many hurdles must be overcome to ultimately ensure the "commercial" success of any instrument.

- The crucial first step is to "sell" to the putative telescope user's community a concept that provides a powerful way (even better unique) to fill the instrumental parameter space. (Field of view and spatial resolution, wavelength range and spectral resolution, object multiplex, etc.) This is an important item as a much increased coverage of the parameter space helps in getting a competitive system, and offers good potential for serendipitous discoveries; this latter point is very difficult to quantify though. The potential impact of the instrument is then quantitatively evaluated by establishing a number of science drivers, important – or at least fashionable – astrophysical conundrums that should be solved, or at least significantly dented by the instrument. This analysis must not be restricted to the instrument per se. On one hand, part of an instrument attraction potential might be to offer some synergy with very different facilities in operation or planned at the time, one prime example in the last two decades being ultra-deep imagery by the Hubble Space Telescope complemented by faint object spectrography by large ground-based telescopes, in particular the Keck Observatory. One must also evaluate potential competition to avoid being entirely scooped before even first light. Even with the best preparation though, there are still a number of pitfalls that might implement ultimately lethal genetic defects.
- With usually close to a decade between initial instrument approval and first light on the sky, even well-thought out original science drivers can turn obsolete, as science priorities worldwide wax and wane. One example is the FORS2 multiobject spectrograph put at the VLT in 1999. With its too low multiplex (19) and too low spectral resolution (~1000), it was unable to jump in the fray for the study of high-redshift galaxies in the reddish wavelength domain, an ultra-hot topic since pioneering work on the Keck 10-m telescope only a few years earlier. The first problem was solved as the FORS2 building team succeeded in retrofitting a much higher multiplex mask unit. Even then, the spectral resolution inadequacy would still have all but killed the instrument's scientific potential without the availability out of the blue of a new grating species, volume-phase holographic gratings. This US defense development had just been adapted to astronomical ventures by an industry-observatory partnership between Kaiser Optical Sciences Inc. and the National Optical Astronomy Observatory, and the bill was footed beautifully. In summary, a near-fatal science case collapse was saved by a combination of work and luck.
- To inflate an instrument science case, and hopefully later its customer base, it is tempting to cram many modes inside the same instrument, a move that on top tends to please the various committees charged with giving the green light for any general-use facility. OASIS, the first ever adaptive optics assisted

integral field spectrograph, installed in 1997 at the CFHT 3.6 m telescope, featured a multimode integral field capability with a number of spatial resolutions (through a battery of enlargers) and spectral resolutions (through a grating wheel), and even a brand new specie, PYTHEAS, a cross-breed between scanning Fabry–Perot and integral field LeCoarer et al. [63]. Such a multimode approach usually proves attractive and fun at the concept level, a workable additional load at the manufacturing level, a burden at the instrument characterization level, and finally a nightmare at the data reduction level. Arguably, OASIS egregious mode multiplicity (largely due to this book's authors!) played a significant role in this short-lived instrument's rather undistinguished science return. Its innovative Pytheas mode was technically working, but addressed a very narrow science field, namely, simultaneous high resolution stellar spectra in globular clusters. Throwing in the huge data analysis development that would have been needed to exploit scientifically this mode, it is no big wonder that it never took off.

- Overselling an instrument looks like a trivial pitfall, not always avoided though. OASIS also heavily suffered from insufficient evaluation oversight. On paper, with the PUEO CFHT adaptive optics corrector delivering ultra-sharp images, it looked as a sure winner. However, OASIS science case was based on too optimistic assumptions on AO performances. Good correction in its 700–900 nm domain could be obtained only during excellent seeing nights, required a relatively bright "guiding" star very close to the science object, and even then only a small fraction of the light was concentrated in diffraction-limited cores for point-like objects. These limitations strongly impacted the science return and consequently customers' satisfaction. In retrospect, the nail in the coffin was that service-mode observation – real-time tuning of the telescope observational program to specific observing conditions – was not yet implemented in the telescope, so the observer could rarely benefit from AO-friendly conditions (small/slow atmospheric turbulence) to carry out his/her program.
- Innovation is in principle great in opening new ecological grounds, but might well encounter strong community resistance, as the tortuous road of the integral field technique shows. Firstly, it took close to a decade between the integral field initial concept put forth around 1980 by Vanderriest [11] and Courtès [119] and the first two working instruments, SILFID and TIGER. One main reason was the total lack of support by the relevant community. Actually, even this book's authors took a long time before jumping on the wagon, starting the crash development of the TIGER line with G. Courtès and Y. Georgelin nearly 9 years after the concept has been put forward. A couple of years later, both instruments had proved scientifically useful, and a handful of teams worldwide had joined the fray. Yet, selling this new paradigm to the astronomical community at large has been then anything but a breeze. This was to be expected, as lifelong users of the venerable long-slit spectrograph were not particularly enthused to help make their cherished tool obsolete. There was also, to be fair, a perfectly valid rationale in the critical data reduction area: it took years of hard work to achieve routinely that capability, a key step toward acceptance of any observing technique. And, in retrospect, the final key move toward general acceptance happened when the

various instrument builder teams agreed to be "robbed" from their offsprings, putting much effort to deliver turnkey general-use facilities to most large observatories by the mid-2000s, some 25 years after the initial concept had been put forward. In sheer contrast, acceptance of multiobject spectroscopy has been swift and relatively painless, probably because it is just more of the same, spectra of hundreds of galaxies/stars in parallel versus only one or two with their long-slit spectrograph ancestor. This purely quantitative move proved much easier to swallow than the major qualitative change brought by the integral field approach. And, ultimately, both the multiobject and integral field approaches "won" because they turned out to address in an efficient manner important – or at least currently fashionable – science drivers, such as large-scale galaxy mapping for the former and galaxy evolution for the latter.

Starting an instrument with a finely holed gene pool is a good start, but it takes caring relatives for successful nurturing. As elaborated in Section 1.8, a competent, motivated, and well organized builder team is crucial, especially so for general-use instruments given their long timeline and the daunting task ahead. Putative customers expect a full turnkey system, including a comprehensive data reduction pipeline, fully automatic operation with very little overheads, and essentially no technical breakdown. Observatories, for good reasons, always insist on full documentation, easy operation, and little, if any, maintenance needs. And, crucially, about 10 years after project start, the instrument should still be technically competitive and with a strong science case, not necessarily the original one though. This takes good choices at the start, often the ability to steer the instrument capabilities in mid-course (if at possibly some risk of project collapse), and admittedly a most important ingredient, luck. Note that what matters in terms of performance is not that of the instrument alone. What really counts is the global performance of the site, telescope, instrument, and data reduction pipeline as an integrated system. In that respect, the first few semesters of use at the telescope are crucial, as even only one user's negative appreciation might destroy an instrument as swiftly as an earthquake, especially if coming from an opinion leader in the domain.

Arguably, the most efficient way to commit such a suicide is to offer an instrument with a data reduction pipeline that is *almost* ready from the builder's point of view, which translates into *not working at all* by the end user. This is not easy to avoid though, as many data reduction aspects must be finely tuned to the many instrument idiosyncrasies, a number of which are initially poorly known, some even totally unknown, at least before on-sky commissioning. To reduce that major risk, it is essential to test carefully the whole instrument before it is shipped to its home destination and feedback the results to the data reduction scheme. it is also most advisable to build right at the start of the project a comprehensive instrument numerical model that produces realistic simulated raw data, incorporating site and telescope impact, optical aberrations, detector noise, cosmic ray impacts, calibration uncertainties, and so on. Such a tool is very useful to discover manufacturing errors, test the robustness of the data reduction software, and better predict the instrument's observing potential.

To illustrate the many factors playing in the development of general-use instruments, here is an insider view of the MUSE case.

The MUSE project started in 2001 with an answer to the ESO call for idea for the VLT second generation instrumentation. This is a slicer-based wide-field integral-field spectrograph including many innovative concepts (see Section 5.4.1 for a detailed description of the instrument). It was an extremely ambitious project with an order of magnitude more detector pixels than on previous similar instruments. The project has been carried out under the responsibility of a large European Consortium of seven institutes (ESO included) with a staff of ~50 engineers. The program went through the classical road path with its many reviews before being authorized to move to the Paranal Observatory for deployment and commissioning, 13 years after first proposal.

At the time of this writing (Spring 2016), the instrument is open to the community since only 18 months, and it is thus too early to assess its long-term impact. However, given the oversubscription of the proposal from the community and the ~50 papers already published, this development is already successful, at least in its main wide-field mode. Here are some of the main contributing factors:

- MUSE has a comprehensive science case over a wide variety of astrophysical objects, but also with a clear single driver, namely, very high-redshift deep field searches. This has been invaluable to guide the instrument development and perform the right tradeoffs.
- The instrument has only two modes: one main, highest priority, wide-field mode, plus a lower priority narrow field one to be used once the adaptive optics (AO) facility is available, hopefully by early 2018.
- The pipeline was thoroughly tested in advance of first light, using an instrument numerical model as well as laboratory data from the first prototype spectrograph unit. The effort in the data reduction software started very early and did not stop after first light, but continued full speed.
- A large science team (50 staff, postdocs, and PhDs) has been involved in the exploitation of the 255 nights guaranteed time allocation. Its science strategy is focused on a few large programs, benefitting to a large fraction of the science team rather than being split in many small individual observations.
- When the instrument entered operation it offered unique capabilities, in particular superb throughput, a large field of view, and a wide spectral range. Only one other instrument, the soon to be fully deployed VIRUS facility at the Hobby–Eberly Telescope, will provide similar data cubes. However, with quite different science driver and technical tradeoffs, it is not expected to be a direct competitor.
- Early spectacular science results have been obtained. In particular, the discovery of dozens of extended Lyman-α emitters in the Hubble Deep Field South (see figure REF) beyond the limiting magnitude of the HST has demonstrated the MUSE discovery capability and attracted the attention of the community.
- Much care has been taken to get a viable AO-assisted diffraction-limited mode, with in particular extremely low-noise detectors and the use of a cutting-edge multilaser AO facility fully integrated with the telescope. Whether this will be

enough to avoid an OASIS-like ugly demise is however a multimillion euros question mark left for a few years ahead.

13.3 Team-Use Instruments

A team-use instrument is proposed, developed, and operated for at least one major survey, with the observing data fully reduced and interpreted, by a dedicated and enthusiastic team. A major proviso is that the many telescope nights – easily hundreds – required to complete such surveys must absolutely be pre-empted with the project home observatory before project start. The usually much reduced development timeline helps to avoid, or at least mitigate, many of the pitfalls above, even if the same professionalism is still required in all relevant areas to get the required performance. With very few modes, sometimes only a single one, team instruments development is easier and a few minor hiccups in the operation status, while not welcome, are usually more manageable than with a general-use instrument. One big plus is also that the team has only to "sell" the instrument to itself, which leaves still a major trap to avoid at all cost along the way, namely, being ultimately wiped out by the competition: this is not an idle threat; at least one major all-sky imaging/spectroscopic survey has been outrightly killed by the hugely successful SDSS survey. Its demise came from one lethal flaw, developing an ambitious project at the wrong time, with insufficient funds and bodies.

To illustrate these points, let us look at the SAURON project, a detailed 3D survey of 71 nearby galaxies with a dedicated lenslet-based integral field spectrograph. This was a "private" project, hosted at the William Herschel Telescope in the 2000s and operated by its builder team. Instrument specifications ($30'' \times 40''$ field of view, $1''$ sampling, small 470–540 nm spectral range) were finely tuned to its single science goal. The instrument was built in fast track on a shoestring budget in less than 3 years, using most of the technology developed previously for TIGER and OASIS, with very little technical innovation. SAURON has been used at WHT for 13 years, completing the initial survey, and later extended to the Atlas3D survey. All in all, it was a scientific success with more than 60 refereed publications and a couple of international honors. Seen from inside, here are a few factors for this success story:

- A strong, international and enthusiastic science team of ~15 members was set up right at the start between the three institutes involved (Leiden Observatory, at first Durham then Oxford University, and Lyon Observatory). This team was a good mix of instrumentalists, observers, and theoreticians, chaired by the three co-principal investigators.
- With its rather coarse $1''$ sampling, the system was well adapted to most atmospheric conditions at La Palma.
- The instrument has basically a single mode, which helped in reducing development and implementation time, and kept the project under its tight budget.
- A lot of attention was devoted to optimize the data reduction and data analysis. Advanced tools were developed to extract maximum information from the

data cube, for example, Voronoi binning to maximize the signal to noise ratio [120].
- The survey strategy, entirely executed by the team, allowed to maximize the science return and streamline the operation.

13.4 The Bumpy Road to Success

This book covers the life cycle of a number of successful instrumental lines over the last 40 years or so. Each started from a sound concept, that is, one that permitted to get relevant information on a large number and/or a large variety of putative targets in an efficient way. Most importantly, all were ultimately developed into reliable instrumental facilities, easy to operate and offering a full data reduction process delivering science-ready data to the users. While such assets indeed represent necessary conditions for success, they do not fully guarantee it. The ultimate step, as for any sellable goods, is that these instrumental facilities be "bought" by a sizeable customer pool. That requires objective qualities, in particular faster, better data gathering than their competitors, but also and most importantly hearty embracing by the said community, a largely subjective (sometimes even irrational) aspect. As for most human ventures, success takes skills, a lot of hard work, and at least a modicum of luck.

References

1 Fath, E.A. (1909) The spectra of some spiral nebulae and globular star clusters. *Lick Obs. Bull.*, **5**, 71–77, doi: 10.5479/ADS/bib/1909LicOB.5.71F.

2 Buisson, H., Fabry, C., and Bourget, H. (1914) An application of interference to the the study of the Orion nebula. *Astrophys. J.*, **40**, 241, doi: 10.1086/142119.

3 Schroeder, D.J. and Anderson, C.M. (1971) An echelle spectrograph for astronomical use. *Publ. Astron. Soc. Pac.*, **83**, 438, doi: 10.1086/129150.

4 Tully, R.B. (1974) The kinematics and dynamics of M51. 1. The observations. *Astrophys. J. Suppl. Ser.*, **27**, 415, doi: 10.1086/190304.

5 Maillard, J.P. and Simons, D. (1992) Spectro-imaging mode of the CFHT-FTS with a NICMOS camera: first results, in *European Southern Observatory Conference and Workshop Proceedings*, vol. **42** (ed. M.H. Ulrich), p. 733.

6 Hill, J.M., Angel, J.R.P., Scott, J.S., Lindley, D., and Hintzen, P. (1982) Multiple object fiber optic spectroscopy, in Proceedings of SPIE 0331, Instrumentation in Astronomy IV, p. 279.

7 Butcher, H. (1982) Multi-aperture spectroscopy at Kitt-Peak, in Proceedings of SPIE 0331, Instrumentation in Astronomy IV, p. 296.

8 Adam, G., Bacon, R., Courtes, G., Georgelin, Y., Monnet, G., and Pecontal, E. (1989) Observations of the Einstein cross 2237+030 with the TIGER integral field spectrograph. *Astron. Astrophys.*, **208**, L15–L18.

9 Angonin, M.C., Vanderriest, C., and Surdej, J. (1990) Bidimensional spectrography of the "clover leaf" H1413+117 at sub-arcsec. spatial resolution, in *Gravitational Lensing*, Lecture Notes in Physics, vol. **360** (eds Y. Mellier, B. Fort, and G. Soucail), Springer-Verlag, Berlin, p. 124, doi: 10.1007/BFb0009246.

10 Cameron, M., Weitzel, L., Krabbe, A., Genzel, R., and Drapatz, S. (1993) 3D: the new MPE near-infrared field imaging spectrometer, in *American Astronomical Society Meeting Abstracts*, Bulletin of the American Astronomical Society, vol. **25**, p. 1468.

11 Vanderriest, C. (1980) A fiber-optics dissector for spectroscopy of nebulosities around quasars and similar objects. *Publ. Astron. Soc. Pac.*, **92**, 858–862, doi: 10.1086/130764.

12 Krabbe, A., Genzel, R., Eckart, A., Najarro, F., Lutz, D., Cameron, M., Kroker, H., Tacconi-Garman, L.E., Thatte, N., Weitzel, L., Drapatz, S.,

Geballe, T., Sternberg, A., and Kudritzki, R. (1995) The nuclear cluster of the Milky Way: star formation and velocity dispersion in the central 0.5 parsec. *Astrophys. J.*, **447**, L95, doi: 10.1086/309579.

13 Amelio, G.F., Tompsett, M.F., and Smith, G.E. (1970) Experimental verification of the charge coupled device concept. *Bell Syst. Tech. J.*, **49**, 593–600.

14 Jorden, P.R., Jordan, D., Jerram, P.A., Pratlong, J., and Swindells, I. (2014) e2v new CCD and CMOS technology developments for astronomical sensors, in Proceedings of SPIE 9154, High Energy, Optical, and Infrared Detectors for Astronomy VI, p. 91540M, doi: 10.1117/12.2069423.

15 Walsh, J.R., Kummel, M., and Kuntschner, H. (2010) Slitless Spectroscopy with HST Instruments, in Hubble after SM4. Preparing JWST, p. 8.

16 Meyer, R.D., Kearney, K.J., Ninkov, Z., Cotton, C.T., Hammond, P., and Statt, B.D. (2004) RITMOS: a micromirror-based multi-object spectrometer, in Proceedings of SPIE 5492, Ground-based Instrumentation for Astronomy (eds A.F.M. Moorwood and M. Iye), pp. 200–219, doi: 10.1117/12.549897.

17 Zamkotsian, F., Spano, P., Lanzoni, P., Ramarijaona, H., Moschetti, M., Riva, M., Bon, W., Nicastro, L., Molinari, E., Cosentino, R., Ghedina, A., Gonzalez, M., Di Marcantonio, P., Coretti, I., Cirami, R., Zerbi, F., and Valenziano, L. (2014) BATMAN: a DMD-based multi-object spectrograph on Galileo telescope, in Proceedings of SPIE 9147, Ground-based and Airborne Instrumentation for Astronomy V, p. 914713, doi: 10.1117/12.2055192.

18 McCarthy, J.K., Cohen, J.G., Butcher, B., Cromer, J., Croner, E., Douglas, W.R., Goeden, R.M., Grewal, T., Lu, B., Petrie, H.L., Weng, T., Weber, B., Koch, D.G., and Rodgers, J.M. (1998) Blue channel of the Keck low-resolution imaging spectrometer, in Proceedings of SPIE 3355, Optical Astronomical Instrumentation (ed. S. D'Odorico), pp. 81–92.

19 Oke, J.B., Cohen, J.G., Carr, M., Cromer, J., Dingizian, A., Harris, F.H., Labrecque, S., Lucinio, R., Schaal, W., Epps, H., and Miller, J. (1995) The Keck low-resolution imaging spectrometer. *Publ. Astron. Soc. Pac.*, **107**, 375, doi: 10.1086/133562.

20 Faber, S.M., Phillips, A.C., Kibrick, R.I., Alcott, B., Allen, S.L., Burrous, J., Cantrall, T., Clarke, D., Coil, A.L., Cowley, D.J., Davis, M., Deich, W.T.S., Dietsch, K., Gilmore, D.K., Harper, C.A., Hilyard, D.F., Lewis, J.P., McVeigh, M., Newman, J., Osborne, J., Schiavon, R., Stover, R.J., Tucker, D., Wallace, V., Wei, M., Wirth, G., and Wright, C.A. (2003) The DEIMOS spectrograph for the Keck II Telescope: integration and testing, in Proceedings of SPIE 4841, Instrument Design and Performance for Optical/Infrared Ground-based Telescopes (eds M. Iye and A.F.M. Moorwood), pp. 1657–1669, doi: 10.1117/12.460346.

21 Seifert, W., Appenzeller, I., Fuertig, W., Stahl, O., Sutorius, E., Xu, W., Gaessler, W., Haefner, R., Hess, H.J., Hummel, W., Mantel, K.H., Meisl, W., Muschielok, B., Tarantik, K., Nicklas, H.E., Rupprecht, G., Cumani, C., Szeifert, T., and Spyromilio, J. (2000) Commissioning of the FORS instruments at the ESO VLT, in Proceedings of SPIE 4008, Optical and IR Telescope Instrumentation and Detectors (eds M. Iye and A.F. Moorwood), pp. 96–103.

22 Le Fevre, O., Saisse, M., Mancini, D., Vettolani, G.P., Maccagni, D., Picat, J.P., Mellier, Y., Mazure, A., Cuby, J.G., Delabre, B., Garilli, B., Hill, L., Prieto, E., Voet, C., Arnold, L., Brau-Nogue, S., Cascone, E., Conconi, P., Finger, G., Huster, G., Laloge, A., Lucuix, C., Mattaini, E., Schipani, P., Waultier, G., Zerbi, F.M., Avila, G., Beletic, J.W., D'Odorico, S., Moorwood, A.F., Monnet, G.J., and Reyes Moreno, J. (2000) VIMOS and NIRMOS multi-object spectrographs for the ESO VLT, in Proceedings of SPIE 4008, Optical and IR Telescope Instrumentation and Detectors (eds M. Iye and A.F. Moorwood), pp. 546–557.

23 Davies, R.L., Allington-Smith, J.R., Bettess, P., Chadwick, E., Content, R., Dodsworth, G.N., Haynes, R., Lee, D., Lewis, I.J., Webster, J., Atad, E., Beard, S.M., Ellis, M., Hastings, P.R., Williams, P.R., Bond, T., Crampton, D., Davidge, T.J., Fletcher, M., Leckie, B., Morbey, C.L., Murowinski, R.G., Roberts, S., Saddlemyer, L.K., Sebesta, J., Stilburn, J.R., and Szeto, K. (1997) GMOS: the GEMINI multiple object spectrographs, in Proceedings of SPIE 2871, Optical Telescopes of Today and Tomorrow (ed. A.L. Ardeberg), pp. 1099–1106.

24 Kashikawa, N., Inata, M., Iye, M., Kawabata, K., Okita, K., Kosugi, G., Ohyama, Y., Sasaki, T., Sekiguchi, K., Takata, T., Shimizu, Y., Yoshida, M., Aoki, K., Saito, Y., Asai, R., Taguchi, H., Ebizuka, N., Ozawa, T., and Yadoumaru, Y. (2000) FOCAS: faint object camera and spectrograph for the Subaru Telescope, in Proceedings of SPIE 4008, Optical and IR Telescope Instrumentation and Detectors (eds M. Iye and A.F. Moorwood), pp. 104–113.

25 Wolf, M.J., Mulligan, M.P., Smith, M.P., Adler, D.P., Bartosz, C.M., Bershady, M.A., Buckley, D.A.H., Burse, M.P., Chordia, P.A., Clemens, J.C., Epps, H.W., Garot, K., Indahl, B.L., Jaehnig, K.P., Koch, R.J., Mason, W.P., Mosby, G., Nordsieck, K.H., Percival, J.W., Punnadi, S., Ramaprakash, A.N., Schier, J.A., Sheinis, A.I., Smee, S.A., Thielman, D.J., Werner, M.W., Williams, T.B., and Wong, J.P. (2014) Project status of the Robert Stobie spectrograph near infrared instrument (RSS-NIR) for SALT, in Proceedings of SPIE 9147, Ground-based and Airborne Instrumentation for Astronomy V, p. 91470B, doi: 10.1117/12.2056736.

26 Byard, P.L. and O'Brien, T.P. (2000) MODS: optical design for a multi-object dual spectrograph, in Proceedings of SPIE 4008, Optical and IR Telescope Instrumentation and Detectors (eds M. Iye and A.F. Moorwood), pp. 934–941.

27 Cepa, J., Aguiar-Gonzalez, M., Bland-Hawthorn, J., Castaneda, H., Cobos, F.J., Correa, S., Espejo, C., Fragoso-Lopez, A.B., Fuentes, F.J., Gigante, J.V., Gonzalez, J.J., Gonzalez-Escalera, V., Gonzalez-Serrano, J.I., Joven-Alvarez, E., Lopez-Ruiz, J.C., Militello, C., Cano, L.P., Perez, A., Perez, J., Rasilla, J.L., Sanchez, B., and Tejada, C. (2003) OSIRIS tunable imager and spectrograph for the GTC. Instrument status, in Proceedings of SPIE 4841, Instrument Design and Performance for Optical/Infrared Ground-based Telescopes (eds M. Iye and A.F.M. Moorwood), pp. 1739–1749, doi: 10.1117/12.460913.

28 Eikenberry, S., Elston, R., Raines, S.N., Julian, J., Hanna, K., Hon, D., Julian, R., Bandyopadhyay, R., Bennett, J.G., Bessoff, A., Branch, M., Corley, R.,

Eriksen, J.D., Frommeyer, S., Gonzalez, A., Herlevich, M., Marin-Franch, A., Marti, J., Murphey, C., Rashkin, D., Warner, C., Leckie, B., Gardhouse, W.R., Fletcher, M., Dunn, J., Wooff, R., and Hardy, T. (2006) FLAMINGOS-2: the facility near-infrared wide-field imager and multi-object spectrograph for Gemini, in Proceedings of SPIE 6269, Society of Photo-Optical Instrumentation Engineers (SPIE) Conference Series, p. 626917, doi: 10.1117/12.672095. astro-ph/0604577.

29 Ichikawa, T., Suzuki, R., Tokoku, C., Uchimoto, Y.K., Konishi, M., Yoshikawa, T., Yamada, T., Tanaka, I., Omata, K., and Nishimura, T. (2006) MOIRCS: multi-object infrared camera and spectrograph for SUBARU, in Proceedings of SPIE 6269, Society of Photo-Optical Instrumentation Engineers (SPIE) Conference Series, p. 626916, doi: 10.1117/12.670078.

30 McLean, I.S., Steidel, C.C., Epps, H.W., Matthews, K., Konidaris, N., Kulas, K., Mace, G.N., Rudie, G.C., and Trainor, R. (2013) MOSFIRE: first light and early performance on the Keck I telescope, in American Astronomical Society Meeting Abstracts, Vol. **221**, id. 345.04.

31 Buschkamp, P., Seifert, W., Polsterer, K., Hofmann, R., Gemperlein, H., Lederer, R., Lehmitz, M., Naranjo, V., Ageorges, N., Kurk, J., Eisenhauer, F., Rabien, S., Honsberg, M., and Genzel, R. (2012) LUCI in the sky: performance and lessons learned in the first two years of near-infrared multi-object spectroscopy at the LBT, in Proceedings of SPIE 8446, Ground-based and Airborne Instrumentation for Astronomy IV, p. 84465L, doi: 10.1117/12.926989.

32 Garzón, F., Abreu, D., Barrera, S., Becerril, S., Cairós, L.M., Díaz, J.J., Fragoso, A.B., Gago, F., Grange, R., González, C., López, P., Patrón, J., Pérez, J., Rasilla, J.L., Redondo, P., Restrepo, R., Saavedra, P., Sánchez, V., Tenegi, F., and Vallbé, M. (2007) EMIR, the GTC NIR multi-object imager-spectrograph, in Revista Mexicana de Astronomia y Astrofisica Conference Series, vol. **29** (ed. R. Guzmán), pp. 12–17.

33 Hill, J.M., Angel, J.R.P., Scott, J.S., Lindley, D., and Hintzen, P. (1980) Multiple object spectroscopy - the Medusa spectrograph. *Astrophys. J.*, **242**, L69–L72, doi: 10.1086/183405.

34 Lewis, I.J., Cannon, R.D., Taylor, K., Glazebrook, K., Bailey, J.A., Baldry, I.K., Barton, J.R., Bridges, T.J., Dalton, G.B., Farrell, T.J., Gray, P.M., Lankshear, A., McCowage, C., Parry, I.R., Sharples, R.M., Shortridge, K., Smith, G.A., Stevenson, J., Straede, J.O., Waller, L.G., Whittard, J.D., Wilcox, J.K., and Willis, K.C. (2002) The Anglo-Australian observatory 2dF facility. *Mon. Not. R. Astron. Soc.*, **333**, 279–299, doi: 10.1046/j.1365-8711.2002.05333.x.

35 Akiyama, M., Smedley, S., Gillingham, P., Brzeski, J., Farrell, T., Kimura, M., Muller, R., Tamura, N., and Takato, N. (2008) Performance of Echidna fiber positioner for FMOS on Subaru, in Proceedings of SPIE 7018, Advanced Optical and Mechanical Technologies in Telescopes and Instrumentation, p. 70182V, doi: 10.1117/12.788968.

36 Fisher, C.D., Braun, D.F., Kaluzny, J.V., Seiffert, M.D., Dekany, R.G., Ellis, R.S., and Smith, R.M. (2012) Developments in high-density Cobra fiber positioners for the Subaru Telescope's Prime Focus Spectrometer, in Proceedings of SPIE 8450, Modern Technologies in Space- and Ground-based Telescopes and Instrumentation II, p. 845017, doi: 10.1117/12.927161. 1210.2734.

37 Gilbert, J., Goodwin, M., Heijmans, J., Muller, R., Miziarski, S., Brzeski, J., Waller, L., Saunders, W., Bennet, A., and Tims, J. (2012) Starbugs: all-singing, all-dancing fibre positioning robots, in Proceedings of SPIE 8450, Modern Technologies in Space- and Ground-based Telescopes and Instrumentation II, p. 84501A, doi: 10.1117/12.924502.
38 Wilkinson, A., Sharples, R.M., Fosbury, R.A.E., and Wallace, P.T. (1986) Stellar dynamics of CEN A. *Mon. Not. R. Astron. Soc.*, **218**, 297–329, doi: 10.1093/mnras/218.2.297.
39 Carranza, G., Courtes, G., Georgelin, Y., Monnet, G., and Pourcelot, A. (1968) Interferometric study of ionized hydrogen in M 33. New kinematical and physical data. *Ann. Astrophys.*, **31**, 63.
40 Taylor, K. and Atherton, P.D. (1980) Seeing-limited radial velocity field mapping of extended emission line sources using a new imaging Fabry-Perot system. *Mon. Not. R. Astron. Soc.*, **191**, 675–684, doi: 10.1093/mnras/191.4.675.
41 Boulesteix, J., Georgelin, Y., Marcelin, M., and Monnet, G. (1984) First results from CIGALE scanning Perot-Fabry interferometer, in Proceedings of SPIE 445, Instrumentation in Astronomy V (eds A. Boksenberg and D.L. Crawford), pp. 37–41.
42 Bland, J. and Tully, R.B. (1989) The Hawaii imaging Fabry-Perot interferometer (HIFI). *Astron. J.*, **98**, 723–735, doi: 10.1086/115173.
43 Cepa, J., Bland-Hawthorn, J., González, J.J., and OSIRIS Consortium (2000) OSIRIS: a tunable filter spectrograph for the GTC 10m telescope (Poster), in *Imaging the Universe in Three Dimensions*, Astronomical Society of the Pacific Conference Series, vol. **195** (eds W. van Breugel and J. Bland-Hawthorn), p. 597.
44 Rangwala, N., Williams, T.B., Pietraszewski, C., and Joseph, C.L. (2008) An imaging FABRY-PÉROT system for the Robert Stobie spectrograph on the Southern African Large Telescope. *Astron. J.*, **135**, 1825–1836, doi: 10.1088/0004-6256/135/5/1825.
45 Drissen, L., Alarie, A., Martin, T., Lagrois, D., Rousseau-Nepton, L., Bilodeau, A., Robert, C., Joncas, G., and Iglesias-Páramo, J. (2012) New scientific results with SpIOMM: a testbed for CFHT's imaging Fourier transform spectrometer SITELLE, in Proceedings of SPIE 8446, Ground-based and Airborne Instrumentation for Astronomy IV, p. 84463S, doi: 10.1117/12.925202. 1209.3929.
46 Grandmont, F., Drissen, L., Mandar, J., Thibault, S., and Baril, M. (2012) Final design of SITELLE: a wide-field imaging Fourier transform spectrometer for the Canada-France-Hawaii Telescope, in Proceedings of SPIE 8446, Ground-based and Airborne Instrumentation for Astronomy IV, p. 84460U, doi: 10.1117/12.926782.
47 Drissen, L., Rousseau-Nepton, L., Lavoie, S., Robert, C., Martin, T., Martin, P., Mandar, J., and Grandmont, F. (2014) Imaging FTS: a different approach to integral field spectroscopy. *Adv. Astron.*, **2014**, 293856, doi: 10.1155/2014/293856.
48 Bacon, R., Adam, G., Baranne, A., Courtes, G., Dubet, D., Dubois, J.P., Emsellem, E., Ferruit, P., Georgelin, Y., Monnet, G., Pecontal, E., Rousset, A., and Say, F. (1995) 3D spectrography at high spatial resolution. I. Concept and realization of the integral field spectrograph TIGER. *Astron. Astrophys. Suppl.*, **113**, 347.

49 Lee, D., Haynes, R., Ren, D., and Allington-Smith, J. (2001) Characterization of lenslet arrays for astronomical spectroscopy. *Publ. Astron. Soc. Pac.*, **113**, 1406–1419.

50 Bacon, R. (1995) The integral field spectrograph TIGER: results and prospects. Tridimensional Optical Spectroscopic Methods in Astronomy. Astronomical Society of the Pacific Conference Series, Vol. 71 (eds G. Comte and M. Marcelin), p. 239.

51 Afanasiev, V.L., Dodonov, S.N., Drabek, S.V., and Vlasiouk, V.V. (1995) Bidimensional spectroscopy with the 6-meter telescope in time resolving mode. Tridimensional Optical Spectroscopic Methods in Astronomy. Astronomical Society of the Pacific Conference Series, Vol. 71 (eds G. Comte and M. Marcelin), p. 276.

52 Bacon, R., Emsellem, E., Combes, F., Copin, Y., Monnet, G., and Martin, P. (2001) The M 31 double nucleus probed with OASIS. A natural m = 1 mode? *Astron. Astrophys.*, **371**, 409.

53 Bacon, R., Copin, Y., Monnet, G., Miller, B.W., Allington-Smith, J.R., Bureau, M., Carollo, C.M., Davies, R.L., Emsellem, E., Kuntschner, H., Peletier, R.F., Verolme, E.K., and de Zeeuw, P.T. (2001) The SAURON project - I. The panoramic integral-field spectrograph. *Mon. Not. R. Astron. Soc.*, **326**, 23–35.

54 McDermid, R., Emsellem, E., Cappellari, M., Kuntschner, H., Bacon, R., Bureau, M., Copin, Y., Davies, R.L., Falcón-Barroso, J., Ferruit, P., Krajnović, D., Peletier, R.F., Shapiro, K., Wernli, F., and de Zeeuw, P.T. (2004) OASIS high-resolution integral field spectroscopy of the SAURON ellipticals and lenticulars. *Astron. Nachr.*, **325**, 100–103.

55 Lantz, B., Aldering, G., Antilogus, P., Bonnaud, C., Capoani, L., Castera, A., Copin, Y., Dubet, D., Gangler, E., Henault, F., Lemonnier, J.P., Pain, R., Pecontal, A., Pécontal, E., and Smadja, G. (2004) SNIFS: a wideband integral field spectrograph with microlens arrays, in Proceedings of SPIE 5249. Optical Design and Engineering (eds L. Mazuray), p. 146.

56 Larkin, J., Barczys, M., Krabbe, A., Adkins, S., Aliado, T., Amico, P., Brims, G., Campbell, R., Canfield, J., Gasaway, T., Honey, A., Iserlohe, C., Johnson, C., Kress, E., Lafreniere, D., Magnone, K., Magnone, N., McElwain, M., Moon, J., Quirrenbach, A., Skulason, G., Song, I., Spencer, M., Weiss, J., and Wright, S. (2006) OSIRIS: a diffraction limited integral field spectrograph for Keck. *New Astron. Rev*, **50**, 362–364.

57 Vanderriest, C. (1995) Integral field spectroscopy with optical fibres, in *IAU Colloq. 149: Tridimensional Optical Spectroscopic Methods in Astrophysics*, Astronomical Society of the Pacific Conference Series, vol. **71** (eds G. Comte and M. Marcelin), p. 209.

58 Dressler, A., Bigelow, B., Hare, T., Sutin, B., Thompson, I., Burley, G., Epps, H., Oemler, A., Bagish, A., Birk, C., Clardy, K., Gunnels, S., Kelson, D., Shectman, S., and Osip, D. (2011) IMACS: the Inamori-Magellan areal camera and spectrograph on Magellan-Baade. *Publ. Astron. Soc. Pac.*, **123**, 288–332, doi: 10.1086/658908.

59 Chonis, T.S., Hill, G.J., Lee, H., Tuttle, S.E., and Vattiat, B.L. (2014) LRS2: the new facility low resolution integral field spectrograph for the Hobby-Eberly

telescope, in Proceedings of SPIE 9147, Ground-based and Airborne Instrumentation for Astronomy V, p. 91470A, doi: 10.1117/12.2056005. 1407.6016.

60 Gil de Paz, A., Gallego, J., Carrasco, E., Iglesias-Páramo, J., Cedazo, R., Vílchez, J.M., García-Vargas, M.L., Arrillaga, X., Carrera, M.A., Castillo-Morales, A., Castillo-Domínguez, E., Eliche-Moral, M.C., Ferrusca, D., González-Guardia, E., Lefort, B., Maldonado, M., Marino, R.A., Martínez-Delgado, I., Morales Durán, I., Mujica, E., Páez, G., Pascual, S., Pérez-Calpena, A., Sánchez-Penim, A., Sánchez-Blanco, E., Tulloch, S., Velázquez, M., Zamorano, J., Aguerri, A.L., Barrado y Naváscues, D., Bertone, E., Cardiel, N., Cava, A., Cenarro, J., Chávez, M., García, M., Guichard, J., Gúzman, R., Herrero, A., Huélamo, N., Hughes, D., Jiménez-Vicente, J., Kehrig, C., Márquez, I., Masegosa, J., Mayya, Y.D., Méndez-Abreu, J., Mollá, M., Muñoz-Tuñón, C., Peimbert, M., Pérez-González, P.G., Pérez Montero, E., Rodríguez, M., Rodríguez-Espinosa, J.M., Rodríguez-Merino, L., Rosa-González, D., Sánchez-Almeida, J., Sánchez Contreras, C., Sánchez-Blázquez, P., Sánchez Moreno, F.M., Sánchez, S.F., Sarajedini, A., Serena, F., Silich, S., Simón-Díaz, S., Tenorio-Tagle, G., Terlevich, E., Terlevich, R., Torres-Peimbert, S., Trujillo, I., Tsamis, Y., Vega, O., and Villar, V. (2014) MEGARA: a new generation optical spectrograph for GTC, in Proceedings of SPIE 9147, Ground-based and Airborne Instrumentation for Astronomy V, p. 91470O, doi: 10.1117/12.2047825.

61 Content, R. (1997) New design for integral field spectroscopy with 8-m telescopes, in Proceedings of SPIE 2871, Optical Telescopes of Today and Tomorrow (ed. A.L. Ardeberg), pp. 1295–1305.

62 Bacon, R., Vernet, J., Borisova, E., Bouché, N., Brinchmann, J., Carollo, M., Carton, D., Caruana, J., Cerda, S., Contini, T., Franx, M., Girard, M., Guerou, A., Haddad, N., Hau, G., Herenz, C., Herrera, J.C., Husemann, B., Husser, T.O., Jarno, A., Kamann, S., Krajnovic, D., Lilly, S., Mainieri, V., Martinsson, T., Palsa, R., Patricio, V., Pécontal, A., Pello, R., Piqueras, L., Richard, J., Sandin, C., Schroetter, I., Selman, F., Shirazi, M., Smette, A., Soto, K., Streicher, O., Urrutia, T., Weilbacher, P., Wisotzki, L., and Zins, G. (2014) MUSE commissioning. *The Messenger*, **157**, 13–16.

63 Le Coarer, E., Georgelin, Y., and Bensammar, S. (1992) Pytheas - a multi-channel spectrometer for the analysis of astronomical images, in European Southern Observatory Conference and Workshop Proceedings, vol. 40 (ed. M.H. Ulrich), p. 297.

64 Groff, T.D., Kasdin, N.J., Galvin, M., Peters, M.A., Chilcote, J.K., Brandt, T., Knapp, G.R., Carr, M., Loomis, C., McElwain, M.W., Mede, K., Guyon, O., Jovanovic, N., Takato, N., and Hayashi, M. (2016) Laboratory performance and commissioning of the CHARIS IFS, in American Astronomical Society Meeting Abstracts, vol. **227**, id. 146.07.

65 Sharples, R., Morris, S., and Content, R. (2002) MEIFU - a million element integral field unit for deep Ly-α searches, in *Next Generation Wide-Field Multi-Object Spectroscopy*, Astronomical Society of the Pacific Conference Series, vol. **280** (eds M.J.I. Brown and A. Dey), p. 125.

66 Sharples, R., Bender, R., Agudo Berbel, A., Bennett, R., Bezawada, N., Castillo, R., Cirasuolo, M., Clark, P., Davidson, G., Davies, R., Davies, R., Dubbeldam, M., Fairley, A., Finger, G., Schreiber, N.F., Genzel, R., Haefner, R., Hess, A., Jung, I., Lewis, I., Montgomery, D., Murray, J., Muschielok, B., Pirard, J., Ramsay, S., Rees, P., Richter, J., Robertson, D., Robson, I., Rolt, S., Saglia, R., Saviane, I., Schlichter, J., Schmidtobreik, L., Segovia, A., Smette, A., Tecza, M., Todd, S., Wegner, M., and Wiezorrek, E. (2014) Performance of the K-band multi-object spectrograph (KMOS) on the ESO VLT, in Proceedings of SPIE 9147, Ground-based and Airborne Instrumentation for Astronomy V, p. 91470W, doi: 10.1117/12.2055496.

67 Eikenberry, S.S., Bennett, J.G., Chinn, B., Donoso, H.V., Eikenberry, S.A., Ettedgui, E., Fletcher, A., Frommeyer, R., Garner, A., Herlevich, M., Lasso, N., Miller, P., Mullin, S., Murphey, C., Raines, S.N., Packham, C., Schofield, S., Stelter, R.D., Varosi, F., Vega, C., Warner, C., Garzón, F., Rosich, J., Gomez, J.M., Sabater, J., Vilar, C., Torra, J., Gallego, J., Cardiel, N., Eliche, C., Pascual, S., Ballester, O., Illa, J.M., Jimenez, J., Cardiel-Sas, L., Galipienzo, J., Carrera, M.A., Hammersley, P., and Cuevas, S. (2012) MIRADAS for the Gran Telescopio Canarias: system overview, in Proceedings of SPIE 8446, Ground-based and Airborne Instrumentation for Astronomy IV, p. 844657, doi: 10.1117/12.925686.

68 Pasquini, L., Alonso, J., Avila, G., Barriga, P., Biereichel, P., Buzzoni, B., Cavadore, C., Cumani, C., Dekker, H., Delabre, B., Kaufer, A., Kotzlowski, H., Hill, V., Lizon, J.L., Nees, W., Santin, P., Schmutzer, R., Kesteren, A.V., and Zoccali, M. (2003) Installation and first results of FLAMES, the VLT multifibre facility, in Proceedings of SPIE 4841, Instrument Design and Performance for Optical/Infrared Ground-based Telescopes (eds M. Iye and A.F.M. Moorwood), pp. 1682–1693, doi: 10.1117/12.458915.

69 Croom, S.M., Lawrence, J.S., Bland-Hawthorn, J., Bryant, J.J., Fogarty, L., Richards, S., Goodwin, M., Farrell, T., Miziarski, S., Heald, R., Jones, D.H., Lee, S., Colless, M., Brough, S., Hopkins, A.M., Bauer, A.E., Birchall, M.N., Ellis, S., Horton, A., Leon-Saval, S., Lewis, G., López-Sánchez, A.R., Min, S.S., Trinh, C., and Trowland, H. (2012) The Sydney-AAO multi-object integral field spectrograph. *Mon. Not. R. Astron. Soc.*, **421**, 872–893, doi: 10.1111/j.1365-2966.2011.20365.x.

70 Bundy, K., Bershady, M.A., Law, D.R., Yan, R., Drory, N., MacDonald, N., Wake, D.A., Cherinka, B., Sánchez-Gallego, J.R., Weijmans, A.M., Thomas, D., Tremonti, C., Masters, K., Coccato, L., Diamond-Stanic, A.M., Aragón-Salamanca, A., Avila-Reese, V., Badenes, C., Falcón-Barroso, J., Belfiore, F., Bizyaev, D., Blanc, G.A., Bland-Hawthorn, J., Blanton, M.R., Brownstein, J.R., Byler, N., Cappellari, M., Conroy, C., Dutton, A.A., Emsellem, E., Etherington, J., Frinchaboy, P.M., Fu, H., Gunn, J.E., Harding, P., Johnston, E.J., Kauffmann, G., Kinemuchi, K., Klaene, M.A., Knapen, J.H., Leauthaud, A., Li, C., Lin, L., Maiolino, R., Malanushenko, V., Malanushenko, E., Mao, S., Maraston, C., McDermid, R.M., Merrifield, M.R., Nichol, R.C., Oravetz, D., Pan, K., Parejko, J.K., Sanchez, S.F., Schlegel, D., Simmons, A., Steele, O., Steinmetz, M., Thanjavur, K., Thompson, B.A., Tinker, J.L., van den Bosch, R.C.E., Westfall, K.B., Wilkinson, D., Wright, S.,

Xiao, T., and Zhang, K. (2015) Overview of the SDSS-IV MaNGA Survey: mapping nearby Galaxies at Apache Point Observatory. *Astrophys. J.*, **798**, 7, doi: 10.1088/0004-637X/798/1/7.

71 Smee, S.A., Prochaska, T., Shectman, S.A., Hammond, R.P., Barkhouser, R.H., DePoy, D.L., and Marshall, J.L. (2012) Optomechanical design concept for GMACS: a wide-field multi-object moderate resolution optical spectrograph for the Giant Magellan Telescope (GMT), in Proceedings of SPIE 8446, Ground-based and Airborne Instrumentation for Astronomy IV, p. 84467N, doi: 10.1117/12.926437.

72 McGregor, P.J., Bloxham, G.J., Boz, R., Davies, J., Doolan, M., Ellis, M., Hart, J., Jones, D.J., Luvaul, L., Nielsen, J., Parcell, S., Sharp, R., Stevanovic, D., and Young, P.J. (2012) GMT integral-field spectrograph (GMTIFS) conceptual design, in Proceedings of SPIE 8446, Ground-based and Airborne Instrumentation for Astronomy IV, p. 84461I, doi: 10.1117/12.925259.

73 Goodwin, M., Brzeski, J., Case, S., Colless, M., Farrell, T., Gers, L., Gilbert, J., Heijmans, J., Hopkins, A., Lawrence, J., Miziarski, S., Monnet, G., Muller, R., Saunders, W., Smith, G., Tims, J., and Waller, L. (2012) MANIFEST instrument concept and related technologies, in Proceedings of SPIE 8446, Ground-based and Airborne Instrumentation for Astronomy IV, p. 84467I, doi: 10.1117/12.925125.

74 Bigelow, B.C., Radovan, M.V., Bernstein, R.A., Onaka, P.M., Yamada, H., Isani, S., Miyazaki, S., and Ozaki, S. (2014) Conceptual design of the MOBIE imaging spectrograph for TMT, in Proceedings of SPIE 9147, Ground-based and Airborne Instrumentation for Astronomy V, p. 914728, doi: 10.1117/12.2055297.

75 Moore, A.M., Larkin, J.E., Wright, S.A., Bauman, B., Dunn, J., Ellerbroek, B., Phillips, A.C., Simard, L., Suzuki, R., Zhang, K., Aliado, T., Brims, G., Canfield, J., Chen, S., Dekany, R., Delacroix, A., Do, T., Herriot, G., Ikenoue, B., Johnson, C., Meyer, E., Obuchi, Y., Pazder, J., Reshetov, V., Riddle, R., Saito, S., Smith, R., Sohn, J.M., Uraguchi, F., Usuda, T., Wang, E., Wang, L., Weiss, J., and Wooff, R. (2014) The Infrared Imaging Spectrograph (IRIS) for TMT: instrument overview, in Proceedings of SPIE 9147, Ground-based and Airborne Instrumentation for Astronomy V, p. 914724, doi: 10.1117/12.2055216. 1407.2995.

76 Thatte, N.A., Clarke, F., Bryson, I., Schnetler, H., Tecza, M., Bacon, R.M., Remillieux, A., Mediavilla, E., Herreros Linares, J.M., Arribas, S., Evans, C.J., Lunney, D.W., Fusco, T., O'Brien, K., Tosh, I.A., Ives, D.J., Finger, G., Houghton, R., Davies, R.L., Lynn, J.D., Allen, J.R., Zieleniewski, S.D., Kendrew, S., Ferraro-Wood, V., Pécontal-Rousset, A., Kosmalski, J., Richard, J., Jarno, A., Gallie, A.M., Montgomery, D.M., Henry, D., Zins, G., Freeman, D., García-Lorenzo, B., Rodríguez-Ramos, L.F., Revuelta, J.S.C., Hernandez Suarez, E., Bueno-Bueno, A., Gigante-Ripoll, J.V., Garcia, A., Dohlen, K., and Neichel, B. (2014) HARMONI: the first light integral field spectrograph for the E-ELT, in Proceedings of SPIE 9147, Ground-based and Airborne Instrumentation for Astronomy V, p. 914725, doi: 10.1117/12.2055436.

77 Evans, C., Puech, M., Afonso, J., Almaini, O., Amram, P., Aussel, H., Barbuy, B., Basden, A., Bastian, N., Battaglia, G., Biller, B., Bonifacio, P., Bouché, N., Bunker, A., Caffau, E., Charlot, S., Cirasuolo, M., Clenet, Y., Combes, F., Conselice, C., Contini, T., Cuby, J.G., Dalton, G., Davies, B., de Koter, A., Disseau, K., Dunlop, J., Epinat, B., Fiore, F., Feltzing, S., Ferguson, A., Flores, H., Fontana, A., Fusco, T., Gadotti, D., Gallazzi, A., Gallego, J., Giallongo, E., Gonçalves, T., Gratadour, D., Guenther, E., Hammer, F., Hill, V., Huertas-Company, M., Ibata, R., Kaper, L., Korn, A., Larsen, S., Le Fèvre, O., Lemasle, B., Maraston, C., Mei, S., Mellier, Y., Morris, S., Östlin, G., Paumard, T., Pello, R., Pentericci, L., Peroux, C., Petitjean, P., Rodrigues, M., Rodríguez-Muñoz, L., Rouan, D., Sana, H., Schaerer, D., Telles, E., Trager, S., Tresse, L., Welikala, N., Zibetti, S., and Ziegler, B. (2015) The Science Case for Multi-Object Spectroscopy on the European ELT. *ArXiv e-prints*, 1501.04726.

78 Bland-Hawthorn, J., Ellis, S., Leon-Saval, S., Haynes, R., Roth, M., Lahmannsrben, H.G., Horton, A., Cuby, J.G., Birks, T., Lawrence, J., Gillingham, P., Ryder, S., and Trinh, C. (2011) A complex multi-notch astronomical filter to suppress the bright infrared sky. *Nat. Commun.*, **2**, 581.

79 Le Coarer, E. (2009) Put Detector in your optics, in *EAS Publications Series*, vol. **37** (ed. P. Kern), pp. 31–34, doi: 10.1051/eas/0937004.

80 Le Coarer, E., Blaize, S., Benech, P., Stefanon, I., Morand, A., Lérondel, G., Leblond, G., Kern, P., Fedeli, J.M., and Royer, P. (2007) Wavelength-scale stationary-wave integrated Fourier-transform spectrometry. *Nat. Photon.*, **1**, 473–478.

81 Turner, M.J.L., Abbey, A., Arnaud, M., Balasini, M., Barbera, M., Belsole, E., Bennie, P.J., Bernard, J.P., Bignami, G.F., Boer, M., Briel, U., Butler, I., Cara, C., Chabaud, C., Cole, R., Collura, A., Conte, M., Cros, A., Denby, M., Dhez, P., Di Coco, G., Dowson, J., Ferrando, P., Ghizzardi, S., Gianotti, F., Goodall, C.V., Gretton, L., Griffiths, R.G., Hainaut, O., Hochedez, J.F., Holland, A.D., Jourdain, E., Kendziorra, E., Lagostina, A., Laine, R., La Palombara, N., Lortholary, M., Lumb, D., Marty, P., Molendi, S., Pigot, C., Poindron, E., Pounds, K.A., Reeves, J.N., Reppin, C., Rothenflug, R., Salvetat, P., Sauvageot, J.L., Schmitt, D., Sembay, S., Short, A.D.T., Spragg, J., Stephen, J., Strüder, L., Tiengo, A., Trifoglio, M., Trümper, J., Vercellone, S., Vigroux, L., Villa, G., Ward, M.J., Whitehead, S., and Zonca, E. (2001) The European Photon Imaging Camera on XMM-newton: the MOS cameras. *Astron. Astrophys.*, **365**, L27–L35, doi: 10.1051/0004-6361:20000087.

82 Perryman, M.A.C., Favata, F., Peacock, A., Rando, N., and Taylor, B.G. (1999) Optical STJ observations of the Crab Pulsar. *Astron. Astrophys.*, **346**, L30–L32.

83 Romani, R.W., Miller, A.J., Cabrera, B., Nam, S.W., and Martinis, J.M. (2001) Phase-resolved Crab studies with a cryogenic transition-edge sensor spectrophotometer. *Astrophys. J.*, **563**, 221–228, doi: 10.1086/323874.

84 Keller, C.U., Gschwind, R., Renn, A., Rosselet, A., and Wild, U.P. (1995) The spectral hole-burning device: a 3-dimensional photon detector. *Astron. Astrophys. Suppl.*, **109**, 383–387.

85 Keller, C.U. (2000) 5,000 by 5,000 spatial by 15,000 spectral resolution elements: first astronomical observations with a novel 3-D detector, in *Imaging the Universe in Three Dimensions*, Astronomical Society of the Pacific Conference Series, vol. **195** (eds W. van Breugel and J. Bland-Hawthorn), p. 495.
86 Shannon, C.E. (1949) Communication in the presence of noise. *Proc. Inst. Radio Eng.*, **37**, 10–21.
87 Fixsen, D.J., Offenberg, J.D., Hanisch, R.J., Mather, J.C., Nieto-Santisteban, M.A., Sengupta, R., and Stockman, H.S. (2000) Cosmic-ray rejection and readout efficiency for large-area arrays. *Publ. Astron. Soc. Pac.*, **112**, 1350–1359.
88 Smith, A., Bailey, J., Hough, J.H., and Lee, S. (2009) An investigation of lucky imaging techniques. *Mon. Not. R. Astron. Soc.*, **398**, 2069–2073, doi: 10.1111/j.1365-2966.2009.15249.x.
89 Obereder, A., Ramlau, R., and Fedrigo, E. (2014) Mathematical algorithms and software for ELT adaptive optics. The Austrian In-kind contributions for adaptive optics. *The Messenger*, **156**, 16–20.
90 Foy, R. and Labeyrie, A. (1985) Feasibility of adaptive telescope with laser probe. *Astron. Astrophys.*, **152**, L29–L31.
91 Calia, D.B., Feng, Y., Hackenberg, W., Holzlöhner, R., Taylor, L., and Lewis, S. (2010) Laser development for sodium laser guide stars at ESO. *The Messenger*, **139**, 12–19.
92 Rigaut, F.J., Ellerbroek, B.L., and Flicker, R. (2000) Principles, limitations, and performance of multiconjugate adaptive optics, in Proceedings of SPIE 4007, Adaptive Optical Systems Technology (ed. P.L. Wizinowich), pp. 1022–1031.
93 Beckers, J.M. (1988) Increasing the size of the isoplanatic patch with multiconjugate adaptive optics, in Very Large Telescopes and their Instrumentation, Vol. 2, pp. 693–703.
94 Rigaut, F., Neichel, B., Boccas, M., d'Orgeville, C., Arriagada, G., Fesquet, V., Diggs, S.J., Marchant, C., Gausach, G., Rambold, W.N., Luhrs, J., Walker, S., Carrasco-Damele, E.R., Edwards, M.L., Pessev, P., Galvez, R.L., Vucina, T.B., Araya, C., Gutierrez, A., Ebbers, A.W., Serio, A., Moreno, C., Urrutia, C., Rogers, R., Rojas, R., Trujillo, C., Miller, B., Simons, D.A., Lopez, A., Montes, V., Diaz, H., Daruich, F., Colazo, F., Bec, M., Trancho, G., Sheehan, M., McGregor, P., Young, P.J., Doolan, M.C., van Harmelen, J., Ellerbroek, B.L., Gratadour, D., and Garcia-Rissmann, A. (2012) GeMS: first on-sky results, in Proceedings of SPIE 8447, Adaptive Optics Systems III, p. 84470I, doi: 10.1117/12.927061.
95 Rigaut, F. (2002) Ground conjugate wide field adaptive optics for the ELTs, in European Southern Observatory Conference and Workshop Proceedings, vol. **58** (eds E. Vernet, R. Ragazzoni, S. Esposito, and N. Hubin), p. 11.
96 Puga, E., Feldt, M., Alvarez, C., Henning, T., Apai, D., Le Coarer, E., Chalabaev, A., and Stecklum, B. (2006) Outflows, disks, and stellar content in a region of high-mass star formation: G5.89-0.39 with adaptive optics. *Astrophys. J.*, **641**, 373–382, doi: 10.1086/500389.
97 Mobasher, B., Crampton, D., and Simard, L. (2010) An Infrared Multi-Object Spectrograph (IRMS) with adaptive optics for TMT: the

science case, in Proceedings of SPIE 7735, Ground-based and Airborne Instrumentation for Astronomy III, p. 77355P, doi: 10.1117/12.858061.

98 Puech, M. and Sayede, F. (2004) FALCON: extending adaptive optics corrections to cosmological fields, in Proceedings of SPIE 5492, Ground-based Instrumentation for Astronomy (eds A.F.M. Moorwood and M. Iye), pp. 303–311, doi: 10.1117/12.549805.

99 Patat, F., Moehler, S., O'Brien, K., Pompei, E., Bensby, T., Carraro, G., de Ugarte Postigo, A., Fox, A., Gavignaud, I., James, G., Korhonen, H., Ledoux, C., Randall, S., Sana, H., Smoker, J., Stefl, S., and Szeifert, T. (2011) Optical atmospheric extinction over Cerro Paranal. *Astron. Astrophys.*, **527**, A91, doi: 10.1051/0004-6361/201015537.

100 Le Fèvre, O., Saisse, M., Mancini, D., Brau-Nogue, S., Caputi, O., Castinel, L., D'Odorico, S., Garilli, B., Kissler-Patig, M., Lucuix, C., Mancini, G., Pauget, A., Sciarretta, G., Scodeggio, M., Tresse, L., and Vettolani, G. (2003) Commissioning and performances of the VLT-VIMOS instrument, in Proceedings of SPIE 4841, Instrument Design and Performance for Optical/Infrared Ground-based Telescopes, Society of Photo-Optical Instrumentation Engineers (SPIE) Conference Series (eds M. Iye and A.F.M. Moorwood), pp. 1670–1681, doi: 10.1117/12.460959.

101 Bacon, R., Brinchmann, J., Richard, J., Contini, T., Drake, A., Franx, M., Tacchella, S., Vernet, J., Wisotzki, L., Blaizot, J., Bouché, N., Bouwens, R., Cantalupo, S., Carollo, C.M., Carton, D., Caruana, J., Clément, B., Dreizler, S., Epinat, B., Guiderdoni, B., Herenz, C., Husser, T.O., Kamann, S., Kerutt, J., Kollatschny, W., Krajnovic, D., Lilly, S., Martinsson, T., Michel-Dansac, L., Patricio, V., Schaye, J., Shirazi, M., Soto, K., Soucail, G., Steinmetz, M., Urrutia, T., Weilbacher, P., and de Zeeuw, T. (2015) The MUSE 3D view of the Hubble Deep Field South. *Astron. Astrophys.*, **575**, A75, doi: 10.1051/0004-6361/201425419.

102 Fruchter, A.S. and Hook, R.N. (2002) Drizzle: a method for the linear reconstruction of undersampled images. *Publ. Astron. Soc. Pac.*, **114**, 144–152, doi: 10.1086/338393.

103 Glazebrook, K. and Bland-Hawthorn, J. (2001) Microslit nod-shuffle spectroscopy: a technique for achieving very high densities of spectra. *Publ. Astron. Soc. Pac.*, **113**, 197–214, doi: 10.1086/318625.

104 Roth, M.M., Kelz, A., Fechner, T., Hahn, T., Bauer, S.M., Becker, T., Böhm, P., Christensen, L., Dionies, F., Paschke, J., Popow, E., Wolter, D., Schmoll, J., Laux, U., and Altmann, W. (2005) PMAS: the potsdam multi-aperture spectrophotometer. I. Design, manufacture, and performance. *Publ. Astron. Soc. Pac.*, **117**, 620–642, doi: 10.1086/429877.

105 Roth, M.M., Löhmannsröben, H.G., Dosche, C., Sandin, C., Reich, O., Haynes, R., Leick, L., Chávez Boggio, J.M., and Kelz, A. (2010) Supercontinuum light sources for use in astronomical instrumentation: a test with PMAS, the Potsdam multi-aperture spectrophotometer, in Society of Photo-Optical Instrumentation Engineers (SPIE) Conference Series, vol. **7739**, p. 26, doi: 10.1117/12.857381.

106 Sarazin, M. and Roddier, F. (1990) The ESO differential image motion monitor. *Astron. Astrophys.*, **227**, 294–300.

107 Bertin, E. and Arnouts, S. (1996) SExtractor: software for source extraction. *Astron. Astrophys. Suppl.*, **117**, 393–404, doi: 10.1051/aas:1996164.
108 Weilbacher, P.M., Streicher, O., Urrutia, T., Jarno, A., Pécontal-Rousset, A., Bacon, R., and Böhm, P. (2012) Design and capabilities of the MUSE data reduction software and pipeline, in Society of Photo-Optical Instrumentation Engineers (SPIE) Conference Series, vol. **8451**, p. 84510B, doi: 10.1117/12.925114.
109 Davies, R.I., Agudo Berbel, A., Wiezorrek, E., Cirasuolo, M., Förster Schreiber, N.M., Jung, Y., Muschielok, B., Ott, T., Ramsay, S., Schlichter, J., Sharples, R., and Wegner, M. (2013) The software package for astronomical reductions with KMOS: SPARK. *Astron. Astrophys.*, **558**, A56, doi: 10.1051/0004-6361/201322282.
110 Filippenko, A.V. (1982) The importance of atmospheric differential refraction in spectrophotometry. *Publ. Astron. Soc. Pac.*, **94**, 715–721, doi: 10.1086/131052.
111 Owens, J.C. (1967) Optical refractive index of air: dependence on pressure, temperature, and composition. *Appl. Opt.*, **6**, 51, doi: 10.1364/AO.6.000051.
112 Streicher, O., Weilbacher, P.M., Bacon, R., and Jarno, A. (2011) Sky subtraction for the MUSE data reduction pipeline, in *Astronomical Data Analysis Software and Systems XX*, Astronomical Society of the Pacific Conference Series, vol. **442** (eds I.N. Evans, A. Accomazzi, D.J. Mink, and A.H. Rots), p. 257.
113 van der Loo, M.P.J., Groenenboom, G.C., Jamieson, M.J., and Dalgarno, A. (2006) Raman association of H2 in the early universe. *Faraday Discuss.*, **133**, 43, doi: 10.1039/b516803a.
114 Osterbrock, D.E., Fulbright, J.P., Martel, A.R., Keane, M.J., Trager, S.C., and Basri, G. (1996) Night-sky high-resolution spectral atlas of OH and O2 emission lines for echelle spectrograph wavelength calibration. *Publ. Astron. Soc. Pac.*, **108**, 277, doi: 10.1086/133722.
115 Mink, D.J. and Kurtz, M.J. (1998) RVSAO 2.0 - a radial velocity package for IRAF, in *Astronomical Data Analysis Software and Systems VII*, Astronomical Society of the Pacific Conference Series, vol. **145** (eds R. Albrecht, R.N. Hook, and H.A. Bushouse), p. 93.
116 Smithsonian Astrophysical Observatory (2000) SAOImage DS9: a utility for displaying astronomical images in the X11 window environment, Astrophysics Source Code Library. 0003.002.
117 Centre de Données Astronomiques de Strasbourg (Cds) (2011) Aladin: Interactive Sky Atlas, Astrophysics Source Code Library. 1112.019.
118 Ott, T. (2012) QFitsView: FITS file viewer, Astrophysics Source Code Library. 1210.019.
119 Courtes, G. (1982) An Integral Field Spectrograph (IFS) for large telescopes, in *IAU Colloq. 67: Instrumentation for Astronomy with Large Optical Telescopes*, Astrophysics and Space Science Library, vol. **92** (ed. C.M. Humphries), p. 123, doi: 10.1007/978-94-009-7787-7_17.
120 Cappellari, M. and Copin, Y. (2003) Adaptive spatial binning of integral-field spectroscopic data using Voronoi tessellations. *Mon. Not. R. Astron. Soc.*, **342**, 345–354, doi: 10.1046/j.1365-8711.2003.06541.x.

121 Zehavi, I., Zheng, Z., Weinberg, D.H., Blanton, M.R., Bahcall, N.A., Berlind, A.A., Brinkmann, J., Frieman, J.A., Gunn, J.E., Lupton, R.H., Nichol, R.C., Percival, W.J., Schneider, D.P., Skibba, R.A., Strauss, M.A., Tegmark, M., and York, D.G. (2011) Galaxy clustering in the completed SDSS redshift survey: the dependence on color and luminosity. *Astrophys. J.*, **736**, 59, doi: 10.1088/0004-637X/736/1/59.

122 Le Fèvre, O., Tasca, L.A.M., Cassata, P., Garilli, B., Le Brun, V., Maccagni, D., Pentericci, L., Thomas, R., Vanzella, E., Zamorani, G., Zucca, E., Amorin, R., Bardelli, S., Capak, P., Cassarà, L., Castellano, M., Cimatti, A., Cuby, J.G., Cucciati, O., de la Torre, S., Durkalec, A., Fontana, A., Giavalisco, M., Grazian, A., Hathi, N.P., Ilbert, O., Lemaux, B.C., Moreau, C., Paltani, S., Ribeiro, B., Salvato, M., Schaerer, D., Scodeggio, M., Sommariva, V., Talia, M., Taniguchi, Y., Tresse, L., Vergani, D., Wang, P.W., Charlot, S., Contini, T., Fotopoulou, S., López-Sanjuan, C., Mellier, Y., and Scoville, N. (2015) The VIMOS Ultra-Deep Survey: ~10 000 galaxies with spectroscopic redshifts to study galaxy assembly at early epochs $2 < z < 6$. *Astron. Astrophys.*, **576**, A79, doi: 10.1051/0004-6361/201423829.

123 de Zeeuw, P.T., Bureau, M., Emsellem, E., Bacon, R., Carollo, C.M., Copin, Y., Davies, R.L., Kuntschner, H., Miller, B.W., Monnet, G., Peletier, R.F., and Verolme, E.K. (2002) The SAURON project - II. Sample and early results. *Mon. Not. R. Astron. Soc.*, **329**, 513–530, doi: 10.1046/j.1365-8711.2002.05059.x.

124 Stott, J.P., Swinbank, A.M., Johnson, H.L., Tiley, A., Magdis, G., Bower, R., Bunker, A.J., Bureau, M., Harrison, C.M., Jarvis, M.J., Sharples, R., Smail, I., Sobral, D., Best, P., and Cirasuolo, M. (2016) The KMOS Redshift One Spectroscopic Survey (KROSS): dynamical properties, gas and dark matter fractions of typical z 1 star-forming galaxies. *Mon. Not. R. Astron. Soc.*, **457**, 1888–1904, doi: 10.1093/mnras/stw129.

125 Sánchez, S.F., Kennicutt, R.C., Gil de Paz, A., van de Ven, G., Vílchez, J.M., Wisotzki, L., Walcher, C.J., Mast, D., Aguerri, J.A.L., Albiol-Pérez, S., Alonso-Herrero, A., Alves, J., Bakos, J., Bartáková, T., Bland-Hawthorn, J., Boselli, A., Bomans, D.J., Castillo-Morales, A., Cortijo-Ferrero, C., de Lorenzo-Cáceres, A., Del Olmo, A., Dettmar, R.J., Díaz, A., Ellis, S., Falcón-Barroso, J., Flores, H., Gallazzi, A., García-Lorenzo, B., González Delgado, R., Gruel, N., Haines, T., Hao, C., Husemann, B., Iglésias-Páramo, J., Jahnke, K., Johnson, B., Jungwiert, B., Kalinova, V., Kehrig, C., Kupko, D., López-Sánchez, A.R., Lyubenova, M., Marino, R.A., Mármol-Queraltó, E., Márquez, I., Masegosa, J., Meidt, S., Mendez-Abreu, J., Monreal-Ibero, A., Montijo, C., Mourão, A.M., Palacios-Navarro, G., Papaderos, P., Pasquali, A., Peletier, R., Pérez, E., Pérez, I., Quirrenbach, A., Relaño, M., Rosales-Ortega, F.F., Roth, M.M., Ruiz-Lara, T., Sánchez-Blázquez, P., Sengupta, C., Singh, R., Stanishev, V., Trager, S.C., Vazdekis, A., Viironen, K., Wild, V., Zibetti, S., and Ziegler, B. (2012) CALIFA, the Calar Alto Legacy Integral Field Area survey. I. Survey presentation. *Astron. Astrophys.*, **538**, A8, doi: 10.1051/0004-6361/201117353.

126 Sánchez, S.F., García-Benito, R., Zibetti, S., Walcher, C.J., Husemann, B., Mendoza, M.A., Galbany, L., Falcón-Barroso, J., Mast, D., Aceituno, J.,

Aguerri, J.A.L., Alves, J., Amorim, A.L., Ascasibar, Y., Barrado-Navascues, D., Barrera-Ballesteros, J., Bekeraite, S., Bland-Hawthorn, J., Cano Díaz, M., Cid Fernandes, R., Cavichia, O., Cortijo, C., Dannerbauer, H., Demleitner, M., Díaz, A., Dettmar, R.J., de Lorenzo-Cáceres, A., del Olmo, A., Gallazzi, A., García-Lorenzo, B., Gil de Paz, A., González Delgado, R., Holmes, L., Iglésias-Páramo, J., Kehrig, C., Kelz, A., Kennicutt, R.C., Kleemann, B., Lacerda, E.A.D., López Fernández, R., López Sánchez, A.R., Lyubenova, M., Marino, R., Márquez, I., Mendez-Abreu, J., Mollá, M., Ortega Minakata, R., Torres-Papaqui, J.P., Pérez, E., Rosales-Ortega, F.F., Roth, M.M., Sánchez-Blázquez, P., Schilling, U., Spekkens, K., Vale Asari, N., van den Bosch, R.C.E., van de Ven, G., Vilchez, J.M., Wild, V., Wisotzki, L., Yıldırım, A., and Ziegler, B. (2016) CALIFA, the Calar Alto Legacy Integral Field Area survey: IV. Third Public data release, *ArXiv e-prints* (arXiv:1604.02289).

127 Brammer, G.B., van Dokkum, P.G., Franx, M., Fumagalli, M., Patel, S., Rix, H.W., Skelton, R.E., Kriek, M., Nelson, E., Schmidt, K.B., Bezanson, R., da Cunha, E., Erb, D.K., Fan, X., Förster Schreiber, N., Illingworth, G.D., Labbé, I., Leja, J., Lundgren, B., Magee, D., Marchesini, D., McCarthy, P., Momcheva, I., Muzzin, A., Quadri, R., Steidel, C.C., Tal, T., Wake, D., Whitaker, K.E., and Williams, A. (2012) 3D-HST: a wide-field grism spectroscopic survey with the Hubble Space Telescope. *Astrophys. J. Suppl. Ser.*, **200**, 13, doi: 10.1088/0067-0049/200/2/13.

128 Skelton, R.E., Whitaker, K.E., Momcheva, I.G., Brammer, G.B., van Dokkum, P.G., Labbé, I., Franx, M., van der Wel, A., Bezanson, R., Da Cunha, E., Fumagalli, M., Förster Schreiber, N., Kriek, M., Leja, J., Lundgren, B.F., Magee, D., Marchesini, D., Maseda, M.V., Nelson, E.J., Oesch, P., Pacifici, C., Patel, S.G., Price, S., Rix, H.W., Tal, T., Wake, D.A., and Wuyts, S. (2014) 3D-HST WFC3-selected photometric catalogs in the five CANDELS/3D-HST fields: photometry, photometric redshifts, and stellar masses. *Astrophys. J. Suppl. Ser.*, **214**, 24, doi: 10.1088/0067-0049/214/2/24.

129 Cheung, E., Bundy, K., Cappellari, M., Peirani, S., Rujopakarn, W., Westfall, K., Yan, R., Bershady, M., Greene, J.E., Heckman, T.M., Drory, N., Law, D.R., Masters, K.L., Thomas, D., Wake, D.A., Weijmans, A.M., Rubin, K., Belfiore, F., Vulcani, B., Chen, Y.M., Zhang, K., Gelfand, J.D., Bizyaev, D., Roman-Lopes, A., and Schneider, D.P. (2016) Suppressing star formation in quiescent galaxies with supermassive black hole winds. *Nature*, **533**, 504–508, doi: 10.1038/nature18006.

130 Martins, F., Genzel, R., Paumard, T., Eisenhauer, E., Ott, T., Trippe, S., Abuter, R., Gillessen, S., and Maness, H. (2007) Stellar populations in the Galactic Center with Sinfoni. Science perspectives for 3D Spectroscopy, ESO Astrophysics Symposia. Springer-Verlag Berlin Heidelberg, 2007, p. 229.

131 García-Lorenzo, B., Arribas, S., and Mediavilla, E. (2000) INTEGRAL: A simple and friendly integral field unit available at the WHT. *ING Newsl.*, **3**, 25–28.

132 Kenworthy, M.A., Parry, I.R., and Taylor, K. (2001) Spiral Phase A: A prototype integral field spectrograph for the anglo-australian telescope. *Pub. Astron. Soc. Pac.*, **113**, 215.

133 Beck, T.L., and McGregor, P. (2006) The Near-Infrared Integral Field Spectrograph at Gemini North Observatory. American Astronomical Society Meeting 208, id. 67.04. *Bulletin of the A.A.S.*, **38**, p. 147.

134 Thatte, N., Tecza, M., Clarke, F., Goodsall, T., Lynn, J., Freeman, D., and Davies, R.L. (2006) The Oxford SWIFT Integral field spectrograph, in Proceedings of SPIE 6269, Ground-based and Airborne Instrumentation for Astronomy, 62693L (29 June 2006); doi: 10.1117/12.670859.

Index

a

adaptive optics (AO)   100, 107, 115, 138, 173, 246
   integral field spectroscopy   127, 239
   systems   102
   techniques   193
   telescope   186
ADU-flux conversion   215
airmass   199, 202, 210
   correction for   229
Aladin Array   240
anisotropic crystals   14
anti-reflection coatings   42–43
apochromatic corrections   121
apodization   158
ARCONS camera   145
Arrayed Waveguide Gratings (AWG) devices   141
assembly, integration and testing (AIT)   51
astrometric calibration   232
astrometry   208
astronomical data cubes *vs.* common-sense data cube   237
astronomical efficiency   129
astronomical nights   191
astronomical seeing   167
Astrophotonics   138–144
astrophysical quantity image   238
astrophysical spectral   98
atmospheric curse   82
atmospheric dispersion   226
atmospheric extinction   189
atmospheric turbulence   1, 137, 155, 169, 171, 181, 194
Australian Astrophysical Observatory (AAO)   72, 124

b

bandpass   173
basic spectroscopic principles   18–19
BEAR instrument   90
bias (subtraction)   221
blaze angle   28
Bragg grating   140
bright time   200
BTO telescope   103

c

calibrated flux values   218
calibrations   207
   astrometric   230
   detector   215
   geometric   215, 225
   spectrophotometric   229
   wavelength   215, 224
CALIFA survey   149, 241
canonical plane grating long-slit spectrograph   28
Cassegrain focus   47
catadioptric systems   36
cat's-eye mounting   90
CFHT telescope   102
Chandra X-ray Science Center (CXC)   240

Optical 3D-Spectroscopy for Astronomy, First Edition. Roland Bacon and Guy Monnet.
© 2017 Wiley-VCH Verlag GmbH & Co. KGaA. Published 2017 by Wiley-VCH Verlag GmbH & Co. KGaA.

Charge Coupling Device (CCD)   33, 48, 83, 101, 159
  array   144
  detector   122
  pixels   226
charge injection devices (CIDs)   35
chart (pointing)   200
classical 3-mirror derotator   47
closed-loop control systems   89, 174
coarse sampling   107
coherence radius angle   171
completeness   206
cone effect   179, 184
contiguous spaxels   130
corner-cube   89
coronagraphic grade optics   40
coronagraphic system   116
correlated noise, effect of   232
cosmics rays   161
  impact of   163, 201
covariance (matrix)   218
crowding effects   230
cryogenics
  digital detector, proper temperature   48, 75
  NIR instruments   49
current wide-field projects   117

d

dark (subtraction)   222
dark current   160
dark exposures   222
dark time   200, 202
data analysis
  definition   237
  handling data cubes
    spatial view   239
    spectral view   238–239
    3D view   239
  handling data cubes: 3D spectrographs   237
  multi-object spectrography   237
  viewing data cubes
    CXC   240
    data cube visualisation   240
    monochromatic layers   241
data cubes   81, 216, 217, 229, 232, 238

data cubes handling
  spatial view
    monochromatic images, data cubes   239
    overlap, differential spectral information   239
  spectral view
    Doppler-Fizeau shift   238
    integral field data cubes   238
    non-linear process   238
  3D spectrographs   237
  3D view   239
data gathering   199
data reduction
  schemes   216, 218
  slitless MOS   216
  software system testing   213
  system   213
data reduction systems (DRS)   162
data sampling   153
data visualisation software   237
deformable mirror (DM)   173, 174, 176, 182
deployable multi-integral field systems (DMIFS)   126
detector calibrations   207, 215
detector noise   202
detector pixels   130, 154
diamond-turning   41
dichroic beamsplitters   19
differential atmospheric correction   226–228
differential image motion monitor (DIMM)   208
diffraction effects   154
diffraction limit   16, 169
digital detectors   62
digital multi-mirror device (DMD)   66, 67
digital signal processors   177
direct photonics coupling   141
dispersers
  efficiency   28
  grating principle   27–28
  prisms   25–27
dispersion (atmospheric)   199
dithering   205

Doppler–Fizeau shift   140, 238
Drizzle   232
DS9   240
DS9 data cube view   240

e

earth remote sensing hyper-spectral data cubes   237
Echelle spectroscopy   5
effect non-telecentricity   38
8-m class telescopes   103
electromagnetic field   141
electromagnetic spectrum   167, 189
electromagnetic wave   14
ESO VLT Observatory   120
etendue   15, 23, 26, 30, 53, 71, 88, 108, 137, 143, 169
  concept   16
etendue conservation
  infinitesimal surface element   15
  minimum beam etendue   16
  2D illustration   16
  two dimensions, plane section   17
EUCLID NISP spectrograph   64
European Southern Observatory (ESO)   65
evanescent field   141
exoplanet
  direct detection of   115
  imagers   116
exoplanets   116, 178
exposure time calculator (ETC)   200, 204
extremely large telescopes (ELT)   194

f

Fabry–Pérot (FP)   2, 4, 6, 20, 81, 83, 92, 130, 184, 216, 238
  psf   172–173
feasibility study   51
fiber-based slicer approach   103
fiber-based spectrograph
  MOS approach   75
  Schmidt collimator   75
fiber slicer   103
fiber systems performance   75
field mounting   24

field programmable arrays   177
filter bandpass   84
first order diffraction angle   28
Fishermen pond type systems   73
flat field (super)   215, 218, 222
flat-field calibrations   222
Fourier transform spectrometer (FTS)   6, 81, 88, 92, 130, 153, 217, 238
4 m-class telescopes   103
4-m Mayall telescope   67
FP cavity   20
Fresnel equations   14
full-width at half maximum (FWHM)   21, 154, 170, 211

g

γ rays   192
Gaussian error propagation   226
Gaussian fitting   238
Gaussian function   172
Gaussian model   204
Gaussian noise propagation   232
Gaussian PSF   97, 202
general-use instruments   245
geometrical calibrations   208, 215, 225
geometrical optics formulae   13
GMACS   138
grating   27, 140
grating etendue
  optimum sensitivity   30
  shelf reflective gratings   30
  VPHG   30
grating fundamental equation   27
grating spectrograph
  holographic   29
  long-slit concept   28
  surface relief grating   29
grey time   200
grism   96
ground-based multi-slit systems   68
ground-based telescope   170, 173
ground-layer adaptive optics (GLAO)
  approach   186
  technique   184, 194
ground seeing   170

h

H-band 167
HgCdTe 160
high-altitude dry site 167
high aperture ratio 103
high-contrast integral field spectrometer 116
High Energy Astrophysics Science Archive Center (HEASARC) 240
high-order echelle grating 30
high performance dioptric systems 37
high reflectivity coatings 43–44
high SNR spectra 208
homogeneous data cubes 237
Hubble flow 156
Hubble Space Telescope (HST) 64, 192, 194, 216
hydroxyl radical (OH) 139
hyper-spectral imaging 95
hysteresis effects 86, 188

i

IFU-type instrument 216
image processing software 239
incidence plane 13
infrared detector pixels 131
instrument grasp 129
integral-field spectrographs (IFS) 8, 95, 115, 116, 129, 130, 137, 173, 185, 238
  fiber-based 130
  mirror slicer-based 107
integral field technique 247
integrating 3D detector 145
integration time 201
interference filter 22–24, 83, 86, 87
interferogram 217
IR arrays 35
IRIS2 spectrograph 141

j

James Webb Space Telescope (JWST) 39, 63, 192, 194
J-band 167
JWST micro-shutter waffle-like array 68

k

K-band 167
KMOS instrument 125
KMOS kinematic survey (K2S) 135, 240
Kolmogorov turbulence model 170, 171, 195

l

laser tomography adaptive optics (LTAO) 181, 186
lenslet
  array 99
  fiber hybrid approach 103
  IFS SAURON 113, 238
lenslet-based approach 103
lenslet-based IFS 116, 119
light beams 13
light flux concentration 184
line spread function (LSF) 157, 158, 228
long-slit 2, 4, 25, 185
  spectrograph 63, 92, 98, 122
low-resolution spectra 65
lucky imaging 171

m

mask, properties of 225
McDonald Observatory 144
mechanical design
  alignments 48
  classical kinematic mounting 46
  NIR instruments 49
  optical components 45
MEIFU 119
meters-size telescopes 168
microlens array 95, 96, 100
micropupils 96
micro-shutter array 68
microwave kinetic inductance detectors (MKID) 144
mini-lenses 95
mirror-based image slicers 104, 119
MOAO 187
MOFFAT 156, 234
monolithic poly-methyl-methacrylate lenses 100

mono-mode fiber   140, 142
moon   206
Moon's diffuse light   200
Moore's law   177
Mosaicing   206
MOS-type instrument   216
multi-aperture interferometry   143
multi-arm approach   19
multi-conjugate adaptive optics (MCAO)   182
multi fiber-based slicer   95
multi-fiber positioning systems
   AAO   71
   2dF   71
multi-filter imager   83, 145
multi mirror-based slicer   95
multi-object spectrographic (MOS) instruments   6, 61, 77, 132, 137, 185
   history, digital age
      digital detectors   62
      NGT   62
   history, pioneers
      astronomical technique   61
      'slitless' approach   61
   multi-slit based multi-object spectroscopy
      multi-fiber concept   70–71
      multi-slit concept   64–66
      multi-slit holders   66–69
   slitless based multi-object spectroscopy
      slitless concept   62–64
multiple wavefront sensors   183
multi-pupil imaging   97
multi-slit based multi-object spectroscopy
   multi-fiber spectrograph   70
   multi-slit concept   64–66
multi-slit instruments
   NIR multi-slit spectrographs   70
   optical multi-slit spectrographs   70
multi-slit MOS   130
multi-slit selection softwares   66
multi-slit spectrograph   65
multi-slit systems   83
exchangeable zero-deviation grisms   69
kinematic mountings   69
multi-slit systems patrol field   66
multi-wavelength imaging   19
MUSE   119, 120, 155, 201, 215
   e2v CCD detectors   222
   IFU pipeline   219
   instrument   186
   integral field spectrograph   236

n

narrow-band imaging   129
Nasmyth focus   47
Nasmyth platform   46
Near Infra-Red (NIR) Array   35, 48, 138
new generation telescope (NGT)   62
night-sky emission spectrum   139
nod and shuffle   205
noise (variance)   217
noise probability function   162
noise variance   204, 226
   estimation   217
non-scanning 3D techniques   92
normal distribution   162, 204
numerical aperture (NA)   18
Nyquist   153
Nyquist sampling   186
Nyquist-Shannon sampling theorem   153, 154, 157

o

OH software avoidance technique   139
OH-suppression fibers   138, 141, 143
one-step resampling scheme   226
optical computation
   inherent chromatic aberrations   37
   schmidt mounting   39
optical etendue   17
optical fabrication
   diamond-turning   41
   moulding techniques   41
   substrate grinding   40
optical fibers principle   18
optical imaging systems   17
optical systems   116, 117

optical throughput   15
optimal extraction   204
optimum spectral resolution   31
opto-mechanical slicer   105
opto-mechanical systems   118
orthogonal plane mirrors   89

p
packing efficiency   99, 130
peak wavelength   21
phase correction   216
photometric conditions   230
photometric night   189
photon counting 3D detectors   144
photon counting detector   87, 184
photonics-based spectrograph   138, 146
photonics Fourier transform spectrometer   141
photon noise   202
photoresist lenslet array   100
pick-up noise   160
piezo deformable mirrors   176
piezo-driven technique   85
pixels detector   129
pixel table   226, 232
Planck constant   202
pointing chart   200
point-spread function (PSF)   131, 154
Poisson discrete probability distribution   159
Poisson noise   226
polychromatic light beam   27
polynomial coefficients   234
prescan/overscan pixels   215
principle investigator (P.I.)   50
prism-like effect   185
prism's principle   25
prism transmissions   26
PSF (spatial)   188, 234
pupil mounting   24

q
Qfitsview   241
Quality control   234
quantum efficiency   86, 141, 146, 202
quasi-Gaussian   183

r
radial velocities, spectra   238
Rayleigh scattering   180
real-time computer (RTC)   174, 176, 177
residual opto-mechanical errors   172
residual wavefront distortions   175
resolving power   157
resonant cavity quality factor (Q)   21
RITMOS instrument   68
RVSAO IRAF packages   238

s
SALT telescope   88
sampling   154, 155
SAURON survey   113, 238
Scanning Fabry–Pérot   238
scanning filters
  Fabry–Pérot (FP) interferometer   20
  FWHM   21
  interference filters
    classical etalon   23
    pupil image   24
    sky field image   24
  zero photon flux   21
scanning Fourier transform spectrograph   217
scanning FP technique   87
scanning interferometry   129
scanning long-slit
  spectrometers   129
  spectroscopy   129
  technique   83
scanning long-slit spectrograph (SLSS)   81
scanning slit spectroscopy   157
scanning technics   129
scintillation effect   167
seeing   206
seeing effects   168, 172
seeing principles   168
seeing properties   170
sensitivity curse   82
SExtractor software   239
Shack–Hartmann wavefront sensor   177
Shannon   153

Shannon's theorem   131
signal to noise ratio (SNR)   88, 141, 153, 162, 200, 201, 215
SILFID   9, 102, 247
single conjugate adaptive optics (SCAO)   175, 182
single-mode laser beam   89
SITELLE   91
sky spaxels   130
sky spectrum   93, 238
sky subtraction   228–229
slit effect   97, 157
'slitless' approach   61
slitless spectroscopy concept   62–64
slitless technique   62
sloan digital sky survey   59, 236
SNR spectra   241
solar light   191
solar system   115, 117
spatial dimensions   238
spatial pixels   82, 87, 125, 154, 186
spatial point spread function (PSF)   154, 172
spatial resolution   154
spatial sampling   107
spatio-spectral data cube   96
spaxel   99, 154
spectel   155
spectral flux distribution   200
Spectral Hole Burning Device (SHBD)   145
spectral pixels   82, 116, 155
spectral resolution   18, 157, 230
spectral sampling   156
spectrographic technique   83
spectrophotometry   130, 208
spectroscopic instruments   13
spectroscopic redshift survey   79, 237
spectroscopic toolbox
  basic spectroscopic principles, spectroscopic case   18–19
  etendue conservation
    infinitesimal surface element   15
    geometrical optics   13–15
    optical computation   37

scanning filters
  Fabry–Pérot Filter   24
  Fabry–Pérot (FP) interferometer   20
  interference filters   22–24
SPIFFI   9, 104
SpIOMM instrument   91
splitting exposures   204
Stationary Wave Integrated Fourier Transform Spectroscopy (SWIFTS)   141
STIS instrument   64
Strehl ratio   193
Subaru telescope   188
superconducting tunnel Junctions (STJ)   144
super-flatfieldind   218
SWIFTS principle   141
systematics   162, 203, 218

t

team-use instrument   250–251
telecentric image   101
telescope
  aperture   171
  polishing errors   172
  primary mirror   16
telluric correction spectrum   229
thermal expansion coefficient   20
3D data set   120
3D data visualization   240
3D de-blending software   239
3D detector   144
3D instrumental approaches   245
3D scanning techniques   91
3D spectrographs   240
3D visualization   241
three mirror anastigmat (TMA)   40
TIGER   9, 101, 247
Tiger-type spectrographs   101
tip/tilt isoplanetic field   181
total light flux   92
trace mask   223
transition edge sensors (TES)   144
tuneable filter   86, 88

2D detectors   92
   optical detectors   32–35
   photographic plate   32
   spectro-imagers   31
   2D infrared arrays   35
2D etendue conservation   16
2dF (2-degree field) robotic system   72
2D spatial field   117, 119

v

vacuum deposition   22
variance   241
variance data cube   241
VIMOS   202
   exposure, multi-object mode   65
   ultra deep survey, spectra   79, 237
VIRUS/MUSE approach   117
visibility   206
   curve   206
   source   199
VLT Observatory   181

volume phase holographic gratings (VPHG)   30
volume pixels   129
voxel   239

w

wavefront sensing   36
wavefront sensor (WFS)   173, 176, 187
waveguide arrays   143
wavelength calibration errors   162
wavelength calibrations   208
WFOS-MOBIE   138
wide-band interference filter   90
wide-field
   integral field spectroscopy   117
   MOS systems   76
wide-field: NIR MOS survey   77
William Hershell Telescope   113, 238

z

zero-field mode   143